Moritz Epple / Claus Zittel (Eds.)
Science as Cultural Practice
Vol. I: Cultures and Politics of Research
from the Early Modern Period to the Age of Extremes

WISSENSKULTUR UND GESELLSCHAFTLICHER WANDEL

Herausgegeben vom Forschungskolleg 435
der Deutschen Forschungsgemeinschaft
»Wissenskultur und gesellschaftlicher Wandel«

Band 24

Science as Cultural Practice

Volume I:
Cultures and Politics of Research from the Early Modern Period to the Age of Extremes

Edited by
Moritz Epple and Claus Zittel

Akademie Verlag

Gedruckt mit Unterstützung der Deutschen Forschungsgemeinschaft, des Landes Hessen und der Stiftungsgastprofessur »Wissenschaft und Gesellschaft« der Deutschen Bank AG.

Einbandgestaltung unter Verwendung eines Ausschnittes aus: Certificate of Membership of Emil du Bois-Reymond in the Berlin Physical Society (Bois Reymond, Emil du 1845)

Bibliografische Information der Deutschen Nationalbibliothek
Die Deutsche Nationalbibliothek verzeichnet diese Publikation in der Deutschen Nationalbibliografie; detaillierte bibliografische Daten sind im Internet über http://dnb.d-nb.de abrufbar.

ISBN 978-3-05-004407-1

Das eingesetzte Papier ist alterungsbeständig nach DIN/ISO 9706.

Satz: Amos Schindler, Laurens Schlicht und Dorothea Wolkenhauer; Frankfurt/M.
Druck und Bindung: Druckhaus »Thomas Müntzer«, Bad Langensalza
Einbandgestaltung: Dorén + Köster, Berlin

Printed in the Federal Republic of Germany

Contents

Part II: Science in the Age of Extremes

Moritz Epple and Claus Zittel

Introduction

Eighteen years ago when Andrew Pickering opened his widely read collection *Science as Practice and Culture* with an essay reviewing the various trends in the field of science studies since the formation of a sociology of scientific knowledge in 1970's, he entitled his reflections "From science as knowledge to science as practice", and he summarised the "key advance" inspiring his collection in the sentence:

> "This [advance] is the move toward studying scientific practice, what scientists actually do, and the associated move toward scientific culture, meaning the field of resources that practice operates in and on" (Pickering 1992: 2).

A lengthy footnote explained the sense in which the two terms 'practice' and 'culture' were to be understood: scientific practice was explained as a kind of work composed of "acts of making (and unmaking)", and culture was defined to be the "field of resources that scientists draw upon in their work" (Pickering 1992: 2-3).

If our collection now bears the title "Science *as* cultural practice" one may rightfully wonder whether there is much of a difference in approach in comparison with Pickering's volume, and if so, what that difference would be. Perhaps the simplest way to put it is to say that while most if not all of the studies presented in this volume would subscribe to a view of science as a kind of practice – and in this sense there is no need to reinvent the wheel – there is also a consensus that the practices of science should not be viewed separately from cultural practices but rather as a particular *subset* of these practices. This shift in perspective may seem slight, and indeed we deviate from Pickering's notion of 'culture'[1] quoted above more strongly than from the understanding of science advocated in his volume. The articles presented here argue for, and present examples of, the notion that scientific practices participate in the production of the web of meanings given by a society to 'nature' and to the actions of its members, in other words, to the production of a culture, if one is willing to accept – as many authors in this volume would do – a largely Geertzian notion of culture (Geertz 1973). Pickering's idea of cultural 'resources' can be embedded in such a view in a more symmetrical fashion: which cultural practices play the role of resources for the practices of scientists, or which cultural practices draw on the resources offered by scientific practices, may vary and can be left open to historical investigation – there is nothing that would place scientific practice in an asymmetrical position in this respect.

Understanding science as one cultural practice among others – or rather, as a whole array of cultural practices intertwined with other such practices – offers both methodological and

1 Pickering's understanding of culture in relation with the history of science has been discussed critically by Jan Golinski in chapter 6 "Culture and Construction" of (Golinski 1998), see also (Dear 1995a); (Zittel 2002).

thematic advantages. On the side of methodology the relation between the historiography of science and other fields of cultural study such as anthropology, literary studies, art history, 'Bildwissenschaft', cultural and social history broadly conceived, political and religious studies, becomes more symmetrical as well. Rather than 'applying' the methods of these fields to the purportedly singular subject of science, one finds oneself immersed in a broader field of available methods in which the historical topic rather than a preconceived 'approach' can be given weight in selecting appropriate tools of investigation. There is, then, no longer a fundamental difference between a historian of mathematics, say, using anthropological research techniques and an anthropologist making use of the history of arithmetics in studying the counting and calculating practices of a given population.[2]

On the topical side, viewing scientific practices as part of the larger field of social practices in which cultural meaning is produced enables one to understand the core of scientific practice, i.e., the generation of both ideals of scientific knowledge and fragments of knowledge satisfying these ideals as a process not separated from the social production of knowledge in general and the formation of norms and values regulating social and cultural life. Just like other cultural practices, the activities of scientific research *give sense* to certain areas of perceived reality; the generation of knowledge and the production of cultural meaning are inextricably linked. Even if one might hope that such a remark would no longer be necessary it must be pointed out that from this point of view, the old and – unfortunately – repeatedly recurring opposition between internalist and externalist approaches to science turns out to be just a pair of partial and, in the best of all worlds, complementary perspectives on a subject matter which is more complex than each of these partial perspectives is able to encapsulate alone.[3]

Two further facets of viewing scientific practices as special kinds of cultural practice are that this view invites us to ask *comparative* questions that cut across traditional disciplinary boundaries, and that it offers the possibility to place our efforts in the continuing discourse of what has recently become known under the name of historical epistemology. In which ways can the visual techniques of scientists be compared with the visual techniques of painters or of other groups engaged in the production of visual media?[4] In what ways are the political strategies of scientific collectives similar to or different from the political strategies of other social groups?[5] To ask questions such as these becomes necessary on the premises which the studies in this volume advocate.

Finally, and as indicated at the beginning, we see the emphasis of this collection as reinforcing and hopefully advancing the discussion of a stringently historicized epistemology which began around 1900 (at the latest) and culminated in the contributions of authors such

2 One striking example for the establishment of the cultural approach to science are the entries in the *Neues Kindler Literaturlexikon*, where not only many scientific texts but also books such as Shapin and Schaffer's *Leviathan and the Air-Pump* are now treated as part of the canon of the history of literature.

3 For further discussion of similar points, see also (Biagioli 2009).

4 Here one may point to studies such as (Bredekamp 2007) or (Biagioli 1999a,b, 2006).

5 A political history of science has of course long been in the interest of some contributors to the field, including Paul Forman or Donald MacKenzie, among others, see e.g. (Forman 1971, 1987); (MacKenzie 1981, 1990). In recent years, the interest in such issues has increased. From a wealth of publications let us mention (Forman, Sánchez-Ron 1996); (Kaufmann 2000); (Galison, Gordin, Kaiser 2001); (Pestre 2003); (Dahan, Pestre 2004); (Nye 2004); (Schiebinger, Swan 2004); (Sachse, Walker 2005); (MacKenzie 2006).

as Ludwik Fleck, Gaston Bachelard, Karl Mannheim and Edgar Zilsel in the 1920's and 1930's before the turmoil of World War II interrupted this most productive exchange – an interruption that was only very slowly ended in the second half of the last century. Here viewing science as a cultural practice becomes reflexive – and indeed the historiography of science is a cultural practice as well, intertwined in its own way with general history and cultural trends. For this reason among others several essays in this volume address issues concerning the history of this historiography in the last century.[6]

*

The present volume collects studies whose topics range from the early modern period to very recent times. A recurrent argument is that the modern sciences, which have grown to social networks of substantial size, are not only shaped by the social and cultural conditions under which they exist, but also generate their own patterns of social and cultural behaviour that tie in with the patterns of social life at large. In this way, one or several scientific sub-cultures are formed in any given period, and indeed these sub-cultures immersed in a society generate their own *politics* as well.

Investigating these sub-cultures and the politics of scientific practice in society, determining the ways in which they partake in weaving the cultural web of meanings as well as in the shaping of the material life of society still represents a formidable task for historical research. While this task cannot be coped with in the short run, we hope that the present volume offers a modest contribution to a better understanding of the complex issues involved in such a task. A second volume *Science as Cultural Practice 2: Modernism in the Sciences, ca. 1900-1940* (edited by Moritz Epple and Falk Müller) will address the particular period and topic of scientific modernity and investigate the analogies, interrelations, similarities and differences between it and the cultural trends of the same period.

This first volume is divided into two parts. The first half collects case studies ranging from the 17th to the 19th century which illustrate how the formation of scientific concepts, of experimental practice, of scientific rationality, of research strategies were connected with broader cultural patterns and political issues. The contributions in the second half are grouped under the title *Cultures and politics of research in the 20th century*. Most of these contributions arose from talks in a lecture series organised in Frankfurt in the winter of 2005/2006 under the same title. Some guiding topics of this series were the hybrid nature of research constellations in a period of accelerated modernisation, two World Wars, and extreme political and cultural tensions. How did experimenters and theoreticians, engineers and philosophers of science act and interact with other groups in their respective 'trading zones', in a world of rapid social change (here Galison's fitting notion must be understood in its widest possible sense)? It turns out that a particular strand in these inquiries is devoted to *reflexive* efforts and their history. Scientists, as well as their philosophical, historical and political observers, have been challenged to rethink the idea of science and scientific research in what we have come to call the age of extremes, and 20th century history is marked by significant cultural and even political responses to the changes in scientific practice and its reflexive analysis.

---

6 See (Rheinberger 2007) for a concise introduction to this topic.

Let us briefly survey the different alternatives available for reacting, at a methodological level, to these issues. One option certainly continues to be an analysis of the 'extra scientific' social and educational backgrounds and motives of influential individual scientists, in particular if it is possible to highlight crucial features of a scientist's research activities in this way. In her analysis of Robert Boyle's education Catherine Wilson illustrates how 'romantic' motives came to be important to him that would later shape his scientific practice, prompting him to harmonise scientific evidence and Christian revelation, rather than setting them in opposition to each other.

The complex motives of scientific research can also be considered from a perspective that leans more to social anthropology than to biography. Indeed Simon Schaffer proposes viewing core practices in the 'hard sciences' – such as astronomy and astrophysics – very much like the field work of anthropologists. Just like anthropologists in the field, scientists in their observatories and laboratories have to establish local socio-technical and political networks of support while at the same time acting in larger communication networks; they have to prudently integrate local and global knowledge. Juxtaposing anthropology and astrophysics in India in the 19th and early 20th centuries in this way, Schaffer throws an interesting light on the formation of a colonial science in the British Empire.

Friedrich Steinle's and Moritz Epple's contributions seek to provide an analysis of cultural elements in the generation of experimental and mathematical knowledge, respectively. Steinle's account of explorative inquiries into the nature of electricity at the beginning of the 18th century, and in particular in the work of Charles Dufay, points to the essential role of initial concept formation when making an inroad into a new field of experimental inquiry. Here concept formation is found to be a creative activity, and indeed first and foremost a creation of new language (rather than a refinement of long-standing ideas) that makes it possible to express things differently than before and thus may completely reorder the cosmos that the experimenters see. This linguistic and conceptual creativity participating in the genesis of facts makes the scientist similar to a poet or writer; both change not merely our language but also what it can be used to express and explore.

From a broader perspective, Moritz Epple's essay aims at determining certain patterns in which the production of knowledge may be linked with extra-scientific cultural activities. He argues that while sophisticated histories of experimentation have exhibited such patterns in many interesting cases, room still exists for pursuing a similar analysis in histories of mathematical research practice. Just like experimental practices, research practices in this scientific field draw on a wide variety of resources, they condense in mathematical ways of life, and they experience what Rheinberger has termed 'conjunctures', indeed conjunctures with episodes of cultural activities far beyond mathematics.

The papers of Lorraine Daston, Sven Dierig and Norton Wise, in turn, engage with the aesthetic dimension of research practices, another area in which it turns out to be impossible to separate scientific practice from its cultural embedding. Discussing the travels, visions and research practices of Alexander von Humboldt, Daston shows how a new image of nature became possible on the basis a tightly-knit fabric of cooperation between researchers and travellers, observation skills, means of transport and communication, and by integrating an aesthetic model of the perception of nature with strategies of exact measuring and detailed description. A key role is played here by Humboldt's vision of combining an attention to local variation with the idea of universal laws, and indeed of doing so for both nature and culture. Consequently, Humboldt's integrated *cosmos* required a corresponding integration of practices in which 'scientific' and 'cultural' aspects became inseparable.

Another ideal that linked aesthetics, humanist *Bildung* in Wilhelm v. Humboldt's sense and scientific research was Emil du Bois-Reymond's programme of *Bildung durch Wissenschaft*, located by Sven Dierig in du Bois-Reymond's practice of experimentation. Rather than a creative, aesthetic view of nature, in this programme the beauties of an artistically arranged experiment and the aesthetic experiences it allowed were intended to educate the subject. Dierig interprets Du Bois-Reymond's series of plates that show a youthful Apollo while performing experiments, as '*Bildungsbilder*', because these drawings offer a corrective of Wilhelm v. Humboldt's one-sided humanist outlook by bringing to the fore the education of the senses and of the body in scientific experimentation.

Norton Wise takes up another facet of the topics addressed by Daston and Dierig. His paper analyses a bourgeois fashion of the 1830's and 1840's that ran through art, architecture and, so Wise claims, in the end science: the art of drawing curved lines. Noting occasional earlier instances, Wise still sees in this fashion the reason for a significant surge in the use of a technique that became widespread in all areas of mathematics and mathematised science, the technique of representing processes of nature or of society by means of curve diagrams. Once again, Emil Du Bois-Reymond is the crucial actor, and once again the scientific and bourgeois culture of early 19th century Berlin provides the setting in which this fashionable link could be forged. Wise's essay leads us to a wealth of paintings and drawings, pointing out that these cultural practices obviously informed the cognitive practices of arranging scientific knowledge.

The second part of our volume is opened by Hans-Jörg Rheinberger's survey of his analysis of experimental systems in molecular biology. Viewing the experimental systems of mid-20th century biological research as particular forms of what Paul Rabinow has termed 'assemblages', i.e., constellations of actors, things and institutions that allow certain things to happen (and at times in unexpected ways), Rheinberger draws attention to the extraordinarily contingent and involved processes in which objects of scientific research and experimental techniques as well as laboratory set-ups co-evolve. Of particular importance in the context of this volume is Rheinberger's emphasis on the open boundaries of laboratory research – on the technical level, elements from other experimental and technological domains were integrated including those related to warfare, while on the practical level, scientific activity appears as a 'practice of art', and on a moral or political level, motivations from medical policies were incorporated as well as visions of a reshaping of the living organism – including the human one. Comparing this analysis with the contributions in the first part of our volume the reader will find that many strands in the discussions of early modern and 19th century scientific practices are again found in those of the 20th century, a century, however, in which the consequences of any social activity – including scientific activity – could have repercussions on human life on an unprecedented scale.

The possibility of these repercussions was also felt where 20th century scientists engaged with the epistemology and historiography of their disciplines, acknowledging in various forms and in different social and scientific circumstances that scientific research needs to be understood as a changing *assemblage*. From this perspective, Claus Zittel and Mary Jo Nye discuss some of the most influential theories of the cultural and social production of scientific knowledge and try to show how these theories themselves were both anchored in, and the bearers of, political implications for the convoluted history of the age of extremes. Claus Zittel makes clear how Fleck's theory of thought styles and thought collectives is marked by an inner tension between the relativistic connotations of his account of the social genesis of scientific knowledge – and 'facts' – in situated, collective practices of research and communication on the one hand, and his overarching programme of a comparative, scientific analysis of thought styles on the other. The tendency toward the latter, moving more and more to the foreground of Fleck's epistemological writings in the mid-1930's, clearly has political overtones resonating with a universalism of knowledge that went against some of the more fanatic relativisms of those years.

Similarly, by sketching the social and cultural landscapes in which Karl Popper, Michael Polanyi and Thomas S. Kuhn turned to the history and epistemology of science Mary Jo Nye shows to what extent their endeavours were shaped by political impulses. While for the first two, the experience of emigration and the breakdown of the 'republic of science' (Polanyi) in both National Socialism and Stalinism provided a strong motive for engaging with epistemology, the latter turned to the history of science in the context of post-war general education in the US. It will remain a task for future meta-historians of science to determine the degree in which all these major contributions to our field remained under the spell of the social, cultural and political contexts in which they emerged.

The last two contributions move very close to our present. Both Evelyn Fox Keller and Dominique Pestre try to investigate mutual dependencies of recent developments in the sciences and in social and political life. Looking at the increasing cooperation between physical, biological and engineering sciences, Fox Keller discusses how seemingly reliable and long-standing boundaries such as that between machines and organisms, artificial and natural bodies are gradually given up – not just given up as the simple popular catchwords by which they have long been identified, but right in the involved technical debates of recent research on complex structures and (self-)organization. If she is right, then we are led to a similar point to that in Rheinberger's account of molecular biology – what will, in some not very distant future, be called 'life' or 'human life' may well depend on the strategies and technologies available to scientists and engineers investigating and engineering them.

Dominique Pestre raises the question whether such developments – tied in with hybrid, large scale institutional networks of technoscientific research – mark the advent of a new 'regime of knowledge production'. And – he reflects – would it not be necessary to conceive of perspectives of democratic control, of open access and the pursuit of sensible alternatives (including small scale, distributed research) in our buzzing world of scientific research?

*

The full list of instutions and persons to whom we must address our sincere thanks for help in the making of this volume is too long to be included here. Suffice it to say that we are most grateful to the Sonderforschungsbereich 435 *Wissenskultur und gesellschaflicher Wandel* and the *Stiftungsgastprofessur Wissenschaft und Gesellschaft* of Deutsche Bank AG for financial support, and to Dorothea Wolkenhauer and Laurens Schlicht for their enourmous help in editing the manuscript.

Catherine Wilson

Some Motives and Incentives to the Study of Natural Philosophy

Robert Boyle's thoughtfulness and insightfulness as a metaphysician and epistemologist, as well as a natural philosopher, are increasingly recognized. Many ideas supposedly original with Locke – the doctrine of primary and secondary qualities, nominal and real essences, concepts made by the mind, the bounds of knowledge and about personal identity, are derived from his reading of and sometimes his reaction to, Boyle. Boyle's own moral and philosophical writings are concerned to a large extent with Epicureanism, the ancient philosophy of materialistic atheism, and Epicurism – the dedication to sensory pleasure. References to the philosophy and to the mode of life associated with it appear in virtually every one of Boyle's essays.[1]

Both Epicureanism and Epicurism were topics of some ambivalence where Boyle was concerned. For Boyle subscribed – in a highly qualified way to be sure – to the corpuscularian or mechanical philosophy of Epicurus and Lucretius, and his earlier writings reveal him as a kind of aesthete – though often a self-censoring and even censorious one – *avant la lettre*. To reconcile his scruples with his attractions to both Epicureanism and Epicurism, Boyle invented and commended to his readers the figure of the Christian Virtuoso, and he endowed the Boyle Lectures to confute atheists and libertines. The creation of this model figure, and the subsequent defense of the corpuscularian or mechanical philosophy as intellectually more supportive of the Christian religion than scholastic philosophy, helped to secure legitimacy for the somewhat insecure enterprises of the early Royal Society.

The recovery and interpretation of some of Boyle's many lost and suppressed manuscripts confirms the suggestion that the Christian Virtuoso is an imaginary figure created for a purpose, regardless of the tendency to identify the Christian Virtuoso simplistically with Boyle himself. Michael Hunter – no friend to conspiracy theories – maintains that Oldenburg, Glanvill, Burnett, and Birch contrived an image of Boyle as "a paragon of civility and moderation with a clear and unproblematic strategy for the vindication of a mechanistic view of nature by profuse experimentation, and an accompanying combination of deep piety [...]" (Hunter 2000: 8). Boyle was, Hunter suggests, more "mixed-up" than this, – though in a way that leaves our respect for his intelligence and moral qualities unimpaired. This paper will explore the more and less successful features of this construction in the hope of shedding slight on the larger issue of the revival of Epicureanism and Epicurism in a Christian culture in which Augustinian renunciation of the world remained a powerful motive.

Epicureanism was already recognized as a powerful intellectual force and extensively condemned by Edward Stillingfleet in his *Origines Sacrae* of 1662. "That which makes the most noise in the world [...] is," he says, "the Atomical or Epicurean Hypothesis" (Stil-

---

1 See (MacIntosh 1991).

lingfleet 1662: 282), and he refers suspiciously to "modern philosophers" as its exponent and to the "late sedulous Vindicator" (Hobbes) of the despised philosophy. Boyle's own concerns were engaged at several levels. First, he was troubled by his adoption of a corpuscularian *ontology* associated with atheism and libertinage. Second, Boyle was troubled by the naturalist's evaluation of the material objects of this world as *worthy* of sustained attention and the utilitarian, this-world focus of the experimental philosophy. Third, he was bothered by the discrepancy between the standards of *evidence* employed in the theological realm and those employed in the experimental realm.

1. Corpuscles, Laws of Nature, and Divinity

According to the Epicureans, matter and void had existed from all eternity, *cosmoi* came into being spontaneously and perished, and animals and men alike had material and perishable souls composed of especially fine and mobile atoms. There was no supervisory Providence or creative and directive intelligence in control of the universe. The atheistic atomism of Epicurus and Lucretius was decisively recast by Pierre Gassendi into a theomechanical system that preserved the atomists' ontology and their insistence that the entanglement, motion, and interaction of invisible corpuscles was the basis of all phenomena, while rejecting their denial of divine creation, providence, and the immortality of the human soul. Gassendi adopted the doctrine of atoms but added to it a God, divine creation *ex nihilo*, a system of laws of nature ordained by God, and an incorporeal immortal human soul above and beyond a corporeal animal soul.[2] This theomechanical system, another version of which was presented by Gassendi's rival René Descartes in his *Principles of Philosophy*, was absorbed by Boyle,[3] and is articulated in his more programmatic writings:

> "I plead onely [...] for such a Philosophy as reaches but to things purely Corporeal, and distinguishing between the first original of things, and the subsequent course of Nature, teaches [...] not onely that God gave Motion to Matter, but that in the beginning He so guided the various Motions of the parts of it, as to contrive them into the World. [...] (furnished with the Seminal Principles and Structures or Models of Living Creatures,) and established those Rules of Motion, and that order among things Corporeal, which we are wont to call the Laws of Nature [...]; the Phaenomena of the World thus constituted, are Physically produc'd by the Mechanical affections of the parts of Matter, and [...] operate upon one another according to Mechanical Laws" (Boyle 1674: 104).

Boyle thus evaded Stillingfleet's blanket condemnation[4] of atomistic mechanism and insisted repeatedly that it implied the existence and activity of a God "[A] Machine so Immense, so Beautiful, so well-contrived, and, in a word, so Admirable as the World cannot

2 See, on his substantial emendations (Johnson 2003); also (Osler 2003).

3 Through the mediation of Walter Charleton, whose *Physiologia Epicuro-Gassendo-Charletoniana* appeared in 1654. See (Kargon 1964).

4 "We must wholly alter the present stage of the world, and crumble the whole Universe into little Particles; we must grind the Sun to Powder, and by a new way of Internment turn the earth into Dust and Ashes, before we can so much as imagine how the World could be framed" (Stillingfleet 1662: 238).

have been the effect of mere Chance, or Tumultuous Justlings and Fortuitous Concourse of Atoms, but must have been produced by a Cause exceedingly Powerful, Wise, and Beneficent" (Boyle 1690: 299-300). "God," he said, anticipating the accusation of deism, "hath not Abandoned a Masterpiece so worthy of him, but does still Maintain and Preserve it" (Boyle 1690: 300). Boyle termed his theory of nature 'Anaxagorean' to distinguish it from classical atomism and also from the Cartesian version, which, though it introduced God as the cause and maintainer of corpuscular motions, nevertheless held that the cosmos, and plant and animal life, had emerged spontaneously.[5] According to the doctrine of Anaxagorean mechanism, the frame of the world, and its original plants and animals, or at least their "seeds or seminal principles," had been intelligently and beneficently designed and created, though thereafter, the laws of motion, the structure of objects, and the dispositions of seeds sufficed for the production of all, or almost all effects.[6]

This Anaxagorean system, one might think, reconciled religion and natural philosophy easily, provided one accepted the notion that the laws of nature could in some sense be prescribed to and obeyed by inanimate particles, and provided one was not troubled by the paradoxes of division and composition which militated against material atoms and the void. Yet Boyle was often troubled by his adoption of large parts of a pagan and aggressively anti-theistic system. He believed himself to be living in an exceptionally dissolute age, and he considered the neo-Epicurean threat to religion and morals more serious and less easily defended against than other atheistic and mortalist versions of Aristotelianism and pagan naturalism, such as those represented by Pomponazzi and Vanini. "Libertines," he says, "own themselves to be so upon the account of the *Epicurean* or other *Mechanical Principles of Philosophy*" (Boyle 1675a: 237), and they fail to pay due regard to Aristotle, Scotus, Aquinas and Augustine. He complained of being taken for an Epicurean himself (Boyle 1663: 354). Yet one cannot say Boyle shows much deference to Aristotle or to his scholastic followers. By contrast, there are over one hundred references to Epicurus and Lucretius and quotations of the latter in his writings. If Boyle was sincere in maintaining that he had read little of Lucretius and lacked conversancy with Epicureanism in 1663,[7] he made up for his neglect later.

Two points relentlessly driven home by Lucretius are that the gods – who are conceived as happy immortals existing in the intermundial spaces between *cosmoi* – do not concern themselves with affairs on our earth and that fear of the gods is an oppressive and needless burden. They cannot help us and they cannot harm us either. This view was at odds with the Christian commitment to divine and demonic intervention in historical processes and personal life. According to Boyle's account of his religious history, he had been roused as a boy from religious indifference by a terrifying thunderstorm (Boyle 1744a: 12), and he thereafter perceived the saving intervention of God in several astonishing escapes, when he was nearly killed by horses, collapsing bedroom walls, and poisoned by a careless apothecary.

5 The term 'Anaxagorean' appears in the suppressed or discarded sections of the *Inquiry into the Received Notion of Nature*, HD 14:148.

6 On Boyle's theory of generation, see (Anstey 2002).

7 "If you knew how little Conversant I have been with Epicurean authors," says the Boylean chemist to his interlocutor, and "how great a part of Lucretius himself I have not yet had the curiosity to read, you would be free of this suspicion" (Boyle 1663: 354).

"*Philaretus*," Boyle says, "would not ascribe any of these rescues unto chance, but would be still industrious to perceive the hand of heaven in all these accidents; and he indeed would profess, that in the passages of his life, he had observed so gracious and so peculiar a conduct of providence, that he should be equally blind and ungrateful, should he not both discern and acknowledge it" (Boyle 1744a: 8).

All his life, however, Boyle was troubled by religious doubts. He read many pagan authors, including Seneca, whom he admired.[8] And when the young Boyle visited a Carthusian monastery in the wild mountains near Grenoble, he was nearly driven to suicide by brooding on the fate of St. Bruno.[9] After months of "tedious perplexity," God restored "the withdrawn sense of his favour," and Boyle came to see "these impious suggestions, rather as temptations to be suppressed, than doubts to be resolved; yet never after did these fleeting clouds cease now and then to darken the clearest serenity of his quiet." They were he says, like a toothache – very troublesome, though not fatal. But they were chronic, "a disease to his faith" (Boyle 1744: 12). Hunter has provided transcripts of Boyle's end of life confessions to Gilbert Burnet, Bishop of Salisbury, in which he refers to the "Impious or Blasphemous Suggestions or Injections," which Burnet reassured him were "mere Effects of Distempers of the Body or the Brain [...] Mechanical Effects" (Hunter 2000: 90). To be sure such doubts can be construed as tests of faith for a convinced Christian, or as temporary derangements. Yet in a true condition of doubt, the doubter must necessarily wonder whether he is only suffering from a distemper or being provoked by a demon, or whether his doubt has arisen from a veridical glimpse of things as they are – the *rerum natura*, for example, as described by Lucretius, in which the religious imaginary is a source of needless torment. When a clerical career was proposed to him at the age of thirty-three, he realized that he was not called by the Holy Ghost.[10]

In his tract on "The Christian Virtuoso," Boyle explains that an experimental philosopher is not "a libertine, though ingenious, a sensualist, though curious," or a mere empiric or a "vulgar chemist." His engagement with the material world leads his mind to higher things:

> "[T]he study of the creatures may justly produce in a *virtuoso*, 1st, a profound *admiration* of the majesty, and some of the attributes of God. Secondly, An external *celebration* of him, for them, by hymns and praises. Thirdly, a deep *humility* in the view of his immense greatness and majesty, and the distance of our nature, as we are human creatures, from his gratitude and trust follow" (Boyle 1744b: 481).

But Boyle was concerned to distinguish the physico-theology of the Christian from that of the pagan nature-worshipper. In his famous treatise on the human body, Galen combined detailed descriptions of the functions and uses of its parts with praise for the skill, wisdom, economy, and foresight of the designing intelligence behind them. Galen sometimes directed

---

8 Seneca insists in his *Questiones Naturales* than thunderstorms have no theological significance and are naturally caused.

9 "[T]he devil taking advantage of that deep raving melancholy, so sad a place, his humour and the strange stories and pictures he found there of Bruno, the father of that order, suggested such strange and hideous thoughts, and such distracting doubts of some of the fundamentals of Christianity, that, though his looks did little betray his thoughts, nothing but the forbiddenness of self-dispatch hindered his acting it" (Boyle 1744: 12).

10 (Boyle 1744a: 37).

his worshipful attitude to "the Creator," but more usually to "Nature," whom he apotheosized, as Boyle was fully aware, as a goddess.[11] This worship of divine or semi-divine "Natura," the "filia Dei, semper [...] actuosa," as Comenius calls her,[12] was easily amalgamated with the Venusian paratheology of Lucretius, who, without suggesting that the goddess Venus controls or constructs things out of atoms, nevertheless presents her as a principle of life and renewal (Lucretius 1992: I 1-9; 225-238). Boyle's purpose in writing his *Free Inquiry into the Received Notion of Nature* had been to show that such pagan nature-worship was to be disdained as idolatrous nonsense (Boyle 2000: vol. 14, 446-447), and the Christian Virtuoso makes this point in a positive way.

However, Boyle still needed to provide reasons for accepting the Christian message of the Fall of Man, the Redemption of the World through Christ's martyrdom, and the existence of Heaven and Hell. For reasons to be discussed below, Boyle could not endorse a religion that did not posit an afterlife or recognize what he called "intermeddlings" of the deity with the corporeal world. The revisions of Anaxagorean mechanism were of no help in this task. His ontology had to be carefully fabricated to accommodate these possibilities, and his epistemology enlarged beyond the plain experimental method. There were really three worlds, Boyle decided. Besides the corporeal world, and the present state of things, consisting of "objects proportioned to our unassisted sight," there was a "dioptrical world [...] which consists of all those creatures, that lay concealed, in former ages, from mortal eyes, and are not now discovered, without dioptrical glasses," and also a spiritual world of "good angels, and other intelligent beings, as devils, and separate human minds, that have either no bodies, or none that we can see" (Boyle 1744b: 502). How this world was to be accessed, and how far it could be accessed, without flying in the face of sense and reason – for Boyle disdained fideism – was a problem he addressed in many of his essays, with the results summarized in Section III below.

2. Pleasure and the Material World

Thomas Birch explained Boyle's choice of an occupation as motivated by his desire to divert the upper classes from their customary pursuits and to improve them and the nation:

> "He set himself to phylosophise, and to persuade the nobility and gentry of the nation, who had the means and leisure to pursue such sorts of studies, to follow his example. He was convinced, that it would be of inestimable use to mankind to engage them in these enquiries; it would divert them from those impertinent and criminal amusements, with which most of them busied themselves, and would make them not only better Christians, but likewise more useful members of society" (Birch 1744: 33).

Other studies such as Hooykaas's, have emphasized the physico-theological over the moral and socially altruistic motive (Hooykaas 1997). But the motives and incentives that led

---

11 (Galen 1968). See for example I: 274.

12 (Comenius 1657: 449). Under the heading "What is Nature?" Comenius lists 14 axioms about nature's ways of acting.

Boyle to adopt the experimental life were far more personal and circumstantial than his rationales. The experimental life was, first and foremost, a source of pleasure for him, relatively free of the guilt and distaste that attached to other sources of pleasure. Relatively free – because Boyle, always a moralist, experienced periods of self-doubt and ambivalence in which he was inclined to trivialize his beloved occupation.

The young Boyle was a tender youth with a rich and romantic imagination.[13] His account of his childhood is a surprisingly unguarded account of his experiences of being indulged by his elders, and of gustatory pleasures, and temptations. One of his first memories is of eating half a score of plums. Little Boyle was given balls and tops, fruits and sweetmeats, instructed in an "affable, kind and gentle way," and cajoled into love of study by Mr. Harrison. "He would often as it were cloy him with fruit and sweetmeats, and those little dainties, that age is greedy of [...] He would sometimes give him unasked play-days and oft bestow on him such balls, and tops, and other implements of idleness [...]" (Boyle 1744a: 8). But Boyle always understood the importance of moderation and self-control. He tells a number of stories in which, when Philaretus, the young lover of virtue, is faced with some delicious but wrong prospect, his inner self-restraint triumphs.

> "During his stay at Stalbridge all [one] summer, his father, to oblige him to be temperate, by freely giving him the opportunity to be otherwise, trusted him with the keys of all his gardens and orchards. And indeed Philaretus was very little given to greediness, either in fruits or sweetmeats; in the latter he was almost abstemious, and in the former, he was very moderate: so valuing such niceties and dainties, that although he enjoyed them with delight, he could want them without the least regret" (Boyle 1744a: 10).

His ascetic tendencies again came into play when the sweetmeats of adolescence dangled before him. "[T]hough his boiling youth did often very earnestly solicit to be employed in those culpable delights, that are usual in, and seem so proper for that season; and have repentance adjourned till old age; yet did its importunities ever meet with denials" (Boyle 1744a: 12). He visited brothels – and brothel imagery reappears in his later writings – but remained aloof.

"As a young man," John Clay remarks, "Boyle started avoiding things." He became more "choosy, fastidious, and pernickety" (Clay 1999: 288). As he approached maturity, he began to write clever, misanthropic satires on foppishness, gaming, flirtation, and the vain and foolish boasting of aristocrats.[14] The *Occasional Reflections* are concerned with pleasure and its regulation, and the sweetmeats theme again appears in an entry on "Sweetmeats offer'd him after a Banquet." As the surfeit of sweets nearly turns his stomach, he decides to "use them, that they shall be ever Noveltys to me; & so, tho' Sensualists enjoy more Pleasure (then I) yet Ile enjoy the Greater, & make amends for their Discontinuance with the Freshnes of my Delihts" (Boyle 1664b: 116). Boyle also expressed scruples against the vivi-

13 The pioneering research on this subject was that of (Oster 1993); but see also John T. Harwood's introduction to The Early Essays on Ethics of Robert Boyle, Carbondale IL, Southern Illinois University Press, 1991; and (Principe 1994). An entire issue of the *British Journal for the History of Science* 32 (1999) is devoted to the topic of Boyle and psychoanalysis.

14 (Boyle 2000: vol. 13, 141-145); see also (Hunter 1994: 88).

section of animals, though his practice was later at variance with his declarations (Oster 1989; MacIntosh 1996). His tracts against face-painting and décolletage – Oster refers in this connection to a "theatrical imagination that enacts a polarized view of virtue and vice"[15] – suggest that exaggerated femininity was extremely aversive (Oster 1993: 200 ff.), but the homosexual aggression that he experienced in an episode from his childhood was also recollected with a shudder. The common element in these pleadings is heightened sensitivity. Too much visual and emotional stimulation was experienced as deeply unpleasant, and Boyle even ascribed a delicacy and acuity of the senses to dogs and apes (MacIntosh 1996: 447) in a way that is seemingly unparalleled in 17th century philosophy.

Boyle was scholarly. His meditation on "Scaping into his Study out of a Crowd of extraordinarily vaine Company of both Sexes" to the works of Heraclitus, Democritus, Plato, Hermes, Zeno and other ancient writers hinted at a literary career, and indeed Boyle's career turned out to be largely literary. But he needed a via tertia between the "world" – tiresome visits and parties, love affairs, debauchery, idleness, hunting, gambling, swearing, and politics, all of which he despised, or which bored him – and theology, and the study of nature provided it. "The other humane studies I apply myself to," he reported of his time in Stalbridge in 1646-1647, in addition to ethics, "are natural philosophy, the mechanics, and husbandry, according to the principles of our new philosophical college, that values no knowledge, but as it hath a tendency to use" (Boyle 1744: vol. 1, 20). At the age of twenty-two, he set up a laboratory, and reported to his sister that he was now happy:

> "*Vulcan* has so transported and bewitched me, that as the delights I taste in it, make me fancy my laboratory a kind of Elysium, so, as if the threshold of it possessed the quality the poets ascribe to *Lethe* [...] before their entrance into those seats of bliss, and there forgot my Standish and my books" (Boyle 1744a: 27).

Boyle's involvement with Hartlib's group in London and with Wilkins' in Oxford offered a "scape" from ordinary social life and at the same time for conviviality with like-minded persons. He found the members of the Invisible College very impressive, describing them as

> "men of so capacious and searching spirits, that school-philosophy is but the lowest region of their knowledge; and yet, though ambitious to lead the way to any generous design, of so humble and teachable a genius, as they disdain not to be directed to the meanest, so he can but plead reason for his opinion; persons that endeavour to put narrow-mindedness out of their countenance, by the practice of so extensive a charity, that it reaches unto everything called man, and nothing less than an universall good-will can content it. And indeed they are so apprehensive of the want of good employment, that they take the whole body of mankind for their care" (Boyle 1646-1647: 20).

Malcolm Oster, the first historian to try to construct an account of the young Boyle's motives and incentives to natural philosophy has called attention to the importance of this social incentive for Boyle, participation in a friendly group of men. Though Brett Kahr suggests that "one might surmise that [Boyle] felt safe in the predominantly nonhuman envi-

15 (Oster 1993: 180). The texts are printed as "Letter to Lady Drury" Feb 12 1647 in HD 13:564 and in the following letter on breast-feeding.

ronment of the laboratory" (Kahr 1999: 280), the presence of helpful assistants made it a human environment of a co-operative and often pleasant kind. And the suggestion that scientific pursuits are a substitute for and equivalent to conventional amorous exploits had in any case been laid out very elegantly in Walter Charleton's dialogue on the Epicurean thesis of the mortality of the soul. "Athanasius," the author of an unpopular treatise in natural philosophy corresponding to Charleton's *Physiologia Epicuro-Gassendiana* recalls to the friend of his youth "Lucretius" (John Evelyn) "our ancient Caresses in the days of youth, innocence, and peace." "Since that day I first ventured abroad in to the World," he continues, "I have had no Mistress that held any considerable room in my thoughts but One, and that the very same I have observed you to court with the strongest desires imaginable [...] Her, upon whom women usually transfer the blame for all their imperfections, Nature." "Lucretius" agrees that, although conscious of the vast difference between himself and Nature and his inability to satisfy his desire in the knowledge of the least part of her, still

> "I discover such an infinite variety of fresh beauties and excellencies in her every day, that but to gaze upon them at a distance and view her in the weak and pale reflection made in the glasse of my own Reason, I find the most pleasant and ravishing employment."[16]

Boyle's early moral treatise, the *Aretology*, had not been well-received, dashing his hopes of becoming a major ethical writer. Chemistry enabled Boyle to brush off this disappointment and to forget about the dissonance between his Christian ideals and the behaviour of his social set.[17] It further offered pleasure in producing and experiencing, without the depressing side effects of vulgar hedonism.

Boyle's Epicurist leanings were already evident in his *Aretology* in which he opposed the Platonic doctrine that all pleasure is evil. The Garden of Eden, the original Paradise, he points out, was pleasant, and if pleasure were evil, it would have been no punishment to be thrown out of it. God has "seasoned and as it were sweetened all the Necessary actions of our life with certain pleasures."[18] His hedonism is evident in his *Experiments and Considerations Touching Colours* (1664), written, he says, "to divert and recreate, as well as to excite [...] by the delivery of matters of fact [...]" assisted by experiments that are easy and delightful (Boyle 1664a: 25). The colour essay is a recitation of beautiful experiences and productions, beginning with a description of the glorious colours appearing to persons in a fever or about to suffer a stroke, and lovingly going on to describe the whiteness of beaten eggs, the green of leaves and emeralds, the nap of cornfields and peafields, the colour of heated steel as it changes from yellow, to red, to blue, and the glass that when held at various angles alters from gold to a pale blue or turquoise. There are "experiments" with the juice of blackberries, cochineal and cherries, and the revival of the colour of dried rose leaves fixed with oil of sulphur.

---

16 (Charleton 1657: 3-4). In the *Excellency of Theology*, Boyle refers to natural philosophy as a "Mistris," and to the Virtuoso as "(if not a Passionate) an Assiduous Courter of Nature."

17 On this phase of his life, see (Maddison 1963); (Shapin 1993).

18 *Aretology* in (Harwood 1991: 30). Boyle explored the themes of sensibility, folie d'amour, and sex slavery in the remarkable letter he composed as a personal exercise to Joseph from Potiphar's Wife, and in the early suppressed version of *The Martyrdom of Theodora*.

The world of the chemist is a painted and scented world of graceful and protean forms, very unlike the sub-dioptrical world of atoms, possessed of bulk, figure, and motion alone, that characterized Boyle's official ontology. In his reflections "Upon the sight of N.N. making of Syrup of Violets," Boyle reflects that one might suppose N.N. a very great "Friend to *Epicurism*." His employment seems "wholly design'd to gratifie the senses," as the things "he deals with are Flowers and Sugar" (Boyle 1664b: 148). But regardless of the colour and scent of the one and the sweetness of the other, he is in fact, Boyle insists, making a useful remedy. Flowers, sugar, and sweet-meats – and more generally the colours and textures of things – changeable taffeties, velvet, the blue and golden necks of pigeons, rainbows, little birds and mice – these were the charming components of the experimental life, quite close to the fish, dogs, looking-glasses, clocks, and medicines that Boyle described in his journalistic musings on the trifles of everyday life, the *Occasional Reflections*.

Inevitably, however, guilt attached to these enjoyments. The *Occasional Reflections* often turn to pastoral and natural beauty, but typically in a spirit of self-correction. Describing the paring of an apple ("fresh and lively Vermillion [...] emulation at Rubies themselves [...] the Platonick definition which styles Beauty the *Lustre and Flower of Goodness* [...]"), Boyle reflects that the "gay outside" of the apple is "cut and thrown away" and urges in typical protestant spirit that we "strip and devest [things] of all those flattering Ornaments (or cheating Disguises) which so often conceal or misrepresent their true and genuine nature" (Boyle 1664b: 60-61) – as with the sweetmeats that turned out to be modeled in wax.[19] We must learn to look upon "the curiosest Productions of Nature [...] with a Philosopher's and a Christian's eyes" so that we can gaze on her "bright objects" with pleasure but also turn away our eyes without trouble and disquiet (Boyle 1664b: 60-61). Boyle's appreciation of the conflict between Christian morality and English life was replaced by a more threatening perception of a conflict between Christian theology and the experimental life of involvement not only with small, dirty, and low objects, but with beautiful and transitory ones. This problem had surfaced by 1665 when he composed his paradoxical essay on the *Superiority of Theology over Natural Philosophy*.

Criticism and satire were, meanwhile, severe with regard to the unchristian, frivolous, and self-indulgent character of experiment (Syfret 1950). John Smith, the Cambridge Platonist, maintained in his sermons of the 1640s, that those men who

> "spending themselves about Bodily and Material acts, and conversing only with Sensible things; [...] are apt to acquire such deep Stamps of Material phantasms to themselves, that they cannot imagine their own Being to be any other than Material and Divisible, though of a fine Aetheriall nature: which kind of conceit [...] hath too much prevalencie in Philosophers themselves, their minds not being fully abstracted while they have contemplated the highest Being of all" (Smith 1660: 65).

Thomas Browne's *Religio Medici* (1643) was written to defend doctors from the charge of atheism, but did not defeat the notion that persons who engage too much with the material

19 (Boyle 1664b: 169). Cf. Bach: "The nature of loathsome sins/Is indeed from outside very beautiful/ But you must/Afterwards with sorrow and frustration/Experience much hardship/ From outside it is gold/But if you want to look more closely/It is shown to be only an empty shadow/And whitewashed tomb [...]" Cantata BWV 54 "Widerstehe doch der Sünde."

world as experimentalists come to attach too much respect to the powers of material things. Kenelm Digby pounced on the book on its first appearance:

> "[It cannot] be expected that an excellent Physitian whose fancy is always fraught with the materiall drugs that hee prescribeth his *Apothecary* to compound his Medicines [...] and whose hands are inured to the cutting up, & eies to the inspection of anatomised bodies; should easily, and with successe, flye his thoughts at so towring a *Game,* as a pure intellect, a Separated and unbodyed Soule" (Digby 1643: A7).

Robert South, the prolific and irascible Canon of Christ Church, and Orator of the University of Oxford, in sermons of the 1660s and 1670s described the Royal Society as "sons of Epicurus, both for Voluptuousness and Irreligion."[20] Though Hobbes was in fact excluded from the Royal Society, South saw it as a hotbed of "Hobbism." He availed himself of the religious contrast between knowledge and love of God and the knowledge and love of World: "although there is vanity, a sorrow and dissatisfaction in the knowledge of created, inferior objects, yet we are assured it is life eternal to know God" (Syfret 1950b: 244). In 1669, Meric Casaubon wrote against experimental philosophy as "very destructive to true Religion and Christianity." "High curiosities" such as human flight, the possibility of which had been proposed by John Wilkins in 1638, may be dangerous and the mind is distracted from higher and nobler entities and mysteries.

> "Men that are very much fixed upon matter and secondary causes and sensual objects, if great care not be taken, may in time [...] and by degrees, forget that there be such things in the world as Spirits, substances really existing and of great power, though not visible and palpable by their nature, forget, I say, and consequently discredit supernatural operations: and at last, that there is a God and that their souls are immortal" (Causabon 1669: 30).

Man, Casaubon said, has a soul as well as a body, and we do not live by bread – the promotion of works and trades alone – but need those studies conducive to virtue and holiness. The reading of histories is "better than attending on furnaces or raking into the entrails of men or beasts to find somewhat which it may be will never make them much wiser when they know it nor prove of any great use."

Henry Stubbes's criticism of the Royal Society focused on Thomas Sprat's provocative remark that, where mastering, naming, and looking into the nature of all Creatures were concerned, "This had been the onely Religion if men had continued innocent in Paradise and had not wanted a Redemption" (Sprat 1667: 349). Sprat had insisted in the same breath that there was no just reason "why an *Experimenter* should be prone to deny the essence and properties of *God*, the universal Sovereignty of his *Dominion*, and his providence over the *Creation*" (Sprat 1667: 348). His involvement with material things, he said, does not lead the experimentalist to oppose invisible beings, but on the contrary "puts his thoughts into an excellent good capacity to believe them." Corpuscularianism expands his imagination and leads his thoughts from unseen atoms to unseen angels:

---

20 Quoted in (Syfret 1950b: 234).

> "In every work of nature that he handles, he knows that there is not only a gross substance, which presents itself to all mens eies; but an infinite subtilty of parts, which come not into the sharpest sense. So that what the Scripture relates of the Purity of God, of the Spirituality of his Nature, and that of Angels, and the Souls of men, cannot seem incredible to him, when he perceives the numberless particles that move in everymans Blood, and the prodigious streams that continually flow unseen from every Body" (Sprat 1667: 348).

Nevertheless, Stubbe twisted Sprat's meaning to imply that he thought that Adam's fall was a consequence of his deficiency of natural philosophy and his "not minding the cultivation of the Garden and of natural curiosities" (Stubbe 1670: 38-39). Experimentalism and libertinism remained associated in the minds of critics, enhanced by moral disgust with the behaviour of the King and the aristocracy.

All this put Boyle in a state of conflict, but also afforded an opportunity for him to express the moral-religious side of his personality in a defense of his preferred activities. The difficulty was in finding the language and concepts in which to do so. As Karl Figlio acutely remarks, the problem of the ultimate worth – the destructive potential and resource-consuming wastefulness – of experimental science is one we still face, but financial and institutional power and a developed rhetoric of science have strengthened its position. Boyle had to think out his answers for the first time – and virtually alone (Figlio 1999: 314).

One significant way in which Boyle attempted to resolve the problem presented by the personal and hedonistic incentive to natural philosophy was by insisting that his experimental practice was not accompanied by, but actually served to establish the truth of the Anaxagorean system that he presented in his programmatic works. He was not, he maintained, a mere empirick – even an empirick who happened to be a nobleman, but a philosopher attaining to knowledge of the transempirical world through – somehow – the producing of forms and qualities. The changes of colour and form in the glass had somehow shown him the reality underneath the coloured surface of the world, the latent image of the Creation, and through the vagaries of experiment he had somehow detected the institution of a system of laws of nature.

But was experimentation really worthwhile? In the sketch "Of Naturall Philosophie"[21] of 1649-1653 which rehearses the themes of the later *Usefulness of Natural Philosophy* (1671), Boyle cites as rationales for the pursuit of experimentation, the satisfaction of curiosity, the exciting of devotion, and the applications in physic, husbandry, navigation, and in the useful trades – the latter mainly consisting it seems of solvents and paint-additives (Boyle 2000: vol. 6, 472ff.). He reminds himself that the experimental philosopher must not scorn what seems trivial, or expect too much from experiments, or give way to frustration when they fail. When properly conducted, without excessive fondness and hope, this involvement with mistress Nature is legitimate and productive. But the utilitarian rationale for experimental philosophy was not entirely satisfying to Boyle. His essay on the *Excellency of Theology or the Pre-Eminence of the Study of Divinity above that of Natural Philosophy* written in 1665 is ostensibly addressed to a virtuoso inclined to atheism who, closely involved in experimental practice, sees no need for supplementing his studies with the texts of revealed theol-

21 The text is given in (Hunter 2000: 30-31).

ogy. It is tempting to read this essay not as a reflecting a debate Boyle is having with some unidentified Royal Society colleague, but as a debate he is having with himself. In any case, he acknowledges very strong reasons in favour of the opponent's position, yet comes to the conclusion that the experimentalist needs to and can take on more than a minimal religious commitment to the existence of a wise and provident creator. A remarkable feature of this tract is that every reason Boyle gave for practicing experimental philosophy some years earlier is subverted, and felicity is deemed an effect of theological studies, by contrast with experimental philosophy.

The first part of the treatise takes seriously the naturalist's tendency to skepticism and his lack of concern for Christian doctrine. The philosopher who values the study of nature and its canons of evidence will wonder, Boyle agrees, about the warrant for claims concerning supersensible beings and the events and promises recorded in Scripture. Was Christ's sacrifice necessary? Might not God might have remitted all the penalties of sin? Does the soul remain in suspension until judgement day or go immediately to heaven or hell? Will we know one another and see our mistresses, relatives, princes and subjects? Or be annihilated (Boyle 1671: 27-28)?

"Reason is not much to be trusted when she wanders far from Experiments." Boyle had remarked in "Of Naturall Philosophie" (Hunter 2000: 31). He shrewdly notes that Descartes was unconvinced by his own arguments for the immortality of the soul, those cited so favourably by John Smith, for Descartes himself wrote sarcastically to Elizabeth:

> "my knowledge of [eternal life] is far inferiour to that of Monsieur [he means Sir Kenelm] Digby. For, setting aside that which Religion teaches us of it, I confess, that by meer Natural Reason we may indeed make many conjectures to our own advantage, and have fair *Hopes*, but not any Assurance" (Boyle 1671: 24-25).

The Cartesian argument for the incorporeality of the soul, Boyle notes, is telling against Epicureans and atheists who consider the soul to be a modification of the body, but doesn't prove immortality.

> "Meer Reason," Boyle agrees, "cannot inform us what will become of [the soul] in her separate state, whether she will be vitally united to any other kind of Body or Vehicle; and if to some, of what kind that will be, and upon what terms the Union will be made. For possibly she may be united to an unorganiz'd or very imperfectly organiz'd body, wherein she cannot exercise the same Functions she did in her Humane Body."[22]

Just as the soul's incorporeality does not guarantee her immortality, Boyle recognizes, her immortality does not guarantee the possibility of happiness. What if the soul is later installed in a body that only feels pain and cannot move? How can such a soul experience felicity?

22 (Boyle 1671: 25). For Charleton, as for Stillingfleet, it was not the existence of God, but the immortality of the soul that was the "the grand Base of Religion", the keystone of its arch. "For if the Soul be mortal & subject to utter dissolution with the body: to what purpose doth all Piety and religion serve? What issue can we expect of all our prayers, of all our Adorations, of all our Self-denying acts of obedience, of all our unjust sufferings? Why should we worship God at all? Nay more, why should we consider whether there be a God or no?" (Charleton 1657: 59).

Philosophers from Aristotle to Descartes had maintained that even a disembodied, unsensing soul can be happy because she can always think and contemplate, and those are her most favoured activities. But, says Boyle, men would grow weary of thinking

> "if they received no supply of Objects from without, by Reading, Seeing, or Conversing; and if they also wanted the opportunity of executing their thoughts, by moving the Members of their Bodies, or of imparting them, either by Discoursing, or Writing of Books, or by making of Experiments" (Boyle 1671: 26).

He describes the unfortunate case of a nobleman who was imprisoned for a year in Spain with a diet fitting his station, but without books, light, or company. The prisoner could reminisce and contemplate without interruption but fell a victim to extreme melancholy: his chief impulse was to drink himself into a stupor and fall asleep (Boyle 1671: 26).

Happiness, as proponents of the mechanical philosophy implied or insisted, comes from engagement with material *things* that can be seen, tasted, handled, and from investigating these things and discoursing of them with others. The worry Boyle places in the mouth of the naturalist is a serious one:

> "Among thinking men, whose thoughts run much upon that future state, which they must shortly enter into, but shall never pass out of; there will frequently and naturally arise a distrust, which, though seldom owned, proves oftentimes disquieting enough. For such men are apt to question, how the future condition, which the gospel promises, can afford them so much happiness as it pretends to; since they shall in heaven but contemplate the works of God, and praise him and converse with him; all which they think may, though not immediately be done by men here below, without being happy."

So Boyle suggests that in heaven we must not only meet our relatives and lovers, and converse with God, the saints, and the angels, but also our friends and colleagues. The ordinary naturalist who subscribes to the Epicurean doctrine of universal destruction faces the loss of "the pleasure of his knowledge, by losing those Senses and that World, which are the Instruments and Object of it," as well as facing the threat of eternal misery, while the pious naturalist is not only "freed from the wracking Apprehensions of having his soul reduc'd to a state of Annihilation or cast into Hell [...]" (Boyle 1671: 51) but can look forward to even better working conditions: "Those things, that do here most excite our Desires, and quicken the Curiosity and Industry of our searches, will not onely there Continue, but be Improv'd to a far greater measure of Attractiveness and Influence" (Boyle 1671: 51). We will be free of those interests, and passions, and lusts, that here below impede us in these pursuits. In *Seraphic Love*, Boyle had argued that the Beatific Vision of God would not as was sometimes suspected, bore us with its monotony, but stimulate us with iridescence and sparkle, since it would offer "Variety in further knowledge of the fixed object such as arises from the fixt Beholding of the changing-necks of Doves, or such as we may see in the Diversifie'd Refractions of the same sparkling Diamond" (Boyle 1659: 129).

Part Two of the *Excellency*, however, seems infected by second thoughts. Rather than reinstalling the laboratory in heaven and lending an element of materiality to the beatific vision, Boyle seeks to persuade his imaginary naturalist that the laboratory is not so heavenly.

> "I shall, without Preamble," Boyle says, "begin this Discourse, by considering the Delightfulness of Physicks, as the main thing, that inveigles your Friend, and divers other Virtuosi, from relishing as they ought, and otherwise would, the pleasantness of Theological Discoveries" (Boyle 1674: 55).

The experimental life brings, indeed "sincerer Pleasures, than those the more undiscerning part of Mankind is so fond of," but it is still inferior to a life of devotion and theological study and Boyle now sets out to "weaken the Argument, that is drawn from delightfulness" by showing the inconveniencies of natural philosophy.

The utilitarian argument for experiment and the devotion to the common good that he had cited earlier are considerably discounted in this tract. "The boasted use of Natural Philosophy, by its advancing Trades and Physick," he says, "will still be to serve the Body, which is but the Lodging and Instrument of the Soul" (Boyle 1674: 62). Even if the chemist were to discover a stone capable of transmuting gold, this would not benefit mankind by much; the discovery of the Americas has enriched Spaniards but made many multitudes of men miserable for working in mines (Boyle 1674: 62). Prolongation of life is small in relation to eternity:

> "all the Remedies, and Reliefs, and Pleasures, and Accommodations that Philosophical Improvements can afford a man, will not keep him from the Grave, (which within a few days will make the body of the greatest *Virtuoso* as hideous and loathsome a Carcase, as that of any ordinary man [...])" (Boyle 1674: 64).

The practice is full of frustration, as the naturalist is dependent on "such a variety of mechanick people", as distillers, druggists, smiths turners, &c." He must seek out apparatus, wait upon tradesmen, and repair losses from his own pocket (Boyle 1674: 57). Natural philosophy is not a good way to get fame and reputation either. It is subject to fashion, "wanes and eclipses," vicissitudes and prejudices (Boyle 1674: 85). It is difficult to make new discoveries, hard to be right. The virtuoso is bothered by visits and letters and acquires a bad reputation if he puts people off. It is hard to write books (Boyle 1674: 82-83).

Nevertheless, the choice of the experimental life seems to solve the problem of an innocent and benevolent occupation for humans, diminishing the urgency of the Christian message to repent and reform. That message is more compelling when one is faced with a stark choice between gaming, wenching, and ruthless striving for power and devotional activities. Boyle might have gone down the road evidently traveled by Thomas Sprat, but he did not. Instead, in this context he downgrades the experimental life as useless and uncertain, setting aside both charitable and physico-theological motives, purely in order to correct the view that it is a good and innocent life, one that needs no supplementation through a study of what he will later characterize as the world of bodiless intelligent agents, and the secrets of the future state. The use of the word "mere" is instructive in this respect. Boyle refers repeatedly to "mere Reason" "mere Matter," and to the "merely Corporeal," to "mere Naturalists," and the Phaenomena of "meer Nature." Yet, this downgrading is at the same time an ennobling. In part by seeming to diminish nature and the activities of the naturalist and portraying corporeal nature as supplemented and patterned by spiritual agents and her investigator as necessarily led to appreciate the mysteries of theology, Boyle conferred dignity and worth on these objects and roles.

In the second part of the *Christian Virtuoso*, published in the last years of Boyle's life, the attempt to represent heaven as a superiour form of earth and heavenly intercourse as a superiour form of human sociability are abandoned. Years of trouble with his eyesight and perhaps some degree of social fatigue were perhaps as much responsible as increasing philosophical sophistication. In the future state, "the nature of things corporeal, may be very differing from those that obtain in the present worlds [sic]" and our science may be accordingly useless (Boyle 1744b: 521). God, he says, can give the soul a purely rational form of happiness "though it have no remembrance or knowledge of the world or any corporeal creature." The sweetness of life will have come to seem insipid to it and its former inquiries into nature worthwhile only insofar as they involved "laudable industry" (Boyle 1744b: 521).

3. Evidence and Belief

Boyle insisted that God cannot have made our salvation dependent on our acceptance through faith alone of some manifest irrationality. But he also insisted that we should not reject claims merely because they seem contrary to reason. He also claimed that Scripture contains matters of fact, known by experience and through testimony that carries authority (Boyle 1690: 308-309). This triad of beliefs was not easy to maintain consistently.

For the harmonization of standards of evidence in science and religion, Boyle employed various strategies including concession, ad hominem attacks, prudential reminders, and metaphysical and epistemological arguments. He argued that some doubts were excusable, since the doctrines had not been clearly delivered in Scripture or proved (Boyle 1675a: 284-285), and he charged that Epicurean atheists were biased by their sensuality, lust, and passion (Boyle 1690: 294). He insisted that God can change or neglect to uphold the laws of nature (Boyle 1675a: 251). He pointed out that the dictates of prudence mostly coincide with divine command, and that self-interest might encourage us to require a lower standard of proof when eternal life was at stake (Boyle 1675a: 282-285). He argued that competent testimony can compensate for apparent irrationality; many initially implausible things reported by travellers and experimenters have turned out since to be well grounded in fact (Boyle 1690: 309ff.). He informed readers that the experimental philosophy had revealed many surprising things concerning the magnet, and that instruments like the telescope had revealed much that was previously unknown (Boyle 1675a: 275ff.). He showed that despite powerful objections from Epicureans who believed that the atoms of soul and body dispersed after death, doctrines such as the resurrection were actually supported by observation of plants and animals and experiments involving the redintegration of chemical substances (Boyle 1675b: 305ff).

But since the realm of souls was an incorporeal and invisible world, the experimental method was somewhat difficult to apply to the investigation of the future state, its different corporeal order, and it's possibly nonsensing minds. Boyle tried to turn a minus into a plus: we do not need, he said "anatomical knives, or geometrical globes, or optical telescopes or microscopes, or [...] elaborate instruments" (HD 2000: vol. 8, 305f.) to explore theological questions. But what do we need? Boyle was not very clear on this subject. The experimental life implies the imposition of strict standards for distinguishing competent from the ubiquitous doubtful testimony surrounding it. Hunter rightly points to "the ingenuity Boyle displayed in devising and executing trials" (Hunter 1999: 263), but Boyle could not settle on a method for the investigation of incorporeals. As a kind of epistemological last resort he looked for evidence of angelic, demonic and ghostly activity in the world; if these beings were not to be captured and proved in the laboratory, at least their effects might be witnessed outside it.

One might suppose that Boyle's early religion was replaced over time by a more mechanical view of causation and signification, one in which chance might indeed play a role, and particular, as opposed to general providence could not be so easily inferred – or at least not so many instances of particular providence as he had noted in the course of one small person's early life. The dominant position in his *Free Inquiry* resembles the views of Malebranche and Leibniz. The universe is a perfectly working clock, and God does not need to intervene in the ordinary course of nature since he clearly discerns "what would happen in consequence of the Laws by Him establish'd, in all the possible Combinations of Them, and in all the Junctures of Circumstances wherein the Creatures concerned in them may be found" (Boyle 1686: 567). A secondary theme is however that God Himself causes "many Irregularities and Exorbitances" and extends favour and punishment to individuals (Boyle 1686: 518-529). Anyone of a certain age, he remarks sagely in the *Christian Virtuoso*, who has escaped the perils and misfortunes that befall so many, must realize his indebtedness to God "for particular and personal preservations" (Boyle 1744b).

Jan Wojcik has argued that we ought to read Boyle not as a "confident and single-minded advocate both of the corpuscularian philosophy and rational religion," but as a philosopher who expected to keep running up against the limits of knowledge.[23] Boyle's mature position was perhaps that more typically ascribed to Locke, and deriving from Gassendi via Charleton; the powers of human reason do not extend as far as might be thought, either in natural philosophy or in theology, and our best course of action is "nobly attempting to surmount Difficulties that are superable, and wisely submitting [...] to those that our not so" (Boyle 1675c: 159). We cannot fully understand our own minds, and the Book of Nature is, he said, like a Romance – a mystery novel whose "parts have such a connection and relation to one another, and the things we would discover are so darkly or incompletely knowable by those that precede them, that the mind is not satisfied till it come to the end of the Book" (Boyle 1674: 57) – though we cannot in fact do so. In the last pages of *The Christian Virtuoso,* he appears even to approach to the position of Immanuel Kant or even William James in his references to an "extrinsick Motive," meaningful in the present order of things, for belief. The excellence of religion, he says there, consists in the fact that, regardless of the truth-status of its claims, faith and religious contemplation and study make this life agreeable. If, (however implausibly) "all those *great and precious promises* [...] should prove illusory," if he can only set aside his doubts, the experimentalist will "sweeten all his hardships and sufferings in this life."[24]

23 (Wojcik 1997: 212). Cf. (Boyle 1744b: 475).
24 (Boyle 1744b: 511); see also (Boyle 1675a: 282).

4. Conclusion

The *Christian Virtuoso* followed the pattern of earlier tracts written specifically for the confutation of those given to "Prophane Discourses and Licentious Lives," the "bad men" amongst the Virtuosi, and as a literary scourge to those occupied by "Secular Affairs and Sensual Pleasures" (Boyle 1690: 284-285). Yet Boyle himself was passionately occupied with the secular and sensual things. He insisted that the appreciation of the complexity of natural machines and their beauty was a form of sincere worship, and his discovery of the manifold of qualities producible by his efforts enhanced the naturalist's admiration for God's power, wisdom, and benevolence, though true happiness did not require either sensation or memory – nor life as we know it. The *Christian Virtuoso* constructed by Boyle and presented to the world was at once a reproach to his more light-minded colleagues, the "erring Virtuosi," a figure of the experimental philosopher answering the charge that the experimental philosophy undermined society and morals, and an idealized version of Boyle himself, one free of the doubts and conflicts that are expressed elsewhere in his essays and autobiographical writings. His attempt to reconcile reason and religion, the experimental life with the devotional life, was protracted and intelligent, if not altogether consistent or convincing to the modern eye. He was moreover chronically unable to banish all dangerous ideas from his own mind.

As a young man, Boyle had worried about his tendency to give way to "roving" or "raving,"[25] and it is perhaps significant that his paper on the "Atomicall Philosophy", for all its seemingly innocuous theomechanical content, was amongst the papers Boyle ordered to be destroyed, along with all his papers concerning the romance of Theodora. Boyle wanted to assure us that he was not a friend to Epicureanism – nor to Epicurism. "Corpuscularian Principles," he never tired of insisting, "may not only be admitted *without* Epicurean Errors, but be employ'd *against* them" (Boyle 1675b: 297). Scholasticism, he assured his readers, was closer to pagan naturalism and to materialism than the atomical hypothesis of the moderns. But his own complex reactions to the possibility that Epicurean mortalism was true and to the question of the permissibility of Epicurist sensual enjoyment cannot be read off from his portrayal of the *Christian Virtuoso*, regardless of the degree to which this imaginary figure's beliefs and motives in their certitude, fixity, and purity represented an intellectual and personal ideal to him.

---

25 (Harwood 1991: xlviii ff.) Cf. ibid.: 192.

Friedrich Steinle[1]

Scientific Facts and Empirical Concepts: The Case of Electricity

A widespread image of science is founded upon a basic dichotomy: there are empirical facts, obtained by observation and experiment, on the one hand, and theories and explanations, obtained by reasoning, speculation and creativity, on the other. Whether scientific reasoning should take the inductive path, or the hypothetical-deductive approach, has long been a matter of debate, but the basic dichotomist picture has been left untouched. And there is the concomitant idea that theories may come and go, while facts form the stable, unshaken and ever growing foundation of science. This picture is often advanced in science education, for example in experimental physics courses and their textbooks. Additionally, the use of language reinforces that picture: we say that we "discover" facts (such as the cosmic microwave background, the variability of species), while we create, form and eventually discard theories (like the standard model of elementary particles, string theory, theories of developmental biology and so on). Something that has been "discovered" is attributed a different status than something that has been created and formed.

In recent Science Studies discussion and in more refined public discussions, this picture has been given a critical counterpart. For earlier critique, one could even go back as far as Kant. However, the first author to explicitly address the question of facts was the Polish bacteriologist and philosopher of science, Ludwik Fleck. Being himself a practitioner of experimental science, he gave his principal work the provocative title *Entstehung und Entwicklung einer wissenschaftlichen Tatsache – Genesis and Development of a Scientific Fact.* As indicated by the very title, his aim was to eliminate the rigidity of scientific facts, and ascribe them history ("origin") and flexibility ("development"). In so doing, he strongly emphasized the social aspect: facts are only facts insofar as they are acknowledged as such by a minor or major group, by a community or "Denkkollektiv". What defines a collective, and holds it together, is a common "Denkstil" – "style of thinking" – that exerts a sort of "Denkzwang" (constraint of thought). The latter is mostly interpreted as a sort of sociological enforcement.

Fleck's provocation was deliberate, at his time directed against both the Logical Empiricism of the Vienna Circle, and the too uncritical and unreflected scientific community of his time. But Fleck's book, published in German in 1935, fell under most unfavourable conditions in Nazi Germany and received no recognition during his lifetime. It was rediscovered only after the war, notably by Thomas Kuhn, but even then it would have to wait until the

---

1 Major parts of the work embedded in this paper were conducted with support of the Thyssen Foundation, Cologne, and the Max Planck Institute for the History of Science, Berlin. I am grateful for this support, and also for many discussions in the context of a working group on "Generating Experimental Knowledge", supported by the German-Israeli Foundation for Scientific Research and Development (GIF, grant no. G-767-217.4/2002). Many thanks to Wiebke Herr for stimulating and critical discussion.

late 1970s for an English translation to be made. It has only recently assumed a significant role in science studies.

While Fleck's analysis hits central points and remains pertinent to this day, I also think that it could and should be taken further. On the one hand, it has become clear that reading Fleck as a radical social constructivist or relativist is a mistake.[2] Parallel to that, much remains to be worked out from an epistemological point of view. As many cases in history of science show, scientific facts are often "discovered" not collectively, but individually. Even more, they often definitely transcend the collective "style of thinking," but nevertheless form the basis for strong knowledge claims, even those that become widely accepted. Precisely at the point of creation of novelty and of individual episteme, however, Fleck's analysis becomes somewhat unsharp and leaves us alone. His analytical concepts "Denkstil" and "Denkkollektiv" do no longer take a grip here. At the same time, however, I find the more classical epistemological approaches are not much sharper either. In this situation, I shall propose focussing on the classical notion of a concept, sharpening it somehow, and studying processes of concept formation in empirical research in order to see how they might eventually be turned into facts. My main thesis, then, is *in nuce* that the formation and solidification of concepts are most closely connected to the genesis of facts.

What I shall present in this paper, is not an elaborate scheme, but a snapshot of work in progress.In the first part, I shall present a specific historical case that should serve as an illustration of and provide a material base for the more general discussion in later sections.

Charles Dufay and the two electricities

My historical case deals with electricity.[3] One of the most basic notions we associate with electricity is its bipolarity. Electricity comes either as plus or minus – this notion, or even "fact," is so basic for us that it is usually given no extra emphasis or explication, but merely used as part of the every-day language of dealing with and teaching electricity. Indeed, it is often taken to be a defining feature. Plus and minus, positive and negative, are fundamental categories when we load a new battery in our camera or when we need our neighbour's help to start our old car after a frosty night. We use and remember the concept, and we need it for dealing with electricity day-in and day-out, regardless of whether we know anything about electrical theory or not. This has long been the case, at least since the mid-eighteenth century (when electricity meant what we now call static electricity).[4] Twenty years earlier, however, in 1730, the field was totally different. There was no thought whatsoever of two electricities – let alone a well-defined notion. The concept of two electricities was not proposed until 1734, but, within slightly more than a decade, had led to a fundamental shift on the conceptual level. It is this shift that I shall focus on in this paper.

The main actor is Charles Dufay, active in the early 18th century, a brilliant academician and director of the Paris botanical gardens, who was well-versed in such different fields as botany and astronomy, dyestuffs and fire pumps, fluid dynamics and magnetism. The *Jardin*

---

2 See Zittel, this volume.

3 I have outlined the historical case already in two previous papers (Steinle 2004b) and (Steinle 2006).

4 (Heilbron 1979) provides the best overview of electricity of that period.

Royal offered ample resources for experimental research. In a brilliant experimental series on shining minerals (Bolognese stones, as they were called), Dufay had shown his interest in treating research fields in which even basic classifications were missing, and it was partly this interest that made him take up research in electricity. In the 1730s, the field was in an unstable and incoherent state.[5] More than a hundred years of research throughout Europe had produced a multitude of different and puzzling phenomena, such as:

- Some materials could be electrified by rubbing, others sometimes, others not at all
- Sometimes electricity acted as attraction, sometimes as repulsion.
- Sometimes sudden switches in attractive and repulsive effects were observed.

Dealing with these questions proved difficult, even more so since the experiments were delicate, the effects minute and reproducibility difficult. Dufay conducted extensive experiments, varying the procedure in many ways: he used a great variety of materials, individually or combined with others, and varied their shape, temperature, colour, moisture, air pressure and the experimental setting: two bodies touching, in close proximity, separated by large distance, being connected by a third and so on. His work led to remarkable results and bold claims, such as that *all* materials except metals could be electrified by rubbing and *all* bodies except a flame could be given an electrical charge.

But he was still left with serious questions as to when exactly electrical attraction and repulsion occurred, and when an object sometimes suddenly switched charges. He devoted extensive work to this point. His first step was an attempt to clarify whether or not repulsion existed – Guericke's report of the effect had been contested by some. By varying experimental conditions, he analyzed the conditions under which repulsion occurred and succeeded in demonstrating that the effect existed without question. In a particularly delicate arrangement, Dufay was able to keep a very light leaf of gold hovering at a distance of more than one foot over an electrified glass tube for several minutes, even when he moved through the room with the tube – quite similar to the model of the much-admired Guericke. When the force of the tube weakened, the leaf would sink and finally land on the floor.

Inevitably, Dufay now was confronted with the question of when exactly repulsion occurred in contrast to the usual attraction; after all, attraction was considered the defining feature of the electric virtue! Again, he approached the question experimentally, by varying many parameters of the set-up:

- the manner of electrification (rubbing, contact)
- the degree of electrification
- the size of the electrified body
- the material of the electrified body
- the basis/support of the electrified body (more or less idio-electric material)
- the constellation between the bodies involved: "larger or smaller distance."

5 (Heilbron 1979: chapter 9) gives a brief account of Dufay's research. Dufay's eight Mémoires at the Paris Académie between 1733 and 1737 provide the main source, in particular the fourth one: (Dufay 1733); Cf. also Dufay's own English summary (Dufay 1734).

For a long time, the results were merely puzzling, appeared unstable and did not combine to form anything resembling a regularity or a correlation. Dufay could not formulate the conditions for when attraction and repulsion occurred, let alone when, as sometimes observed, a sudden switch between the two actions took place. This latter effect, however, was ultimately the first to be conceptualized in a rule. For specific constellations Dufay was able to formulate regularities:

- Electrified bodies always *attracted* small unelectrified bodies.
- However, when these small bodies were placed close to the electrified one and became themselves electrified by contact, the previous attraction turned into *repulsion.*

Dufay found this regularity was valid without exception and with it he could account for many of the effects he had obtained. Nevertheless, it was restricted to the interaction of only two bodies when one had been electrified by contact with the other. In other cases, the regularity was not valid and the outcomes still appeared irregular. Dufay did not stop. He found a crucial hint in an experiment with the hovering gold leaf. When he approached this leaf with a third electrified body, strange effects resulted. If this third body was composed of glass, it repelled the leaf strongly, as the tube did. But when it was copal, the gold leaf was strongly attracted! Though this result was most puzzling,[6] it strongly suggested that the material used was crucial here.

In determinedly pursuing that trace, Dufay became increasingly aware of regular behaviour that could not, however, be formulated in accordance with the usual theories of electricity. Sensitized by his former studies, particularly in botany and luminescence, he was more aware than other researchers of how much such work depended on working with appropriate concepts and categories. He was also more prepared to question and revise even the fundamental categories of the field the his colleagues, which he did in a radical way: in order to express the regularities, he proposed that, rather than speaking of electricity in general, one should more accurately speak of two electricities. The regularity then was that electrified bodies repelled all those that had the same electricity, but attracted all those that had the other. As the experiments showed, the electricities maintained their character even when communicated to other bodies. Thus the above regularity of repulsion turned out to be just a special case of the now general regularity. According to Dufay, which of the two electricities a body received by rubbing depended only on its material. Thus the dichotomy of electricities induced the division of all materials into two classes. The electricities could also be characterized according to these classes – Dufay named them vitreous and resinous electricities, due to particularly prominent representatives of the two classes of materials.

We have no explicit notes about what exactly convinced him of the strength of these new concepts. Presumably their immediate "success" was central here: Dufay emphasized that, with these concepts and the regularities they allowed him to formulate, he could not only understand his own quite numerous experiments, but also those reported by others. The concepts enabled him to do exactly what had previously been impossible: to order the whole field in such a way that stable regularities could be formulated. This was a fundamental achievement indeed, and Dufay was well aware of that. He immediately announced his find-

---

6 "[...] me déconcerta prodigieusement" (Dufay 1733: 464).

ings to the Paris academy and, even more, he wrote a direct letter to the Royal Society in London to make his findings known there – a communication behaviour that contrasted strikingly from his previous habits.

Concepts and theories

Turning from the historical case to more general observations, I want to first note that Dufay pursued a type of experimental activity that fits neither the standard view of experimentation nor the naïve inductivistic picture. His work was neither a test of predictions, hypotheses or even theories, nor was it a process of innocently collecting data and then making generalizations about it. What we see instead is a specific type of experiment that I have labelled "exploratory" and discussed in detail elsewhere (Steinle 1997; Steinle 2003; Steinle 2005). For my purpose here, it is important to point out that in exploratory experiments the systematic variation of experimental parameters is in the forefront, with the goal being to formulate regularities and laws, and that, in order to do so, sometimes even the fundamental concepts in the domain of research are questioned and revised.

Indeed, Dufay addressed a specific level of knowledge. He was perfectly acquainted with the various theories of his time that attempted to explain electric effects by air currents, by humorous components of matter, by specific electric effluvia, by a vacuum between electric bodies and so on. However, those theories did not appear in his research. His aim was not to consider the hidden causes of electric effects, but to order them on a phenomenological level, to establish regularities and to discover the appropriate concepts that would allow him to do so. He did not altogether discard the search for hidden causes as inappropriate: rather, he regarded it as premature as long as the other issues had not been resolved. Searching for causes without having previously discovered a large number of phenomena, and having ordered them according to a "few basic principles,"[7] would mean "attempting the impossible," he stated explicitly. From what he actually did, I interpret this as the task of establishing regularities. And in that context, there was no space for theories about hidden processes. The search for correlations and regular dependencies dominated his enterprise, and what he eventually achieved was simultaneously both a new conceptualization and a law. This was definitely not an explanatory theory in the same sense as existing theories were.

What is crucial here is the distinction between concepts and theories. Indeed, the non-distinction of these two has often led to confusion. A particularly sharp and still instructive example was provided by Popper, when, in his reaction to logical positivism, he emphasized that (protocol-)sentences like "here red now" presuppose already a theory of space, of colour and of time.[8] This would mean that everything we think is already theory. There is no space left for a specific role of concepts: we either think in theories or we do not think at all! Such a lack of differentiation is not particularly helpful, neither for understanding science nor everyday life. It would be impossible, for example, to differentiate between our concept of red and the theory of colour. However, once I teach my five-year old boy how to pass a crossroads with traffic lights, I trust (and with good reason) he has a concept of red and

7 (Dufay 1733: 476) my translation.
8 (Popper 1934: appendix to § 30 in the 1968 edition).

green, while at the same time I am pretty sure he has no theory of these colours. Popper certainly made a crucial point in emphasizing that there is no thinking without presupposition, and this was probably the intended thrust of his statement. But equating every such presupposition with theory deprives us of further possibilities of differentiating and hence understanding. Much of the general discussion about theory-ladenness suffers from this point.

How to differentiate between concept and theory has not received too much attention so far, even in recent approaches towards concept formation (often shaped by cognitive science).[9] It is certainly not always possible to draw a sharp distinction, but some essential points are clearly visible. Despite ongoing debates about the exact notion of concepts (Smith, Medin 1981; Prinz 2002; Gurova 2003), there is a shared basic understanding that concepts are *elements* of our thoughts, rather than extended *systems* as theories are. As elements of thought, concepts are indispensable for expressing anything, be it an everyday remark, an individual observation or experimental result, a law, a model, or a theory. Concepts are also foundational for all conscious and intentional acting, whereas theories are not. Moreover, while theories typically do have an explanatory task, concepts are not explanatory as such – nevertheless they are (necessarily) used in explanations. As a sort of corollary to those features, one might add that concepts are usually just used and taken for granted but not discussed or reflected upon. They are generally implicit, whereas theories are normally explicit. Of course, it is exactly in periods of concept formation that concepts are explicitly discussed – but these processes, I claim, have a different character from those of discussions of theories. Finally, and highly important, concepts and categories cannot be right or wrong, true or false, but rather appropriate or inappropriate, useful or useless. They cannot be proved, or confirmed or disconfirmed, but they have to prove themselves against certain goals. What exactly those goals are, what exactly "appropriate" means, is then a matter of the specific historical situation. But the point regarding specific goals is crucial for all attempts at understanding processes of concept formation.

Lining up these differences between concepts and theories, even in such an admittedly provisional manner, helps to understand historical developments, at least in the specific case I have sketched. Indeed, the Dufay case is highly illustrative for all the features mentioned. Before Dufay, electrical effects were conceived in terms of an electric virtue, defined by attracting small bodies nearby. This basic concept was constitutive of the background of the language in which electrical effects were described and handled, recipes given, laws formulated and theories conceived – it formed the basis for handling all electrical apparatuses and machinery. With Dufay's new concepts, the very language changed and took on new elements. In handling electrical effects, in designing, conducting and evaluating experiments, new considerations came to the fore. New questions were required and raised, questions that previously had simply been inconceivable because the very concepts they were based on did not exist. All these points are indicative of an epistemic level that is different from explanatory theories and addresses the more fundamental issues of language. At least for Dufay, and for everyone who then adopted the new conceptualization, the field of electricity appeared to have been transformed. Whether or not one calls this transformation a "Gestalt-switch" in Kuhn's sense, it is clear that it has an essential connection to the concept level, to the language on which all handling and thinking of electricity was based.

9 Among others (Giere 1988); (Nersessian 2000).

Forming a concept

In order to get a more detailed understanding of the process of concept formation, I shall now analyze in some detail what led Dufay to form and propose a new concept. Dufay's research was driven by the goal of formulating a law of attraction and repulsion. To do so, his experimental activity focused on varying many parameters of the experimental arrangement, one by one, while leaving all others unchanged. The goal was to analyze the impact of a single change on the effect, in this case attraction and repulsion. Such an experimental procedure was systematically described and characterized much later, by John Stuart Mill, as the systematic search for causal relations employing four "experimental methods," and Dufay's procedure fits fairly well with such an understanding.[10]

What is striking to observe then is that Dufay's various attempts at establishing such correlations failed again and again: the ever-increasing range of experimental results on electrical attraction and repulsion did not coalesce to form a regularity. He could not identify experimental parameters whose variation consistently led to a regular change of the experimental result. This was indeed a puzzling situation, and exactly the type of situation that Mill could *not* have envisaged or analysed: the method did not succeed. Situations like these do occur time and again in science and, as history tells us, there are various ways in which researchers can react to such a situation. One is to abandon the question, i.e. leaving it at a 'dead end', and turning to other questions. Another one would be to renounce the belief that such a thing as a law exists, as stable correlations for the specific domain, i.e. to admit that the domain in question is profoundly irregular. This option is hard to accept for experimental researchers. A third option is sometimes taken: considering the possibility that the basic concepts by which experimental outcomes are expressed, and in terms of which the sought-after regularities are formulated, are not quite appropriate.

This last is the option Dufay chose. He was ready to question the very notion of electricity as a (uniform) virtue, a notion that had guided all research at least since William Gilbert had identified and named electricity as an research field in its own right some 130 years earlier. In sum, it is obvious that it was a very specific epistemic situation in which Dufay found himself, and in which he was ready to question fundamental concepts and formulate new ones. His goal to establish laws and his repeated failure to do so formed the essential elements of the epistemic situation.

While this situation constituted the background and motivation for the process of conceiving new concepts, it certainly did not determine its outcome. Even when such openness existed at the conceptual level, how exactly to proceed, what resources to look toward, how to identify promising candidates for new concepts, how to revise them in order to make them fit might all still be open questions. As to resources, it is obvious that no concept comes from out of the blue. At any point in time, we are dealing with a great variety of concepts from other scientific fields, from other cultural resources (such as art, religion or tech-

---

10 It should be noted that Mill's understanding of causality was not an essentialist one, where causes are sought in deeper lying processes behind the level of phenomena. Causal relations were rather seen as a set of network of conditions, each of which can experimentally been realized and changed – such an understanding of causality comes close to what has been elaborated as the concept of INUS conditions (J. L. Mackie).

nology) and from everyday life. In considering new concepts, analogies and the appropriation of what the researcher finds around him will play an important role.

Of course, there is no guarantee that any of these resources will be suitable. The central question is what does it mean to be suitable in this context? In other words, what criteria must new concepts meet in order to be regarded as appropriate? Here the goal of the whole enterprise finally comes into view. For Dufay, the goal was to formulate stable correlations and laws, and this provided a criterion that any new concept had to meet: it had to make possible what existing concepts did not – the formulation of a law. This criterion limited freedom of choice considerably. Moreover, it was by no means clear that any of the existing concepts, never mind the field, would fit – this would broaden the perspective to form entirely new ones or to at least modify existing ones profoundly.

We do not know in detail what Dufay's considerations were at this stage of his research (and this might be a typical situation for the historian: in such a state of conceptual uncertainty, where words are lacking, researchers would typically not take many notes). At some point, however, he conceived the idea that there might be more than one electricity. This certainly did not constitute a small step; indeed, it formed a profound break with the tradition in which electricity had always been conceived of as having a uniform nature. Dufay does not appear to have arrived at such an idea by taking a straight route . Analogies can of course be imagined: from magnetism, for example, with its long-accepted bipolarity,[11] or from colour theory, with a multitude of colours, or from shining stones (called Bolognese stones) with a variety of types of luminescence (as we would call it) – all these were domains in which Dufay had previously been active. One could even go farther and take the attraction (and eventually repulsion) of the two sexes as something that might have inspired Dufay's approach. The lack of sources here leaves us with only speculation, but in any case it is obvious that framing such an idea is a creative process, involving decisions, a process of which we can easily surmise that it could also have gone in a different direction.

Once such an approach was taken, however, the epistemic situation changed, and the problem became more specific. Much less freedom of choice was now left. The question at this point became how far the multiplication should be driven. Dufay himself states that he had conceived of the possibility of three electricities (Dufay 1733: 469). That he eventually ended with only two was due to both his goal – that the law should be as general as possible – and to the requirement that his very numerous experimental results could, without exception, be explained by the proposed law, formulated in terms of multiple electricities. The concept of two electricities fulfilled these criteria, and hence Dufay adopted it.

As this case illustrates, processes in the formation of empirical concepts are shaped by heterogeneous aspects – those of creativity, of choice and decisions taken[12] – but also those of empirical input which are not free to be disposed of but have to be taken as is. The process of concept formation is only partly determined by experimental outcomes; it is also a process that involves choice, a process in which the scientific and even the everyday culture

11 As we know, this again was by no means just a "natural" concept, but developed in 13th century by Peregrinus with a strong analogy to astronomical concepts.

12 Speaking of choice might even make the historical situation appear more open than it was: for someone in that particular period, there might have been less openness to revision and choice than it appears for us.

visibly and irrevocably come in.[13] Processes of concept formation provide nice cases in which neither a traditional empiricist nor a radical constructionist picture can really take hold. Perhaps this has something to do with the specific epistemic level at stake here, the level of concepts and language.

Such an account may give us an idea how the process of forming and accepting his new concept might have gone for Dufay himself. However, asking how new concepts come into being is only one of the issues. The other major issue is how new concepts find their way from the individual researcher to the group (not to mention the community) of researchers that deal with the area in question. It is by no means clear that a fundamental modification like the one Dufay proposed will find any resonance with other researchers, which takes us to our next subject.

From concept to fact

Viewing Dufay's achievement essentially as the formation of a new concept contrasts with how it is usually presented. For the most part Dufay is said to have "discovered" the two electricities, or, to be more precise, the *fact* of two electricities or bipolarity. All histories of electricity speak of "discovery": those written by Dufay's near contemporaries and those who have the benefit of a modern, reflective distance.[14]

But this talk of "discovering" the two electricities is ambiguous – it is true and misleading at the same time. What exactly did Dufay discover? Certainly not something like a previously unknown island, a new species, a new physical process or chemical reaction. What he had done was to tentatively form and eventually propose a new concept. His goal or, perhaps better, the purpose which the concept had to serve was to grasp the ever-growing field of electrical experiments in a few regularities or laws. And what he might really be said to have "discovered" is that the concept of two electricities allowed him to do just this: formulate a law. But this is quite different from what we usually understand by "discovering." If we speak of the "discovery" of the fact of two electricities, both the notion of discovery and the notion of facts are far from the straightforward notions we usually have in mind. What plays a decisive role is rather the formation and stabilization of empirical concepts.

What we see in Dufay, and what is highlighted by the common talk of his "discovery," is a process that can be characterized precisely as a progression from scientific concepts to scientific facts. The two central elements are concept formation and concept stabilization or solidification. As to the relation of concepts and facts, it is somehow trivial to say that there are no facts without concepts: in order to conceive and to formulate a fact, we need concepts. However, we can get more specific and say: we cannot express facts without *solidified* concepts. Many concepts are conceivable and indeed conceived, but only some make it to the stage of solidification and can express facts. Hence we have to deal with the question, how does this process work?

---

13 This result resembles a result that Stephanides and Pickering have obtained in analyzing Hamilton's formation of the new mathematical concept of quaternions, and what made them differentiate between "free moves" and "forced moves" in such processes: (Pickering, Stephanides 1992). What we have in Dufay's case, however, is the formation of an empirical concept, i.e. a concept that aims to grasp the empirical world.
14 See, e.g. (Fontenelle 1733); (Priestley 1767: 48) or (Heilbron 1979).

Again it is instructive to look at the historical case. How did the other "electricians" of Dufay's time react to his proposal? Here, the first and most striking historical observation is the nearly complete absence of an explicit discussion of the topic.[15] This is all the more striking, as the proposal did not disappear, quite the contrary: it soon showed up in a very specific manner. Just five years after Dufay's publication, the Leyden professor Pieter van Musschenbroek, writer of an important physics textbook, presented his proposal as unproblematic: "Experience has taught us that electrical power is of a two-fold manner [...]", he told the reader.[16] Note that he did not say "*Dufay* has put forward this claim," but "*Experience* has taught us". This was certainly not meant to diminish Dufay's effort and achievement, but rather served as a significant indication of the specific level of knowledge that his proposal had been granted here: the two electricities were presented as undisputable, as a "fact" of nature, as we would say – and indeed, this quickly became the common view.[17] Another five years later, and in the first textbook ever exclusively devoted to electricity, the Petersburg academician J. Gabriel Doppelmayr went even further and took the twofold nature of electricity as a part of its definition (Doppelmayr 1744: 1). And this is characteristic for the general attitude. In late 18th century textbooks, the twofold nature of electricity would often be introduced as a defining feature, derived from experience, and only later in the text wouldthe specific laws of attraction and repulsion be presented. The concept had become detached from the context in which it had been framed, and presented as an ontological feature, as a fact that was no longer subject to possible dispute.

Such a reception, or more precisely such a pathway of 'filtering in', is strikingly different from the way in which proper theories are received, both due to the absence of explicit discussion and the quick introduction of the concept not as something to be problematised, but as a fact, taught by experience. The peculiar way in which the two electricities found their way to acceptance is owed precisely to the specific epistemic level at stake here: the level of concepts and language. It is instructive that the above-mentioned textbooks, typically divided into a "historical", supposedly purely descriptive, part and a "philosophical" part, dealing with (explanatory) theories, presented Dufay's two electricities in the formerrather than the latter part (Ibid, Frercks 2004).

As mentioned above, concepts (in contrast to theories), cannot be not said to be right or wrong, true of false, but rather appropriate or not. And appropriate always means: appropriate for something – the aim or purpose involved comes into the focus here. As we saw, for Dufay himself the goal was to formulate a law of electrical attraction and repulsion, and here he found the new concept highly appropriate. But solidification for one person does not necessarily lead to an accepted fact. What were the goals of his contemporaries in respect to which the new concepts were such a great success that they were almost immediately adopted?

---

15 This observation holds for the German speaking countries. I am not yet sure about other European countries; a project concerning English and French researchers is currently being conducted at Wuppertal.

16 "Die Erfahrung hat uns gelehrt, dass die electrische Kraft von einer zwiefachen Art sey [...]": (Musschenbroek 1739: 242), in the German translation (Musschenbroek, Gottscheden 1747).

17 (Winkler 1745: 5); (Musschenbroek, Gottscheden 1747: 242); (Gralath 1747: 208). The latter work is a huge and comprehensive (natural and chronological) "History of Electricity," which most later historians of electricity, from Priestley to our time, take as a source.

Certainly Dufay's law of electrical attraction and repulsion was a nice point for everyone dealing with electricity. At the same time, however, Dufay's emphasis on searching for laws was not universally shared: to many the new law might not have appeared as central as it was for Dufay. Others were more concerned with theories proper on the one hand, or with practical machinery and performance on the other. What did the new concepts do in these two areas? As regards theories, the new concepts certainly were not particularly attractive, they did not help sort out the great variety of existing explanatory schemes in any way. It is not likely that electricians adopted the new concepts for theoretical reasons, and indeed I have not been able to find any remarks in that direction.

What seems to have been much more important, in contrast, was success on a practical level. The new concepts allowed for an unprecedented stability in dealing practically with electrical effects. In 1744, Wittenberg professor Georg M. Bose redesigned the older electrical machine in a way that enabled much higher reliability of electrical experiments than before. Before becoming a professor, Bose had made a living by giving public lectures on experimental findings, and hence the goal of high reliability for experiments in constantly changing environments and conditions had been central for him – probably it provided his motive for working on the electric machine. While we do not yet know many details of what exactly motivated his various improvements to the machine, it is indicative that he praised Dufay highly and explicitly made reference to his two electricities (Bose 1744). These new concepts might have played a central role in his design of the new machine, for example, in such core issues as choice of materials and preparation techniques.

As is well known, Bose's invention provided the major impetus for the explosion-like spread of public performances of electricity across Europe within a short period. Beginning in the mid 1740s, electricity became ubiquitous and reached a completely new status, both as a respectable academic field and a source of public entertainment. The news of the Leyden jar shocked and shook electricians throughout Europe and marked another starting event for what is sometimes called the 'golden age' of electricity, with strong electric machines, spectacular displays at public events and the first commercial uses of electric effects. It was in this context that the discourse of two electricities became unproblematic. Students of electricity took it from their textbooks; users of electrical machinery learned about it in the workshop where they bought their instruments. My guess is that it was just this successful use of the new concept that both enabled the unprecedented stability of electric apparatuses and experimentation, and at the same time contributed essentially to the solidification of the concept, to such a degree that it came to be seen as simply expressing one of most basic facts of the domain.

Expanding the perspective

What we see in this historical case is how a new concept was created in a specific context, by an author with specific, if not idiosyncratic interests, and in the course of broad and systematic experimental exploration. We see then how this concept was stabilized and found, in a rather short period of time, its way into the very language of the research field. It has become a solid and unquestioned element of all thinking regarding electricity, expressing a basic fact.

This brings us back to Fleck. If we reread him while keeping in mind concept formation and concept solidification, things fall right into place. What he describes, for example in the case of infectious disease, is a highly instructive case of concept formation and solidification. He shows the very fragility of the process and its specific nature that inseparably combines elements of decision and enforcement. While he focuses primarily on cultural settings and social dimensions, he pays less attention to the analysis of the various epistemic levels involved. Rereading the case with special attention those levels brings the central role of concepts markedly to the fore. And the role of purpose, of the actors' goals, comes much clearer into view than we find it in Fleck.

Moreover, once alerted, one finds such processes again and again, in very different fields of research, from chemistry to physiology to physics to genetics, and in very different periods, from the early modern period up to present day. Just to mention a few cases from the history of electricity, one might think of the concept of a current circuit covering the battery and its external wire at the same time and treating these two with the same language, a concept that Ampère framed to formulate a law of electromagnetic action (Steinle 2000). Or one may think of the concept of magnetic and electric lines of force that Faraday framed over the course of 20 years of intense experimental research and that finally allowed him to express a vast realm of experimental results with very few laws (Steinle 1996; Steinle 2004a). In mechanics, Newton's concept of mechanical force comes to the mind.[18] In quite a different field, the concept of chemical reaction was shaped and developed in the seventeenth century on the basis of broad experimentation, not by any single prominent individual researcher, but rather within a community with a tight communication structure (Klein 1994; Klein 1996). A further example, quite visible these days, is the concept of a gene that, despite its admitted unclarity, plays an absolutely central role in modern research.[19] In many of these cases, the concepts became solidified parts of the language through which facts are expressed, but some retained a more precarious status, like the concept of mechanical force that both Hertz and Mach heavily criticized as being anthropomorphic – Hertz even proposed a mechanics that would completely do without this concept. In any case, the fundamental character and far-reaching impact of these concepts is obvious: to a greater or lesser degree all of them have become essential elements of the language of those research fields and express basic facts.

How deeply concepts like these have become part of our everyday life might be illustrated by looking at newspapers, when finding reports about AIDS, with notions like virus, populations, risk, or about international train traffic (different electric systems, voltage, frequency, polarity), or climate change (radiation balance, average temperature, atmospheric layers, ozone hole) or even economic change (gross national product, productivity, market cycles, buying power) – the list could go on and on. In trying to understand our world, we use scientific concepts that are not themselves discussed. There is no way it could be otherwise: the high specialization of modern science and technology requires specialized concepts. But it easily escapes notice that all those concepts were at one time created and formed in the highly specific context of scientific research and from there found their way into the scientific community and eventually into common language. At the site of their ori-

18 Among many others (Westfall 1971); (Steinle 1991).
19 See, for example (Beurton, Falk, Rheinberger 2000).

gin, they were open, tentative, unstable and flexible, while later on they appeared solidified, stable, as "natural" categories or in some cases simply as "facts".

Facts have a history. Not only the simple one that at some point in time they became known, became "discovered." Rather one has the impression that if they had been "discovered" in different settings, at different times and places, they might have taken different forms. Their history is inextricably connected to the processes of forming and stabilising concepts of the era of their "discovery."

Lorraine Daston

The Humboldtian Gaze

Introduction: Local Nature, Global Vision

In 1807 the Prussian naturalist and explorer Alexander von Humboldt published *Ansichten der Natur*, a book full of fertile contradictions. It was a popular book with detailed scientific footnotes; it combined a personal travelogue with grand generalizations about the whole of nature and human history; it juxtaposed painstaking measurements to freehand landscape drawings, both made by the author himself. Above all, *Ansichten der Natur*, true to its title, created a whole new way of seeing nature, simultaneously local and global:

> "Wer demnach die Natur mit Einem Blicke zu umfassen, und von Local-Phänomenen zu abstrahiren weiß, der sieht, wie mit Zunahme der belebenden Wärme, von den Polen zum Äquator hin, sich auch allmählich organische Kraft und Lebensfülle vermehren. Aber bei dieser Vermehrung sind doch jedem Erdstriche besondere Schönheiten vorbehalten: den Tropen Mannigfaltigkeit und Größe der Pflanzenformen; dem Norden der Anblick der Wiesen, und das periodische Wiederwachen der Natur beim ersten Wehen der Frühlingslüfte" (Humboldt 2004a: 245).

With one god-like glance, the Humboldtian naturalist surveyed the whole planet, from the poles to the equator. Yet this was not the Apollonian vision of the sun god of ancient Greek mythology or the moon-bound modern astronaut: the earth as seen from some remote point in outer space, all its features melted into a glittering blue globe streaked with clouds. To encompass all of nature "mit Einem Blicke" did not, for Alexander von Humboldt, mean that the distinctive features of particular regions would thereby be erased. On the contrary, what Humboldt called the "physiognomy" of each region – Alpine meadows, tropical rainforests, barren deserts – remains visible to the naturalist whose gaze spans hemispheres. Each landscape is not only individualized; each is also beautiful in its own way, even the forbidding arctic: "Auch der eisige Norden erfreut sich Monate lang der krautbedeckten Erde, großblüthiger Alpenpflanzen und milder Himmelsbläue" (Humboldt 2004a: 383-384).

This is what I shall call the Humboldtian gaze: a way of seeing that was at once morphological and numerical, aesthetic and scientific, local and global. We understand these pairs as irreconcilable opposites, and many of Humboldt's contemporaries did as well. It was Humboldt's achievement to forge a scientific research program in which precise measuring instruments were as essential as the eye of a landscape painter. Only a vast collection of carefully coordinated measurements of air pressure, temperature, and magnetic declination could reveal global patterns; only a highly developed sense of color, form, and above all composition could fuse impressions of flora, fauna, and terrain into a landscape, a recognizable 'face' of a region. For Humboldt's colleagues, especially for the French savants from whom he had received so much of his scientific training, the greatest paradox of all was the idea of a natural law that implied regularity, universality, but *not* uniformity. In contrast to Newton's law of gravitation and its Enlightenment elaborations in the work of Daniel Bernoulli, Leonhard Euler, Pierre Simon Laplace, and others, Humboldtian natural laws did not assume that nature was, in its essentials, everywhere and always the same. Humboldtian nature was "bunt", both literally and figuratively. Varying patterns of vegetation and geological formations colored each region differently and distinctively; Humboldt's endless series of world maps were ingeniously colored to show how an order emerged from this patchwork.

Consider the many maps that illustrated the atlas to Humboldt's magnum opus *Kosmos*, a huge compendium of scientific data, literary history, and artistic reflection about the entire universe published at the very end of his long life. This one shows the worldwide geographic distribution of lines of constant temperature, or 'isotherms', computed on the basis of annual averages. Note that the isotherms follow latitude lines only very approximately and that there are some striking deviations in the arctic regions. The idea of tracing iso-lines of constant measured average values was Humboldt's own: his maps display lines of constant temperature, barometric pressure, rainfall, magnetic declination, flood waves, etc. Previously, such measurements had been presented in the form of synoptic tables, insofar as any attempts had been made to compare values for locales distributed in space, rather than the same locale over time. But the global patterns that leap to the eye in the map of isotherms are all but invisible in the table of annual average temperatures. Humboldt's discovery of new regularities in, for example, the global distribution of atmospheric warmth was also a discovery of how to make these regularities visible to the scientific eye.

The isotherms showed unity in the diversity of climate. However, the quantity measured, temperature in this case, was primarily of interest to Humboldt because it was strongly correlated with the distribution of flora and fauna. Here is a Humboldtian map of the global distribution of the major plant families, with special attention to three groups Humboldt believed to be especially important in determining a landscape physiognomy: palms, pines, and heather. A clear north-south pattern emerges, but modulated by elevation above sea-level: for example, the Himmalayas are a pine zone, despite the fact that they are otherwise located in a palm belt. The major mountain peaks of the world, from the Andes to the Himmalayas to the Alps, are displayed side-by-side to highlight the differences in their elevations, but also the similarity in their bands of vegetation as a function of heights. The isolines, the colors, the sketches all served the purpose of comparison: every region was different, but different in a recognizable way that related in previously unsuspected ways to other distant regions. As Humboldt remarked apropos of the vegetation growing on cliffs by a South American waterfall:

> "Mit Melastonen und Droseren, mit kleinen silberblättrigen Mimosen und Farnkräutern geschmückt, bilden sie Blumenbeete auf dem öden Gestein. Sie rufen bei dem Europäer das Andenken an jene Pflanzengruppen zurück, welche die Alpenbewohner *Courtils* nennen: Granitblöcke, mit Blüthen bedeckt, die einsam aus den savoyischen Gletschern hervorragen" (Humboldt 2004a: 186-187).

Analogies both qualitative and quantitative knit together the flower-spangled rocks of the Andes and the Alps, the palm groves of Brazil and India. The art of Humboldt's science lay in learning to see these analogies and in inventing ways to represent them in maps, charts, graphs, and drawings.

Humboldt invented a new kind of natural law; moreover, he invented a new way of visualizing it. Humboldt's science was remarkable for empirical discoveries (e.g., the influence of ocean currents on climate), for its syntheses of huge amounts of empirical data from many disciplines, for its historical and anthropological erudition, and for its striking presentation in both word and image. But the most extraordinary of his contributions was what I shall call the Humboldtian natural law: not just new examples of regularities, but a new *form* of regularity. The aim of my lecture is to explain just what was so new about these Humboldtian natural laws. I shall proceed in four steps: first, an overview of how Enlightenment savants understood the laws of universal nature; second, the alternative vision of local nature, especially in the long-lived Hippocratic medical tradition that survived from antiquity until well into the eighteenth century; third, the Humboldtian innovation of local/global nature; and finally, a few concluding reflections on the vision of the scientific community that made Humboldtian science possible.

Universal Nature under Law

The most spectacular advances of the seventeenth-century Scientific Revolution were almost all made by idealizing highly complex and variable real-world situations into abstract mathematical models. Take the case of Galileo's brilliant analysis of how bodies fall. Galileo was perfectly aware that the size, shape, and weight of bodies, as well as air currents, do in fact influence the speed and trajectory of falling bodies. A leaf falling from a tree in a high autumnal wind will not fall the same way as a cannon ball dropped from the same height. But under ideal conditions, Galileo argued, the leaf and the cannon ball would hit the ground at the same moment. These ideal consitions, which included a perfect vacuum, could only be realized (at least as far as Galileo was concerned) in the mathematical imagination, not in experience. Yet cleverly designed experiments, sometimes performed for the eye of the body and sometimes for the eye of the mind, could approximate these ideal conditions. For example, imagine the leaf and the cannon ball first dropped through honey, then through water, and finally through air. Although they would never hit the ground at the same instant, the divergence between the times of impact would decrease as the density of the medium decreased: the cannon ball's leading margin over the leaf would be greatest in viscous honey and smallest in thin air. Extrapolating from these results to the case of no medium – the vacuum – Galileo concluded that in the limiting case the margin would disappear altogether (Galilei 1974: 70-75).

Implicit in this sort of reasoning was a model of what kinds of causes were at work in a phenomenon like free fall and how they interacted with one another. Galileo assumed that there was one major cause, the heaviness of bodies (we would now call this main cause gravitation), and numerous minor causes, such as the density and turbulence of the medium and their effects on the size and shape of the bodies. In such cases, the mathematically inclined natural philosopher was justified in ignoring all but the major cause and giving a mathematical account of its operation. This process of leaving out, of abstracting from everything but the essential, is what is meant by idealization. In order to be successful, idealization must be applied with a shrewd eye for the essentials and to a domain of phenomena where in fact one major cause dominates many minor ones. It is most striking in mathematical sciences like terrestrial and celestial mechanics, but it also played a central role in the observational sciences of the Enlightenment, such as botany and anatomy, as we shall see. Idealization depends on a metaphysics of underlying regularities or (in the case of organic nature) of types that are obscured but not overwhelmed by contingent and variable circumstances. At least as far as Galileo was concerned, not all aspects of even inorganic nature conformed to this metaphysics. For example, he asserted that there would never be a true science of air currents, which were, in his opinion, all variability and irregularity (Galilei 1974: 224-225). Nature, it seemed, was regular only in patches.

But circa fifty years later, after the publication of Isaac Newton's *Principia* (*Philosophiae naturalis principia mathematica*) in 1687, natural philosophers became intoxicated by the possibility of subsuming *all* of nature under regularities (now increasingly called "laws") of the Galilean sort. The inspiration and prototype for these universal laws was Newton's law of gravitation, formulated as a mathematical relationship between mass, distance, and force of attraction, which was masterfully deployed by Newton in the *Principia* to explain not only the fall of heavy bodies, but also the orbits of the planets, the trajectories of projectiles, and the ocean tides. Only the aura of reverence and awe that surrounds Einstein in our own day can in any way compare with the demigod status Newton was granted for this achievement in the Enlightenment. This triumphal expansion and unification of the domain to which idealized mathematical models were applicable transcended anything Galileo had imagined and emboldened Newton's successors to claim that it was only a matter of time before all natural phenomena, however unpredictable and variable. Newton's colleagues in the Royal Society of London promptly turned to the problem Galileo had given up on, namely a science of air currents and, more broadly, of the weather. From Florence to Leipzig, Upminster to Zurich, Paris to Peking, far-flung observers rose daily at dawn to read their thermometers and barometers (deliberately placed in northern exposures and unheated rooms), note the wind direction, squint at cloud formations, and jot down the results in diaries or in multi-columned tables. They posted their data to scientific societies, which sifted the columns of entries in search of some rule or correlation – be it with the configurations of planets, sunspots, epidemics, the tides, or simply instrument readings in other locales. The early Enlightenment search for the natural laws of weather was as futile and interminable as the task of Sisyphus, but it was apparently undertaken with enthusiasm and pursued with tenacity.

This stubborn persistence in the face of repeated failure indicates how strong the belief in universal natural laws, even for the notoriously changeable weather, had become after Newton's achievements. The belief had theological as well as metaphysical and scientific grounds. In the late seventeenth and early eighteenth centuries, it became a cliché in works of natural philosophy and, especially, of natural theology, to hold up inquiry into the laws of nature as a model for all knowledge. Natural theological considerations reinforced natural philosophical confidence in the orderliness of the atmosphere, and indeed, everything else: "The Atmosphere, or Mass of Air, Vapours and Clouds, which surrounds our Globe, will appear to be a matter of Design, and the infinitely wise Creator's Work, if we consider its Nature and Make and its Use to the World," (Derham 1716: 4) wrote natural theologian William Derham, who encouraged other Fellows of the Royal Society to compile their own tables of weather observations in the early decades of the eighteenth century. Natural laws, in both jurisprudence and natural philosophy, laid claim to universality and uniformity; positive laws (and religious doctrines) varied widely as to time and place. The English natural philosopher Robert Boyle thought God preferred uniformity even to benevolence: God would not make exceptions to "the general laws he at first most wisely established," (Boyle 1966: vol. 5, 252) even to save his creatures from earthquakes, floods, and famines – an argument that Voltaire was bitterly to parody in *Candide* apropos of the thousands of innocent lives lost in the Lisbon earthquake. By the mid-eighteenth century, Galileo's pockets of disorder had all but disappeared, at least officially; according to leading Enlightenment philosophers and savants, nature was uniformly and universally governed by inexorable natural laws.

The mathematical sciences of mechanics, optics, and astronomy were the showcase examples for these laws, as in the works of Newton and Laplace. It is easy to see how a mathematical model of a fluid must subtract out all the details of swirling, flowing, eddying water and flatten it into a geometric diagram. But even for the descriptive sciences of natural history and anatomy, belief in the existence of uniform species types underlying the variability and idiosyncracies of this or that individual specimen was an article of faith among most Enlightenment naturalists, who worked closely with their illustrators to guarantee that the realism of these types was represented in images, rather than the naturalism of what visible to the eye. Every exemplar of the same plant or animal species looked slightly different when examined closely, but it was the task of the experienced naturalist to abstract from such accidents, just as the mathematical natural philosopher abstracted from the accidents of fluids or heavy bodies. Just as a Linnaean botanical description singled out those features common to the entire species (the *descriptio*) and those that also differentiate this species from all others in the genus (the *differentia*), but at all costs avoided features peculiar to this or that individual member of the species, so the Linnaean illustration also aspired to generality – a generality that could transcend the species or even the genus to reflect a never seen but nonetheless real plant archetype.[1]

This vision of nature had legal, moral, and political as well as scientific resonances. In seventeenth- and eighteenth-century jurisprudence, absolutist monarchs attempted to enforce throughout their realms uniform laws that would supersede the sovereignty of local customs. In Enlightenment ethics, it was Kant who most powerfully articulated the ideal of moral law, binding upon all rational beings – a category that embraced not only the entire human species but perhaps Martians as well (Kant once said he would bet his life on the existence of life on other planets). In politics, the famous declarations of the American and French Revolutions proclaiming the universal rights of man grounded them explicitly in the authority of natural laws. Law, ethics, and politics all invoked the order of the non-human universe, defined by unexceptionable regularities that are everywhere and always the same. The American Declaration of Independence of 1775 secured the "inalienable rights of man" in

1 See (Eriksson 1994: 83-84) concerning Linnaeus's idea of the "most natural form" or "archetype" of flowers.

"Nature and Nature's God". Kant seems to have been in the thrall of this analogy when he coupled "the starry heavens above and the moral law within" as the two phenomena that filled him with awe (Kant 1993a: 186), and the English novelist George Eliot also drew moral comfort from the "great conception of the universal regular sequence, without partiality and without caprice" (Eliot in Levine 2002: 109). Although the very determinism of such a universal, regular sequence precluded genuine moral agency in the voluntarist ethical systems to which both Kant and Eliot subscribed, its relentless uniformity could nonetheless serve as a luminous figure for a universal moral order purged of the arbitrary.

This ideal of a universal natural and moral order also acted powerfully upon Alexander von Humboldt. Some of the most rapturous moments in his writing occurred when he invoked universal natural law, in grand tones reverberating with Newtonian and Kantian echoes. Humboldt claimed that even children respond unconsciuosly to the "geheimnißvollen Zusammenhang aller organischen Gestaltung"; eventually such "dunkle Gefühle" lead to knowledge of "einem gemeinsamen, gesetzlichen und darum ewigen Band [das] die ganze lebendige Natur umschlinge" (Humboldt 2004b: 11). Not only the laws themselves, but also the very act of searching for them had something noble about it for Humboldt. When he sadly contemplated the likely fate of all he had written, doomed like most empirical science to be forgotten in a generation or two, he comforted himself with the hope that at least his quest for natural laws would be remembered with honor:

> "Ein Versuch, die Natur lebendig und in ihrer erhabenen Größe zu schildern, in dem wellenartig wiederkehren den Wechsel physischer Veränderlichkeit das Beharrliche aufzuspüren, wird daher in späteren Zeit nicht ganz unbeachtet bleiben" (Humboldt 2004b: 7).

Yet Humboldt's understanding of the general laws he sought diverged significantly from the Newtonian laws of the Enlightenment. His laws were universal, but his nature was not uniform. In order to grasp the difference, we must first return briefly to an ancient way of describing nature's regularities that long antedates the language of universal laws. This was the Hippocratic tradition of local topographies, correlated with the salubrity and insalubrity of various regions, as well as with the character and appearance of their inhabitants. In this tradition, it was common to speak of nature's customs rather than nature's laws, that is, of regularities that applied only locally and then only most of the time, not necessarily and not always. Although this tradition had been ousted from natural philosophy by the mid-eighteenth century, it remained influential in medicine throughout the Enlightenment. As in the case of universal natural laws, the model of nature's local customs had its counterpart in moral philosophy. The young Humboldt would have been familiar with both medical and moral versions; together, they provided an alternative to Newtonian laws that Humboldt adapted to his own purposes.

Local Nature under Custom

What kind of natural order could be analogized to custom, as opposed to law? The local, variable, more-or-less character of custom seems to accord ill with universal, uniform, and determinate nature. This apparent mismatch is however the result of an anachronism: of imposing a post-seventeenth-century description of the natural order as governed by law upon an earlier vision of nature's regularities as rules or customs. Before the metaphor of nature's laws replaced that of nature's customs, the analogy between natural and moral orders would have been patent, and active at several levels. Not only were nature's customs structurally similar to human ones in being bound to local conditions and admitting of the occasional exception; natural and human customs could also interact and modify one another, thereby dovetailing the features of a given locale into a more coherent and distinctive physiognomy. It is, for example, the nature of water everywhere to seek its own level, but its form (babbling brook or surging ocean?), its taste (sweet or brackish?), its color (limpid or muddy?), its medicinal properties (good for ulcers or for rheumatism?), and numerous other features depend on local conditions. And whether water takes on the aspect of swamp or torrent, abundant or scarce resource, molds the collective habitus of the people in that locale.

Although the interplay of local nature and custom lives on not only in poetry but in whole schools of geography and *Annales*-style history, the metaphysics of custom that united both within a common framework from Antiquity through the Enlightenment has been submerged, though never entirely lost. The salient aspects of this doctrine were: first, the specificity of local nature and customs; second, the concord between them; third, the plasticity of both, often modified in concert; and fourth, a model of interaction that was integrative rather than mutually exclusive. Its locus classicus was the Hippocratic text on *Airs Waters Places* (5th ct. B.C.E.), in which itinerant physicians were advised on how to treat the inhabitants of various topographies and climates; its best-known political expression was Montesquieu's *De l'Esprit des lois* (1748), on the harmony between peoples and their laws; and it became the nucleus for a vast nineteenth-century scientific research program inspired by Humboldt's *Ansichten der Natur* (1807), which classified landscapes into "physiognomies". A few examples may serve to render this now submerged metaphysics of nature's customs vivid.

In *Airs Waters Places*, traveling doctors are enjoined to study the effects of winds, seasons, stellar aspects, water, soil, and mode of life on inhabitants of different places in order to diagnose and treat typical local diseases. Medical knowledge gleaned from one locale cannot be readily generalized to another, unless they resemble one another along these crucial dimensions. Although specific natures are uniform in time and space, their complex combination and mutual modification produce distinctive local gestalts. Asia differs from Europe for example "in the nature of all its inhabitants and of all its vegetation. For everything in Asia grows to far greater beauty and size; the one region is less wild than the other, the character of the inhabitants is milder and more gentle" (Hippocrates 1995: 105-107). The ancient Greek historian Herodotus (whom Humboldt read attentively and approvingly) also draws attention to the natural specificity of different locales, and to providential compensations among and within them. Gold may be more plentiful in India than in Greece, but Greece is blessed with a more temperate climate; Arabia abounds with poisonous snakes, but since the female of these species bites off the head of the male during mating, and is in turn killed by her offspring to revenge their father (a kind of serpentine *Oresteia*), their numbers are kept in bounds (Herodotus 1995: 133-137). Although the Hippocratic author invokes neither the gods nor providence, a similar compensatory and complementary logic can be detected in some descriptions: whereas Asian peoples and cattle are typically more well-grown and handsome than their European counterparts, courage and diligence cannot flourish under Asian conditions.

Natural milieu and custom work hand-in-hand. According to Herodotus, the Asians lack spirit, endurance, and industry both because their climate is mild and equable *and* because they are ruled by despots, who offer no incentive to take military risks or work hard (Hippocrates 1995: 115). A similar logic governs the Hippocratic account of the tribe of Longheads who stretch the heads of their young children with bandages and all manner of other devices. This custom was eventually supplemented by nature, *physis* and *nomos* cooperating with one another: "Custom originally so acted that through force such a nature came into being; but as time went on the process became natural, so that custom no longer exercised compulsion" (Hippocrates 1995: 111). Although the Sophists later hardened the distinction between *nomos* and *physis* into a stark opposition (and one steeply slanted towards the side of *physis*), the cooperative model endured well into the early modern period, with significant medieval and Renaissance elaborations.[2] Nature referred to spontaneous qualities or conduct;[3] custom, to the cultivation or modification of nature through training and habit. "A man's nature runs either to herbs or weeds; therefore let him seasonably water the one, and destroy the other," Francis Bacon recommended (Bacon 1825-1834: vol. 1, 133). The garden evoked by Bacon's imagery of herbs and weeds might serve as the emblem of integrated nature and custom: pruned shrubs, grafted fruit trees, and cultivated flowers were all examples of nature bent, but not extinguished by custom.

This cooperative model of the customs of nature and culture working together was on the wane by the end of the seventeenth century. By the middle decades of the eighteenth century the metaphysics of universal natural law had become an orthodoxy, and was taken to imply, against all appearances, natural uniformity as well. Universal and uniform natural laws were held up for emulation by moral philosophy, which sought its own Newton. It was in this spirit that Montesquieu bemoaned the disparity between the laws of the natural and moral realms:

> "But the intelligent world cannot be so well-governed as the physical world. For even though the former also has its laws, it does not follow them as constantly as the physical world follows its [laws] [...]. Man, as a physical being, is, just like other bodies governed by invariable laws; as an intelligent being, he violates incessantly the laws established by God, and changes those that he has established himself."[4]

Here is the by-now familiar dichotomy between the universal laws of nature and the variable customs of culture, clearly set out. Montesquieu's treatise was the last full-dress defense in political theory of the sovereignty of local custom against the standardizing ambitions of absolute monarchies, and he rooted local custom in local nature, on grounds of fitness:

---

2 (Heinemann 1965: 15-28; 83-125). Medieval treatments of the "power of place" emphasized celestial as well as geographical and climatic influences: see for example (Tilmann 1971); also (Fauser 1982); (Park 1997: 134-147).

3 (Collingwood 1960: 43-44); (Soper 1995: 16). Conversely, nature under constraint, e.g. a heavy object thrown upwards, might be described as "against nature [παρα φυσιν]" (Festa, Roux 2001: 237-253).

4 (Montesquieu 1845: I.i, 5) my translation.

> "If it is true that the character of the mind and passions of the heart are extremely different in different climates, laws must be relative both to the difference of these passions, and the differences of these characters."[5]

To use one of Montesquieu's own examples, it would be a scandal in a hot country like Spain for a man to appear drunk in public, but not in cold Scandinavia, where high alcohol consumption was necessary to dilate constricted nerve endings, in order to be able to feel anything at all. When Condorcet and other Enlightenment *philosophes* criticized Montesquieu, it was not because they rejected his neo-Hippocratic theory of the inter-relations of climates and characters, but because any moral standard not as universal and uniform as natural law to their minds risked anything-goes relativism.[6]

Hence it is interesting to note that while Humboldt returned to the Hippocratic model of local topographies – although he modernized it in important ways, as we shall see – he asserted the fundamental unity of mankind at every opportunity and emphasized the universal character hidden beneath the apparent diversity of custom. He contradicted the racial theories of his Göttingen teacher Blumenbach, arguing that continuous gradations in skin color and skull shape pointed to "die Einheit des Menschengeschlechts" (Humboldt 2004b: 184). What held for the body also held for the mind, however diverse and diversely developed various cultures might be. Describing the ornamentation of the burial urns used by South American natives, he remarked on their similarity to rhythmic patterns used by other cultures, from the ancient Greeks and Romans to South Sea islanders. The universality of such decorations was once again proof, Humboldt thought, of the psychic unity of all human beings, in all times and places: "Die Ursachen dieser Ähnlichkeiten beruhen [...] mehr auf psychischen Gründen, auf der innern Natur unserer Geistesanlagen, als daß sie Gleichheit der Abstammung und alten Verkehr der Völker beweisen" (Humboldt 2004a: 191).

Thus Humboldt gave the Enlightenment opposition between uniform nature and variable culture a new twist. While seeking general laws in nature, he nonetheless underscored the variety of natural locales and imbued that variety with new significance. And while he acknowledged the dizzying variety of human cultures, he was forever on the lookout for evidence of human universals. What distinguished Humboldt from both the Hippocratic topographers of natural locales and the Enlightenment advocates of human universality was the way in which he managed to combine both into a new and coherent research program that employed the most up-to-date techniques: instruments that measured everything from the height of mountains to the intensity of blue in the sky; ingenious graphing and mapping methods that replaced indigestible tables of data with a visual language of comparisons; and, above all, quantitative and qualitative averages that smoothed measurements and landscapes into physiognomic wholes. It is to Humboldt's ingenious use of averages to merge local and global that I now turn.

---

5 (Montesquieu 1845: XIV.i, 187) my translation.
6 (Condorcet 1828 [1817]: 380-381) my translation.

The Cult of the Average

Naturalists in the Hippocratic tradition had catalogued the variety and distinctiveness of local natures, but they had rarely if ever remarked on their holistic character, on the way in which their flora, fauna, and topography fit together like the features of a face. It was this precisely this element of composition, of treating natural landscapes as if they were landscape paintings, that distinguishes the Humboldtian "Ansicht" from the ancient doctrines of local nature. The naturalist must learn to see like the artist, to integrate the blooming, buzzing confusion of individual sensations into a whole, what Humboldt in the opening chapters of *Kosmos* called a "Naturgemälde":

> "Durch Trennung und Unterordnung der Erscheinungen, durch ahnungsvolles Eindringen in das Spiel dunkel waltender Mächte, durch eine Lebendigkeit des Ausdrucks, in dem die sinnliche Anschauung sich naturwahr spiegelt, können wir versuchen das All (*το παν*) zu umfassen und zu beschreiben, wie es die Würde des großartigen Wortes *Kosmos*, als *Universum*, als *Weltordnung*, als *Schmuck* des geordneten erheischt. Möge dann die unermeßliche Verschiedenartigkeit der Elemente, die in ein Naturbild sich zusammendrängen, dem harmonischen Eindruck von Ruhe und Einheit nicht schaden, welcher der letzte Zweck einer jeden litterarischen oder künstlerischen Composition ist" (Humboldt 2004b: 38).

To our modern ears, these appeals to the "Ruhe und Einheit" of a work of art or literature sound at least complementary, if not opposed to the exactitude and sobriety of a work of science. Humboldt was well aware that he thereby risked being classed with the champions of speculative 'Naturphilosophie', both in its ancient and more modern (i.e. early nineteenth-century Romantic) versions. He firmly repudiated both traditions. In his opinion, they had produced nothing but "Luftgebilde", shimmering but insubstantial illusions (Humboldt 2004: 3). He insisted over and over again that his brand of science would be firmly anchored in scrupulous empirical observation and measurement, and the many tables, maps, and fact-stuffed footnotes strewn through his works show that he was true to his word. But he was equally insistent that science, at least his brand of science, could not rest content with a great heap of unrelated facts. This was what distinguished his *Kosmos* from a mere "*Encyclopädie der Naturwissenschaften*": "In der Lehre vom *Kosmos* wird das Einzelne nur in seinem Verhältniß zum Ganzen, als Theil der Welterscheinungen betrachtet; [...]" (Humboldt 2004b: 25). The new science of *Kosmos* would require a new way of scientific seeing, both qualitative and quantitative.

At the qualitative level, the naturalist must master the synthetic overview, the panoramic coup d'oeil (Bigg 2007). As Humboldt knew, this ran directly counter to the analytic, microscopic gaze cultivated by the best naturalists of the eighteenth century. Instead of peering at the stamens and pistils of an individual plant, or counting the number of an animal's teeth, as descriptive naturalists since Linnaeus had been taught to do, the new-style distributive botanist must pay attention to the "räumliche und klimatische Verbreitung organischer Typen", what Humboldt called geognosy (Humboldt 2004b: 18, 25). Moreover, it was not enough to inventory the flora and fauna of a particular region, as eighteenth-century explorers had done; the naturalist must now take note of which plants and animals generally cluster together and in which proportions. This ecological perspective had its quantitative as well as its quantitative side for Humboldt; he attempted for example to estimate the proportion of each of the major plant families to the entire vegetable kingdom, for example in various regions of the world (Humboldt 2004b: 183). The prevailing Linnaean viewpoint had in contrast ignored ecological relationships – to the point that Linnaeus himself had been convinced that, with time and patience, it would be possible to transplant successfully any plant to any locale (e.g., tea to Sweden) (Koerner 1999).

It was perhaps in reaction to such universalist fantasies that Humboldt dismissed hothouses and botanical gardens, with their transplanted exotica, as poor substitutes for the actual experience of the teeming tropics or the frozen steppes, although substitutes were far better than nothing. He praised the "Palmenhaus" on the Pfaueninsel for conveying at least the dense, multicolored "illusion" of the tropics, even if the azure skies and brilliant light were missing (Humboldt 2004b: 236). Failing a voyage to distant climes, the naturalist was encouraged to study the "Totaleindruck" achieved by a landscape painting. Humboldt also was hopeful that the new technology of the panorama, which attracted flocks of visitors in metropolises like Berlin and London in the 1830s and '40s with their 360-degree cityscapes, could be converted to the purpose of studying landscapes as well. One could, of course, concentrate on this or that meticulously painted detail, as many viewers did, but it was also possible to spin around and take in the entire panorama in one vertiginous glance.

These qualitative ways of grasping the physiognomy of nature had to be supplemented by quantitative methods. Humboldt rarely opposed qualitative to quantitative observations; nor did he oppose instrument measurements to unaided sensation. He thought of them as natural extensions of one another, different tools employed to the same end. For Humboldt, the quantitative equivalent of the qualitative physiognomy of nature was the average. Let us return for a moment to his map of global isotherms. The temperatures connected by the imaginary lines are annual averages, arrived at through the combination of a long series of individual measurements that often showed considerable seasonal variation. Taking an average was a way of smoothing out variability, just as encompassing a landscape with a glance integrated its parts into a whole. Both were examples of what Humboldt called "das Werk des combinirenden Verstandes" in science (Humboldt 2004b: 29).

Humboldt did not invent the method of averages. Throughout the eighteenth century, astronomers had argued over how best to combine series of observations of the same objects, for example a comet. Every astronomer knew from experience that even observations made by the same observer with the same telescope on the same evening could diverge from one another; the divergence was still greater when observations from other astronomers in different locales were taken into account. Some, like the English Astronomer Royal John Flamsteed, argued that one should select the 'best' observations – that is, the ones taken when both observers and instruments were known to be functioning optimally (e.g. without head colds, choppy air, scratched lenses, or other disturbances).

Others, like the French mathematician and cosmologist Laplace, devised mathematical methods to approximate the 'error curve' – i.e., the distribution of probabilities that the observation erred in one or another direction. If the sources of the error were random rather than systematic, the errors could be presumed to be symmetrically distributed around a mean or average, which was the most probable 'true' value of the observation. These error

Figure 1: Karte von Europa zur Übersicht der Wärme-Verbreitung in diesem Erdtheile.
In (Humboldt 2004b: 1. Abteilung, No. 3).

curves in astronomy were the source of what came to be known as the Gaussian or normal distribution, now used to graph everything from the distribution of heights to that of IQ in a population (Stigler 1986: 88-98; Porter 1986: 91-146). By the end of the eighteenth century, it was standard practice among astronomers to combine their observations into averages, in the hopes of thereby minimizing errors and coming closest to the true value.

Humboldt would have learned such averaging practices from the Parisian savants, and also about the remarkable regularities that emerged if once collected a large enough sample of measurements. Laplace had famously announced in his *Essai philosophique des probabilités* (1812) that the number of dead letters in the Parisian postal system was, on average, startlingly constant from year to year; eighteenth-century demographers like Johann Süssmilch and John Arbuthnot had discerned the hand of divine providence in the remarkable constancy, year after year, of the ratio of male to female births (Daston 1988: 127-138). Perhaps encouraged by these surprising regularities, Humboldt took averages into the realm of the most stubborn variability, namely barometric and temperature measurements. Suddenly, the chaotic data assembled by the local eighteenth-century meteorological observers began to cohere into a global pattern;

> "Die Einsicht in die Wärmevertheilung im Luftkreise hat einigermaßen an Klarheit gewonnen, seitdem man versucht hat die Punkte, in welchen die mittleren Temperaturen des Jahres, des Sommers und des Winters genau ergründet worden sind, durch Linien mit einander zu verbinden" (Humboldt 2004b: 166).

Humboldt applied the same technique to the problem of why the west coast of Africa was so much hotter and drier than the east coast of South America at the same latitude. Of course, there were local exceptions: the coasts of Cumana and Coro, for example, were just as hot as any part of Africa. And of course many concurrent causes interacted in complicated ways to produce contrasting climatic conditions. Nonetheless, by comparing annual temperature averages at the same latitude on the South American and African coasts, Humboldt was led to take measurements of ocean temperatures, revealing the existence of a cold water current that swept along the South American coast. Similar techniques revealed the systematic climatic contrasts between the east coast of North America and west coast of Europe at the same latitude. In a typical turn of phrase, he announced his findings: "Ich habe versucht die mannigfaltigen Ursachen der Nässe und mindern Wärme Amerika's in Ein Bild zusammenzudrängen" (Humboldt 2004a: 109). Here, even the orthography is significant: "Ein" as well as "Bild" is capitalized, fusing them into a single, monolithic entity, no longer a catalogue of miscellaneous data. This was the quantitative analogy to the all-at-once panoramic gaze, both Humboldtian ways of seeing. The link between them was the synoptic global map, which turned the local numerical averages into a picture that could be comprehended at a glance, just as the "Naturgemälde" revealed the physiognomy of a landscape.

Conclusion: The Cosmic Community

Humboldt's "Lehre vom Kosmos" was made possible by a new kind of scientific inquiry that emerged in the nineteenth century and of which he was on the pioneers: the sustained international collaboration. The eighteenth century had witnessed a few scientific collaborations on an international scale, most notably the expeditions to observe the transits of Venus in 1761 and 1769, which had involved the French, Swedes, British, Germans, Dutch, Italians, Russians, Danes, and Portuguese. However, these cooperative efforts were few and episodic, with no binding consequences for the participants (Woolf 1959; McClellan 1985). Enterprises undertaken in common in the latter half of the nineteenth century, like the 'Internationale Gradmessung' project or the standardization of electrical units,[7] were in contrast long-term and quasi-contractual, obligating participants to reform their procedures and techniques. Improved means of transportation and communication were the preconditions for, but not the causes of these ever wider and more densely woven networks of observers. Rather, a new awareness of natural phenomena that dwarfed merely human scales of time and space prompted the organization of vast observational systems with standardized methods and instruments. The shape and path of a storm,[8] the contour of an isotherm,[9] the physiognomy of a landscape (Nicolson 1990: 169-185), the geographic distribution of a species (Camerini 1987) – these were phenomena whose regularities or very existence was hidden to the observer located too firmly in the here and now, whose view was circumscribed by the humanly meaningful co-ordinates of a locale or a lifetime. Humboldt was among the first to conceptualize these new scientific objects and also among the first to create an international network of correspondents to supply him with data from all over the world.

The sine qua non of these scientific networks was commensurability. Data supplied by farflung observers had to be made comparable in order to make Humboldt's syntheses possible: observers as well as instruments and protocols had to be standardized and coordinated. The presupposition of scientific community was communicability. Humboldt had assumed that both his quantitative and qualitative methods of encompassing all of nature in one view could be widely disseminated; one of the aims of works like *Ansichten der Natur* and *Kosmos* was exactly this. Readers would be taught how to see in a new way. Moreover, Humboldt never doubted that what he called the "objective" and "subjective" aspects of his cosmic science were both communicable. If anything, he seemed to think that the subjective response to nature was more deeply rooted in universal human nature, whereas the objective description of nature was a historically recent and culturally localized achievement. His subsumption of the "objective" parts of *Kosmos* under the rubric "Naturgemälde" and his many parallels between landscape painting and plant physiognomy suggest that he also regarded the relationship between art and science as a cooperative one. Not just the visual arts, but also language and literature might yield valuable insights into the constitution of nature. Humboldt commented for example on how older languages boasted rich vocabularies with which to pick out characteristic landscape features – for example, the

> "alt-castilianischen Idiomen die vielen Ausdrücke für die Physiognomik der Gebirgsmassen, für diejenigen ihrer Gestaltungen, welche unter allen Himmelsstrichen wiederkehren und schon in weiter Ferne die Natur des Gesteins offenbaren" (Humboldt 2004a: 215).

7 (Kommission der europäischen Gradmessung 1881); (Schaffer 1992: 23-56); (Smith, Wise 1989: 694-698).
8 (Fleming 1990) chapters 2, 4, 6, on the storm controversy and observing systems.
9 (Cannon 1978) chapter 3, on "Humboldtian science.

But within a decade or so following Humboldt's death, this union of the objective and subjective was no longer conceivable. Ever starker oppositions of art and science came to define and refine the meanings of subjectivity and objectivity. The crux of these oppositions was communicability. The physicist and physiologist Hermann von Helmholtz makes a particularly instructive witness as to how ineluctable these oppositions had become, for he was both a sensitive observer of the arts (especially music) and a seminal contributor to several sciences. In lecture after lecture, Helmholtz deepened the contrast between the unconscious "Taktgefühl" that guided artistic inductions about human character, and the "scharf ausgesprochenen allgemeinen Regeln und Gesetzen" characteristic of the sciences. Only mediocre artists worked "mit Bewusstsein nach allgemeinen Regeln und Abstractionen", but scientists could not do without them. Art welled up spontaneously from the depths of unconscious memory; science was the "bewusste logische Thätigkeit unseres Geistes in ihrer reinsten und vollendetsten Form", accompanied by the unspontaneous virtues of painstaking care [*Mühe*], exactitude [*Genauigkeit*], and caution [*Vorsicht*]. Science was "eiserne Arbeit" that demanded "grosse Hartnäckigkeit und Vorsicht" (Helmholtz 1903a: 171-178); art was "Anschauung [...] ohne Besinnen, ohne geistige Anstrengung, augenblicklich, sowie der entsprechende sinnliche Eindruck auf uns wirkt" (Helmholtz 1903b: 341).

Humboldt would have been astonished to see his lifelong scientific labors consigned to the realm of effortless, intuitive, and, worst of all, incommunicable artistic inspiration. He would have been all the more astonished because Helmholtz purported to share the same goals that had guided the work of Ansichten der Natur and Kosmos: an ever-better ordering of data so that "die Anhäufung der Einzelheiten" did not overwhelm "den Zusammenhang"; and, above all, achieved the aim to unite "die Einzelfälle der Beobachtung und Erfahrung zu allgemeinen Gesetzen von unbedingter Gültigkeit und ausserordentlich umfassenden Umfange [...]" (Helmholtz 1903a: 168, 175). Humboldt would have nodded emphatically, only to be taken aback by Helmholtz's further claim that these goals could only be achieved through "ein gewisses Misstrauen gegen den sinnlichen Schein" (Helmholtz 1903a: 179). In the less than twenty years that had passed between the publication of *Kosmos* and Helmholtz's 1862 lecture in Heidelberg, Humboldtian ways of seeing had been transformed from science into art, and art and science had become polarized into incompatible opposites.

We still live mostly in Helmholtz's world, not Humboldt's. Some aspects of the Humboldtian way of seeing have indeed become part of the permanent apparatus of science: graphs and global maps, iso-values and the distribution of species – these visual tools still bear witness to the Humboldtian vision of natural laws that are both local and global. One might go further and claim that the latest scientific images, from magnetic resonance images of the human brain to the digital images of distant galaxies generated by the Hubbel telescope are the direct ancestors of Humboldt's campaign to make the invisible visible. And no one, least of all the scientists (and perhaps not even Helmholtz, with his dour distinction between laborious science and ludic art), could deny their hypnotic beauty. But if the letter, as opposed to the spirit, of Humboldt's works has survived, if we still read *Kosmos* and *Ansichten der Natur* in their magnificent new editions as something other than as sources for the history of science, it is because they have become great works of literature, no longer living works of science. It has turned out just as Humboldt predicted:

> "Man hat es oft eine nicht erfreuliche Betrachtung genannt, daß, indem rein literarische Geistesprodukte gewurzelt sind in den Tiefen der Gefühle und der schöpferischen Einbildungskraft, alles was mit der Empirie, mit Ergründung von Naturerscheinungen und physischer Gesetze zusammenhängt, in wenigen Jahrzehenden, bei zunehmender Schärfe der Instrumente und allmäliger Erweiterung des Horizonts der Beobachtung, eine andere Gestaltung annimmt; ja daß, wie man sich auszudrücken pflegt, veraltete naturwissenschaftliche Schriften als unlesbar der Vergessenheit übergeben sind" (Humboldt 2004b: 7).

Yet even if scientists have forgotten all of Humboldt's carefully collected and synthesized observations, even if his maps and graphs have long since been surpassed by more comprehensive and precise representations, he has left a lasting imprint on the form if not the content of science with his vision of a local/ global natural law.

M. Norton Wise[1]

What's in a Line?

It is a curious fact of the history of science that only toward the middle of the 19th century did scientists of every stripe acquire the habit of trying to represent natural processes in terms of curves, or better, to read nature as curves. Of course Kepler's ellipses for planetary motion and Galileo's parabolas for cannonballs had been available for two hundred years. But there were no magnetic lines of force, no supply and demand curves, no bell-shaped curves for the propensity to commit suicide with a knife, no Carnot diagrams for steam engines, no curves of rotating storm systems, and no blood pressure recorders, to say nothing of ekg's, seismographs, and plots of the Dow-Jones Industrial Average over the past week. So I want to return to the time before we all came to think of curves as the most mundane and utterly common-sensical way to describe myriad processes going on in the world. All of this took off between about 1830 and 1850. It occurred all over Europe, but one of the places of intense local activity was Berlin. Even more particularly, the ambitious young men who founded the Berlin Physical Society in 1845 made the curve their emblem of progress in scientific knowledge. Figure 1 shows their certificate of membership; actually it is the certificate of Emil du Bois-Reymond, physiologist and charismatic leader of the group, who did the drawing himself. Looking a bit more closely at the tree on the left, it reveals itself as a tree of knowledge, where youthful heroes of physical science are carrying out their investigations. Many of these athletic experimenters are using instruments that seem to produce curves representing nature's laws.

I aim in the present chapter to put a historical foundation under this elaborate image, to which I will return in some detail at the end. It is apparent immediately that du Bois-Reymond had considerable artistic talent. In fact, after finally taking up medicine and physiology, he acknowleged to his fiance that "die Kunst mein eigentliches Talent war", and regreted that he had not made art his career, like his aunt, his grandmother, and his renowned great grandfather Daniel Chodowiecki, who had directed the Berlin Academy of Art at the turn of the 18th century. Du Bois could study Chowdowiecki's engravings at any time in his mother's collection.[2] Similar, if less grand, family histories could be given for several others among the founders of the Berlin Physical Society.

---

1 Department of History, UCLA, Los Angeles, CA. nortonw@history.ucla.edu This paper was originally presented as the Rothschild Lecture at Harvard University, April 2001 and is adapted from *Bourgeois Berlin and Laboratory Science* (in preparation), chapter 5. For valuable discussions and sources I am indebted especially to Sven Dierig, Mechtild Fend, Myles Jackson, and Elaine Wise as well as to participants in colloquia at Harvard University; Internationales Forschungszentrum Kulturwissenschaften, Vienna; Max Planck Institute for History of Science, Berlin; Pennsylvania State University; University of California, Berkeley; and UCLA.

2 (Finckelstein 1996: 54); (Du Bois-Reymond 1853); for Wilhelmine (Minette) Du Bois-Reymond's collection, see *Daniel Chodowiecki's sämmtliche Kupferstiche*, described by (Engelmann 1969: xviii, xxx).

Figure 1: Certificate of membership of Emil du Bois-Reymond in the Berlin Physical Society (du Bois-Reymond 1845).

Figure 2: Several founders of the Berlin Physical Society in 1845. Left to right, back: G. Karsten, W. Heintz, H. Knoblauch; front: E. Brücke, E. du Bois-Reymond, W. Beetz. Archiv der Deutschen Physikalischen Gesellschaft. Signatur 60001.

Figure 2 shows them in a daguerreotype from 1845. Du Bois-Reymond is in the front in the middle. Next to him on the left is another latent artist, Ernst Brücke, whose father, two uncles, and a stepbrother were Berlin artists and who had considered following their lead. Another early member, not shown, was Hermann Helmholtz, whose aunt was a member of the Academy of Art and whose superior while he served as a military doctor in Potsdam was Wilhelm Puhlmann, a family friend and founder of the Potsdam Society of Art. Helmholtz married Puhlmann's niece Olga von Velten and through Puhlmann he became friends with the great painter Adolph Menzel. These three: Brücke, Helmholtz, and du Bois-Reymond, successively taught anatomy to the students of the Academy of Art in the 1840s. Brücke and Helmholtz played active roles in the theory and practice of art thoughout their lives while du Bois, as late as 1890 in a lecture on "Natural Science and Pictorial Art", still conceived of the practice of experimental science in aesthetic terms.[3] Finally, a key figure for the entire group was Gustav Magnus (figure 3a).

3 For perceptive discussions see Sven Dierig, "Apollo's Tragedy: Laboratory Science between Classicism and Industrial Modernism," this volume; (Dierig 2006b: 10-16, 122-144); (Dierig 2006a); (Dierig 2004); (Lenoir 1997) and (Hatfield 1993). (Brücke 1928: 2-4, 137); (Brücke 1892); (du Bois-Reymond 1912).

Figure 3: (a) *Gustav Magnus* by Eduard Magnus (painting hanging in Magnus' former house, now occupied by the Deutsche Physikalische Gesellschaft). Author's photo.
(b) *Eduard Magnus* by Adolph Menzel (Lammel 1993: 28)
(c) *Adolph Menzel* by Eduard Magnus (Lammel 1993: 27).

It was in Magnus's house, located just across an arm of the Spree River from the Royal Museum of Art (*Altes Museum*), that the members of the Society got their start in experimental science. The painting is by Magnus's brother Eduard, a highly successful portrait painter. Figures 3b, c show the elegant Eduard drawn by his friend Menzel and Menzel by Eduard.

So what now was the relation between these close circles of involvement in the world of art and the pursuit of curves as the expression of nature's truths? The following account will depend on developing some interrelationships between the 19th century revival of Albrecht Dürer's art; the analytic mathematics of Gustav Lejeune-Dirichlet and the meteorological representations of Heinrich Wilhelm Dove, both directly dependent on Alexander von Humboldt; and the physiological investigations of du Bois-Reymond and Helmholtz. It will focus on the local culture of art, industry, and science that gave meaning to the curve.

Berlin Realism

On the 18th of April 1828, the 300th anniversary of Dürer's death, a great commemoration was held in Berlin, as well as in Dürer's native city of Nürnberg. The iconic status that Goethe was already acquiring by that date for German culture is well known; less familiar may be the fact that Dürer had become the personification of Germanness at a time when the Germans had discovered the gothic as their own, a wellspring of their unifying national character even though political unity eluded them. Dürer *was* the gothic, incorporating the romantic, the rational, and the Christian in one figure. This gothic identity can be extracted from any number of his celebrated drawings, woodcuts, and engravings (e.g., "Knight,

Death, and the Devil"). But the degree to which Dürer was himself very nearly worshipped as a christlike figure is difficult to comprehend, even after seeing many christomorph portrayals, such as a series of episodes from his life imitating scenes from the life of Christ, commissioned for the celebration in Nürnberg.[4] The theme appears also in the altar wall that graced the auditorium of the *Singakademie* in Berlin for the celebration there (figure 4).

Figure 4: K. F. Schinkel, design for alterlike decoration, Dürer celebration, 1828 (Rave 1962: p. 360, Fig. 399).

The altar was designed by Karl Friedrich Schinkel, already celebrated for the neo-classical architecture that was shaping the modern face of the city.[5] Dürer's larger-than-life statue, modeled after the most famous of his christomorph self-portraits, stands in the middle beneath a large painting of the ascent of Christ into heaven. The painting is essentially an enlargement of one of Dürer's woodcuts, "The Trinity". The seated female figures to his right and left recall his work in the areas of "scenographia" (perspective), "painting", "sculpture", and "military architecture", the interrelation of which thematizes the present chapter. Importantly also, Schinkel had no difficulty incorporating the gothic Dürer into his own modernizing neo-classical statuary and frame, for he himself had recently made that transition in his

4 (Bialostocki 1986: 91-144). (Koerner 1993) gives a penetrating interpretation of Dürer's christomorph self-portraits.

5 (Bialostocki 1986: 121-123); (Mende, Hebecker 1973: 113-115).

architecture following the War of Independence (1813-1815).[6] Schinkel's neo-classicism looked forward toward an age of technology and industry, and it brought aspects of the Dürer renaissance with it.

One such aspect was what Berlin artists admired as Dürer's realism, which provided a point of reference for their own realist tradition. That tradition has often been referred back to du Bois's great-grandfather Chodowiecki, whose woodcuts recall Dürer's. Another canonical reference is a sharp critique from Goethe in 1800, who complained that Berlin artists had lost sight of the universal ideals of classical Greece and become mired in provincialism, both historically and geographically:

> "In Berlin [...] naturalism seems to be at home with the demand for realism and utility and generally to manifest the prosaic Zeitgeist. Poetry is suppressed by history, character and the Ideal by portraiture, [...] and the universally human by nationalism" (Goethe 1965: 809).

In response, Gottfried Schadow, sculptor, painter, and soon to be director of the Academy of Art, called on the memory of Chodowiecki and Dürer in defense of a naturalism that mirrored real people with real emotions living in particular locations. In Berlin, he said, "one gives priority to those artworks that truly and honestly depict an existing model; every work of art is treated here as a portrait, a reflection of nature [*Konterfei*]". An ironic example could be the realistic bust of Goethe himself that Schadow sculpted in 1822-1823, after his former student Christian Daniel Rauch had done one in 1820 that Schadow regarded as too heroic. More representative is his famous double sculpture of the two princesses (*Prinzessinnen von Preussen*, 1795-1797), the future Queen Luise and her sister Fredericke. Regarded as an epitome for German neo-classicism, it presents their teenage beauty in lifelike individual portraits, expressed through the pure lines of universalizing classical purity. The universal lies within the particular, he insisted, and not the particular within the universal.[7]

To digress briefly, Schadow's view can serve as the motto for my own conception of the cultural history of science, as presented here. Within the local circumstances in Berlin I seek the meanings and resources through which the curve came to seem nature's own means of expression for a particularly significant group of scientists. This is not to suggest that they did not continually make use of developments elsewhere but that the conditions of possibility and many of the motivations for their own work had deep roots in the local culture, particularly in the attempts of its educated elite [*Bildungsbürgertum*] to interrelate neo-humanist ideals with new industrial realities. Out of this mix, the self-consciously progressive and reforming young men of the Berlin Physical Society produced curves that made claims on universality. Whether or not the claims would stick depended on the ongoing interaction of the Berlin group with people in other centers of curve production whose motivations and resources were substantially different. For example, in Britain, one thinks of the

---

6 The best place to see this transition is in the Schinkel paintings in the Nationalgalerie Berlin. *Nationalgalerie Berlin: Das XIX. Jahrhundert: Katalog der ausgestellten Werke* (Berlin; E.A. Seemann, 2001). A perceptive account of Schinkel's architectural work is (Bergdoll 1994) with photographs by Erich Lessing.

7 (Schadow 1890), both in (Busch, Beyrodt 1982: 91-100). Schadow's sculptures are depicted and described in (Staatliche Museen zu Berlin 2001: 359-363), *Prinzessinnen Luise u. Friedericke von Preussen*, 1795-1797, Inv.-Nr. B II 34; *Johann Wolfgang von Goethe*, 1822-1823, Inv.-Nr. BI 59.

development of Faraday's lines of force for electromagnetic theory by William Thomson and James Clerk Maxwell. In France, Jean-Victor Poncelet and Arthur Morin at the Ecole d'application for artillery and engineering at Metz were major contributors to developing self-recording instruments. The present discussion, however, following Schadow, seeks to locate the potentially universalizable accomplishments of the Berlin Physical Society within a local portrait.

It was Schadow who organized the Dürer celebration in 1828. By that time he had himself become the representative of established academic art. A rather rebellious younger generation had emerged with full self-consciousness, having organized themselves in 1825 into the Union of Younger Artists [*Verein der Jüngeren Künstler*]. But even in their desire for greater freedom, they maintained Schadow's realist principles along with his pursuit of a truly national art, though with less reverence. Dürer remained their spiritual referent and the focus of a raucous yearly party (figure 5).

Adolph Menzel joined them in 1834. This invitation card for the Dürerfest of 1836 is one of a series of six he produced from 1834 to 1837 (Börsch-Supan 1997: 386). Godfather Dürer frowns benevolently down from the clouds on his drunken disciples.

Figure 5: Adolph Menzel, 1836 invitation card for annual Dürerfest of the Verein der Jüngeren Künstler (Busch 1985: 80, figure 23).

The Line

Let me now leave Dürermania to focus on Dürer's line. The educated public probably had their best opportunity to actually own and study his works through his famous illustrations for the Prayerbook of Emperor Maximillian I, the Holy Roman Emperor in Nürnberg whose patronage he enjoyed. A lithographic, and thus relatively inexpensive, reproduction appeared in 1808 with regular republications afterwards (Dürer 1808).

Figure 6: A page from Dürer's, *Prayerbook of Emperor Maximillian I.* In (Teja-Bach 1996: 172).

It is Dürer's drawings in the margins, rather than the text, that attracted attention. In these drawings, exemplified in figure 6, the lines of the gothic images metamorphose from one form to another and thence into snorkel-like lines and intricate arabesques, sometimes attached to the text itself. The style found numerous imitators in the 19th century, especially following the 1828 celebrations. The most elaborate was a set of drawings done by Eugen Neureuther, with Goethe's enthusiastic support, to illustrate a collection of Goethe's poems and ballads (figure 7).[8]

8 (Neureuther 1977); (Busch 1985) gives a thorough discussion of the arabesque genre, taken in its broadest sense to characterize an era of complexly interwoven modes of literary as well as graphic representation.

Figure 7: Title page (Neureuther 1977: first issue, 1829).

A more typical imitation was another of Menzel's invitation cards for the annual Dürerfest of the Union of Younger Artists [*Verein der Jüngeren Künstler*] in 1837 (figure 8), depicting Dürer's funeral with his long-time friend Peuckheimer giving a farewell blessing.

Notice how Menzel's line moves smoothly between the plant forms, the written message, and the arabesque at the bottom, which symbolically ties the whole together. With the Dürer, Goethe, and Menzel illustrations in mind, it will be apparent that du Bois's certificate for the Physical Society (figure 1) also cites Dürer's drawings for the *Prayerbook* in its basic style, organized by the line that metamorphoses from form to form: from the tree with its society of experimenting youth to the arabesque at bottom center, to the writing of du Bois's name, "Emil Bois" at bottom center, to the name of the engraver, H. R. Heidel at bottom right, who would become an associate member of the Society.

Figure 8: Adolph Menzel, 1837 invitation card for annual Dürerfest of the Verein der Jüngeren Künstler. [Inscription: "Alles schweige, jeder neige ernsten Tönen nun sein Ohr!!! Er wendete die Blüthe höchsten Strebens das Leben selbst an dieses Bild des Lebens."] (Staatliche Museen zu Berlin – Preussischer Kulturbesitz, Kupferstichkabinett).

So far, then, I have wanted only to establish that du Bois conceived his iconography and employed his own line within what had become a popular genre in the Dürer revival, carrying considerable symbolic significance for hopes of rejuvenation of the German nation. It may be indicative of a more direct parallel between Menzel's and du Bois's images that in 1841 du Bois founded a similarly progressive group calling itself the Union of Younger Natural Scientists [*Jüngere Naturforscherverein*], whose five members would form a nucleus for the Physical Society four years later.[9] Like the Younger Artists and the Younger Natural Scientists, the Physical Society presented itself as a vanguard for this movement into the future. I will return to their means of achieving it below.

But I want to pursue more deeply the function of Dürer's line. The art historian Friedrich Teja-Bach has given an illuminating analysis. He shows the lines and arabesques to be integral to Dürer's theory and practice of drawing. They interpret, so to speak, the naturalistic images of the drawings. Compare in figure 6 the arabesque in the top border with the camel at the bottom. As can be seen by superposition (figure 9), the arabesque provides a kind of paraphrase or epitome of the camel. It consists of a line which gives the basic shape of the camel and then returns to play rhythmically on its own forms in a suggestion of the organic unity of the actual animal and perhaps its rhythmic movement (Teja-Bach 1996: 165-193, about Dürer's camel, 172-173, 177).

9 (du Bois Reymond 1918: 29 March 1841: 86); (Finckelstein 1996: 213); (Schwarz, Wenig 1997: 36).

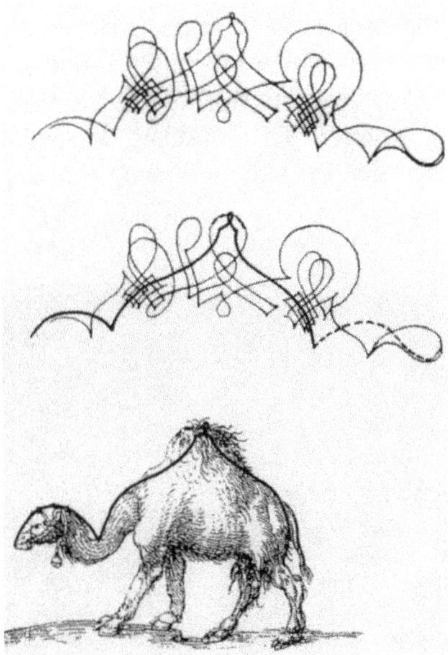

Figure 9: Superposition of arabesque and camel from Figure 7 (Teja-Bach 1996: 173).

This example suggests that Dürer's arabesque provides an abstract essence of naturalistic objects and processes. That is, the abstract line represents an ideal form, in the sense of a Platonic idea. That Dürer intended this Platonic reading seems to be unproblematic among art historians. It attains more depth through Teja-Bach's discussion of how Dürer treated his line as a form of writing (Teja-Bach 1996: 282-297). While the pictures continue the text allegorically, the snorkel-lines and arabesques write out the pictures in an ideal symbolic form.

A similar relation between object, arabesque, and writing is apparent in the drawings of du Bois and Menzel above (figures 1 and 8). The effectiveness of these depictions, however, seems to have depended on other, much more widespread, expressions of the relation between objects and curves.

Consider a depiction of "The Origin of Drawing" (figure 10) done in the 1830s by a Professor at the Academy of Art, Johann Erdmann Hummel, who taught architecture, projection, and optics. Hummel's picture continues an origin myth, also often labelled "the origin of painting", which goes back to Pliny the Elder and continued as a literary tradition into the modern period. But it came to be widely represented in drawings and paintings only from about 1770, in close association with neo-classical ideals, as well as with the popular art of the silhouette and Johann Caspar Lavater's *Physiognomische Fragmente*, illustrated by Chodowiecki.[10] At least six of these allegorical depictions were produced by a lineage of Berlin artists: Christian Bernhard Rode (1790), Schadow (1804), Franz Ludwig Catel (1806),

10 (Rosenblum 1957) discusses both linearity and silhouettes; (Muecke 1999). I thank Claudia Swan for discussion and references. (Lavater 1968-1969).

Schinkel (1830), Hummel (1830s), and Wilhelm Eduard Daege (1834).[11] As the story goes, a Corinthian maid named Dibutades, whose young lover had to depart on a long journey the following day, was inspired to outline his shadow on the wall in order to keep his image clearly before her during his absence. Thus drawing and painting originated in love. Her father Butades, being a potter, filled the silhouette with clay and fired it in his kiln, producing a permanent image.

Figure 10: Johann Erdmann Hummel, *Die Entstehung der Zeichenkunst*, c. 1834 (Hummel 1834).

In the neo-classical aesthetics of the late Enlightenment and Romanticism the story had particular relevance because it gave such prominence to the firmly drawn line, as opposed to color, as the foundation of art. This emphasis was appropiately figured in all of the "origin" drawings and paintings as the line of the silhouette obtained by linear projection. Sharp outlines and smooth surfaces symbolized definiteness, unity, and above all, rationality, seen in contrast to an over-embellished, sentimental Rococo style. Hummel's depiction of Dibutades' lover epitomizes as well the ideal of beauty so widely represented in neo-classicism. He is a youth on the verge of manhood. He exhibits the supple and well-defined body of a

11 (Wille 1960) who shows a different version of Hummel's drawing, dated 1834, and does not mention Daege. (Börsch-Supan, Grisebach 1981: catalogue no. 207a, 267). Wilhelm Eduard Daege in (Staatliche Museen zu Berlin 2001: Inv.-Nr. A I 216).

young athlete, typified as a gymnast, while still recalling the naive innocence and smooth contours of an ambiguously feminine or androgynous adolescence. Quite typically, therefore, he avoids both the insensitivity of a crudely muscular masculinity and the danger of falling into the amorphous, watery softness of feminine emasculation.[12]

Hummel's rendering, while maintaining the ideals of neo-classicism, transforms both the story of Dibutades and the genre of depictions based on it. Normally the potter Butades did not actually appear at all. And if he had, he would have been producing a flat clay model of the silhouette of Dibutades' lover. Here his role is both prominent and different. He is engaged in his everyday work of producing large numbers of vases, all with the same classical form, which we see his assistant arranging on drying racks in the background. The origin of drawing is now manifested in the potter's sharp-eyed concentration on the relation of his daughter's drawing hand to his own shaping hand, the relation of the artist to the craftsman.

Thus Dibutades is the muse who inspires Butades with the beauty of form that she transposes from the idealized beauty of her lover. Her mediating inspiration is reinforced in Hummel's composition through the gentle touch of her lovers' fingers, which unites his arm with hers and passes their complimentary virtues into the line of the silhouette. Their relation reinforces also the view that the firm smooth contour of masculine beauty should maintain an affinity with the softer lines of a more feminine form. These neo-classical ideals then pass through Butades attentive eye to his work in clay. Just as Dibutades' line captures the visual essence of her lover, so a similar line becomes the materialized essence of Butades' vase, whose classical silhouette he shapes in the clay as it rotates on the potter's wheel.

Hummel thus closely juxtaposes the work of art with craft manufacture and connects them through the classical line. His metaphorical picture also seems to depict his teaching of drawing at the Academy of Art, where he promoted the training of the mind through the hand and eye. A grasp of the basic principles of geometrical projection acquired through "numerous examples and drawings" lay behind the capacity to render correctly, as though by second nature, the realistic effects of light and shadow. "Through industrious exercise", he said, "the mind as well as the eye becomes practiced in correctly conceiving the appearances in nature and in making the *laws* on which they rest more intuitively apparent [*anschaulicher*]".[13] This view of *anschaulich* representation of laws as curves begins to get to heart of the present paper.

To see Hummel's conception of the potter's curve in practical use, one need only look at the drawings that Schinkel and Beuth published in the *Prototypes for Manufacturers and Craftsmen* for students at Beuth's *Gewerbe-Institut* (figure 11).

---

12 (Fend 2003) brightly illuminates the neo-classical line and its gender-coding in *Grenzen der Männlichkeit*; esp. ch. 3, "Körpergrenzen in Fluß: Das Ideal und seine Ambivalenzen in der Kunsttheorie des Neoklassizismus". For a succinct survey see (Mosse 1996: 24-36), on Johann Joachim Winckelmann's canonical articulation of the neo-classical ideal of masculine beauty, and (Mosse 1996: 40-45), on the perceived role of gymnastics in sculpting this ideal.

13 (Hummel 1833: vol. 1, vii-viii). See also (Hummel 1842).

Figure 11: Classical vases
(Technische Deputation für Gewerbe 1821).

The collection constituted a kind of canon of aesthetic forms, all classical, for the consumer goods of bourgeois life: tableware, wallpaper, fences, furniture, and architectural ornamentation. It formed part of a widespread attempt to elevate public taste and civic virtue through the artistic quality of the material environment within which the citizens of a modern state would live their lives. As shown for the elegantly simple vases, the *Prototypes* made quite explicit the sought-after relation between ideal curves and manufactured objects. Much more complex prototypes, such as a Corinthian column with its capital of acanthus leaves and decorative objects like rosettes received similar treatment in terms of lines essential to their form.[14] The collection as a whole emphasizes not only the smoothly flowing lines but also the familiar harmonic and periodic curves employed for cornices, decorative borders, fences, and wallpaper.

Importantly, Schinkel and Beuth make no distinction in their *Prototypes* between craft and machine manufacture, even though Beuth served as the most prominent promoter of industrial machinery in Prussia.[15] Indeed, with few exceptions, manufacture remained craft work, even when carried out with machines. More generally, the age of machine manufacture had not yet come to be seen as a tasteless era of "mechanical reproduction" but rather as

14 (Technische Deputation für Gewerbe [P. C. W. Beuth and K. F. Schinkel] 1821). (Matschoss 1921) gives extensive discussion of Beuth's activities in promoting industry. See also (Matschoss 1911: esp. 239-250).
15 Cf., Dierig, this volume. The students trained in the crafts at the *Gewerbe-Institut*, with courses in drawing, modeling, and the natural sciences, fit more nearly the model of J. G. Halske, partner of Werner Siemens than of traditional craftsmen. They were groomed to play an entrepreneurial role.

an era in which a broader cross-section of society could share in the great neo-humanist project of personal self-realization [*Bildung*] and cultivation through the universal forms of classical art and architecture. With his training in state administration [*Staatswissenschaft*] grounded in classical philology, like most members of the Prussian civil service, Beuth joined Schinkel in the attempt to realize these goals through their embodiment in buildings and furnishings of every kind.

Figure 12: Johann Erdmann Hummel, *Das Schleifen der Granitschale*, 1831 (Hummel 1831).

The project did not stop with students at the *Gewerbeinstitut* and *Bau-Akademie* but extended directly to industry itself through such organizations as Beuth's Union for the Advancement of Industry [*Verein zur Beförderung des Gewerbefleisses*], the pendent to the *Gewerbeinstitut*. Its membership placed craftsmen and entrepreneurs alongside professors, artists, and state administrators. Specifically, while Schadow, Schinkel, and Rauch all participated on the administrative committee for Architecture and Fine Art, Hummel's brother Caspar, a mechanic and founder of a machine factory in Berlin, served with other shopowners, professors, and state administrators on the corresponding committee for Mechanics and Mathematics (*Verhandlungen des Vereins zur Beförderung des Gewerbefleisses* 1822: 13). Hummel himself effectively captured this interrelationship on canvas in a well-known series of paintings depicting the machinery – perhaps supplied by Caspar Hummel – required to polish and to manipulate a huge granite bowl, which Schinkel designed to stand before one of his most famous buildings, the Royal Museum [*Königliches Museum*, since 1845 *Altes Museum*].[16]

16 Two of Hummel's paintings of the granite dish are in the Nationalgalerie Berlin, Inv.-Nr. A I 884 (polishing) and Inv. Nr. A I 843 (installed). The third (turning over) has been lost but is reproduced in (Wyss 1985), who also discusses the relation to Caspar (Wyss 1985: 127). (Hummel 1954).

Its sinuous reflecting surface, emerging under the rotative power of a steam engine rather than the potter's wheel, provided an opportunity for Hummel to display his virtuosity in the techniques of optical projection. Thus we may imagine the design by Schinkel, the machinery of Caspar Hummel, and the painting by Johann Hummel all under the control of the classical line of the great bowl.

Geometrical Realism

Having observed some of the ways in which curves were seen to capture essences in both theoretical and practical terms, I want to return to Hummel's "Origin of Drawing" (figure 10) to raise a closely related subject, to which Dürer's name had been attached since the 16th century: geometrical drawing and perspective, but in the new 19th century form of descriptive geometry. It will be apparent that Hummel places the origin of the classically curved but otherwise arbitrary lines of drawing within a highly mathematized space, ruled by linear perspective and by the linear projection of shadows cast by the oil lamp of enlightening antiquity. As professor of architecture, projection, and optics at the Academy of Art, Hummel specialized in producing such constructions in ever more intricate detail, using multiple lighting sources, multiple mirrors, and multiple perspective.

Hummel received high praise from the critics for the extraordinary optical effects that he was able to incorporate in a fully natural manner. For this period, artistic sensibilities in Berlin cohered rather well with geometrical precision in drawing. Architectural realism in painting, for example, characterized not only the works of Schinkel but of well-known painters like Franz Krüger and Eduard Gaertner. Gaertner's "Klosterstrasse" (figure 13), almost photographic in detail, appeared in an engraved version for a great collection in Nürnberg, contributed by artists from all over Germany to honor Dürer.[17]

With its depiction of Beuth, Schinkel, Gaertner, Krüger, and Rauch in the street in front of the *Gewerbeinstitut*, it suggests how closely related were the fine and manual arts in Berlin. One specific vehicle for this (partial) convergence in technique was the teaching of descriptive geometry – Hummel's subject – to students of art, engineering, and technology alike, from the *Kunstakademie*, to the *Bauschule*, to the military schools.

The subject came to Berlin largely as an import from France during and after Napoleon's occupation and followed the mathematical theory and practices of the engineers, Monge, Dupin, and Poncelet (though Hummel preferred the techniques of the earlier Berlin mathematician, Johann Heinrich Lambert). At the technical schools, both civilian and military, descriptive geometry provided part of the foundation for subsequent courses in mathematics and mechanics, as well as drawing.

A good example comes from du Bois Reymond's best friend during his youth, Anton Hallman. Figures 14a,b, suggest the transition from student exercises in descriptive geometry to a realistic anatomical drawing of the architecture of muscle and bone. Interestingly, Hallman learned his descriptive geometry at the *Artillerieschule* in Hannover.[18]

---

17 Eduard Gaertner, *Klosterstraße*, engraving, 1830, in (Mende, Hebecker 1973: 152). The *Stammbuch* continued to grow for several years after 1828.

18 (Fehlemann 1974); Dierig, this volume, and (Dierig, Schnalke 2005: 39-64).

Figure 13: Eduard Gaertner, *Die Klosterstrasse*, 1830. Staatliche Museen zu Berlin Preußischer Kulturbesitz, Nationalgalerie, Inv.-Nr. A II 736.

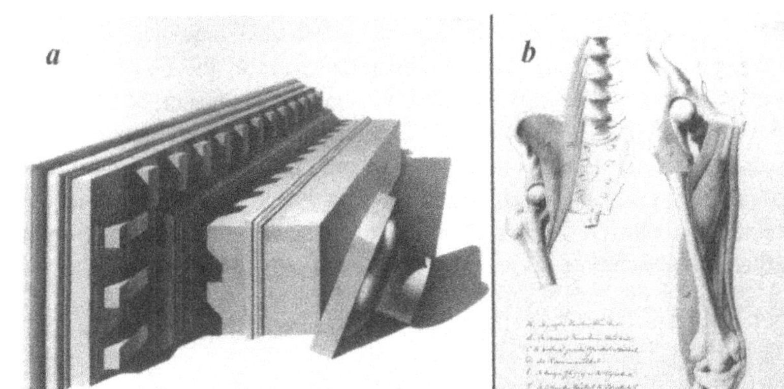

Figure 14: Anton Hallman, (a) "Schattenwürfe eines dorischen Gebälks und eines Kapitells". (b) "*Beckengelenk und Oberschenkelmuskeln*". (Hallman 1830). Both in (Dierig, Schnalke 2005: Fig. 53, Fig. 49).

Werner Siemens provides another marker for this movement in his study of descriptive geometry with Captain Meno Burg at the Artillery and Engineering School in Berlin, which he attended from 1835 to 1838. Like Hummel at the *Kunstakademie*, Burg presented descriptive geometry as the vehicle for learning to produce correct representations according to "mathematical laws". Again like Hummel, Burg couched this mathematical ideal in the all-important language of neo-humanist *Bildung* – self-realization, creative action, truth to nature, reaching the inner form of thing – things all expressed through the properly expressive line:

> "The draftsman must create out of himself [...] and must be capable, using the forms and measurements given to him, of allowing the image, in its outlines and in its intrinsic organization, gradually to emerge in lines."[19]

Burg's students at the Artillery and Engineering School, like Hummel's at the Academy of Art, could reach beyond mechanical reproduction to an authentic creative work only through extensive theoretical and practical exercise with the mathematical laws of projection, until these laws became expressions of the self, even in the depiction of a gun emplacement by Lieutenant Siemens, shown at the annual exhibition of the Academy of Art in 1838 with the title "Theil eines Walles mit einer Erdwinde und 12 pfündigen Kanone."[20] Here was an aesthetics for a particular time and place. What may look today like "mechanical drawing" was in the eyes of the Berlin drawing instructors a path toward attaining *Bildung* and an aesthetics for the modern world.

Had Werner Siemens had the financial means to study at the School of Civil Engineering and Architecture [*Bauschule*], as he had wished, he might have learned his descriptive geometry from none other than the precocious young mathematician Gustav Lejeune Dirichlet, already a member of the Academy of Sciences and associate professor at the University, although his primary teaching duties were at the War College [*Kriegschule*], where he included descriptive geometry in the first year of a three year sequence. Dirichlet also taught descriptive geometry at the *Bauschule*, with classes held at the *Gewerbeinstitut*, in 1835. Earlier, the instructor would have been another university professor, Martin Ohm, brother of Georg Simon Ohm of Ohm's-law fame.[21] That such high-powered analysts were teaching descriptive geometry to architects and civil engineers, military officers, and future technological entrepreneurs speaks once again to the perceived centrality of the subject and to its role as a crucial link – both aesthetically and practically – between the fine arts, modern industry, and the mathematical sciences in a culture obsessed with neo-humanist and neo-classical ideals.

---

19 (Burg 1916: 71-75), paraphrasing his original memorandum of 1816. On Burg and his geometrical drawing, see also, Kathryn Olesko, *Precision in German Society, 1648-1989* (in preparation), chap. 6, "Aesthetic Precision".

20 Siemens' drawing, is listed in (Börsch-Supan 1971: 1838, no. 908).

21 (Lampe 1906); (Dobbert, Meyer 1899: 43, 48).

The Mathematics of Curves

The widespread teaching of descriptive geometry begins to suggest why the progressive young men of the Berlin Physical Society might have been particularly interested in the role of curves in the sciences. But it does not yet suggest how they learned to connect the *irregular* curves of nature's reality with the highly *regular* idealized forms of geometry. It was a longstanding problem. Dürer himself had worked on it with little success. In one of his most famous etchings, "Melancholy I" (figure 15), the celestial figure of astronomy, and more generally of mathematics in both theory and practice, sits in deep contemplation with her book, compass, ruler, saw, plane, and other instruments of analysis and construction strewn about. The figure seems to be reflecting on the relation between human capacity for mathematical and material construction of artificial objects (the sphere and regular solid) and the godlike capacity to create natural objects.

Figure 15: Albrecht Dürer, *Melencolia I*, 1514 (Schuster1991: frontispiece).

Whether her mood is one of frustrated incapacity or of optimistic striving to close the gap is debated among art historians. But the comet in the night sky, still a symbol of frightening events and of God's arbitrary creative power, suggests the degree to which natural phenomena have so far eluded her rationalizing aims. Nature's reason, or God's reason, as Dürer would have preferred, far exceeded geometry.[22]

This is not the place to take up the long history of the problem, nor even the partially successful means that Dürer's geometry and 19th century descriptive geometry offered for dealing with it. To do so would be to investigate, for example, the advantages of projecting an arbitrary three-dimensional curve into two perpendicular planes, yielding a decomposition into complementary two-dimensional curves, which are much easier to manipulate conceptually or to cut into stone. Rather than follow such techniques further, it will be more illuminating to jump instead to the new approach followed by Dirichlet and Ohm.

While Dirichlet was teaching at the military and technical schools, as well as the University, he was also developing the methods of mathematical analysis that initially won him his fame. Most important for physical scientists was a rigorous proof, first presented in 1829, of the generality of the recent discovery by the French engineer and mathematician Joseph Fourier that many mathematical functions could be represented as an infinite sum of sine and cosine functions, or "Fourier series", of which the harmonic vibrations of a violin string are a familiar example (Dirichlet 1829). More formally, in Dirichlet's language, if y is an arbitrary function of x,

$$y = f(x),$$

given over a definite interval of x, say $-\pi$ to $+\pi$, then its general representation as a Fourier series is,

$$y = f(x) = a_o/2 + a_1 \cos x + a_2 \cos 2x + \ldots + a_n \cos nx + \ldots$$
$$+ b_1 \sin x + b_2 \sin 2x + \ldots + b_n \sin nx + \ldots,$$

where the a's and b's are coefficients determined by integrals of the function y over the interval,

$$a_n = (1/\pi) \int y \cos nx \, dx$$
$$b_n = (1/\pi) \int y \sin nx \, dx.$$

Fourier used such series primarily to solve problems in heat conduction, published in 1822 as the *Analytical Theory of Heat*. Dirichlet studied in Paris from 1822-1826, where Fourier became an important mentor and encouraged his extraordinary young friend to carry his own analysis further. This Dirichlet did after his arrival in Berlin in 1828. Without entering on the technical treatment, it is correct to say that Dirichlet established the validity of the Fourier series representation for a very broad class of functions of interest in the physical world, so broad that they exceeded the confines of functions that could be expressed in terms of an analytic law. The term analytic function, as formalized by Euler in 1748, refered in the first instance to algebraic expressions but extended to trigonometric, exponential, logarithmic, and other lawlike forms (Jahnke 2003: esp. 114, 121, 127).

22 (Teja-Bach 1996: 279-282); (Panofsky 1945: 156-171) makes the figure "geometry" and gives the negative reading. In an extended reinterpretation, (Schuster 1991: esp. 107-137), forcefully argues that the figure is "astronomy" and that her mood is one of striving, not depression.

Throughout the great flowering of analytic mathematics, from D'Alembert and Euler through Lagrange, Cauchy, and Fourier, mathematicians continued in practice to treat functions as analytic functions even when they recognized that greater generality was necessary. Thus Euler, with respect to the famous problem of the vibrating string, had stated by 1749 that the string could initially take on the shape of an arbitrary curve, which he sometimes called "mechanical" and which he would later describe as "not determined by any definite equation, of the kind wont to be traced by a free stroke of the hand" (Youschkevitch 1976: esp. 64-69, on 68). By 1755 he had generalized his formal definition of a function to such a degree that it corresponds to present-day views: "If, therefore, x denotes a variable quantity, then all quantities which depend upon x in any way or are determined by it are called functions of it" (Youschkevitch 1976: 70). And yet Euler actually considered only functions of analytic form in his many subsequent works (Lützen 2003: 156-158). Even much later, Cauchy in his *Cours d'analyse* of 1821 and Fourier in his *Analytical Theory of Heat*, continued this implicit reliance on analytic functions, or at least continuous functions, while giving formal definitions that rejected those limitations. Fourier in particular insisted on the arbitrary nature of a function f(x) as a succession of ordinates, remarking that "We do not suppose these ordinates to be subject to a common law; they succeed each other in any manner whatever, and each of them is given as if it were a single quantity." And yet in his proof of the convergence of the Fourier series of a function f(x) he assumed that the function was continuous, with each value differing infinitely little from those immediately adjacent.[23] The Fourier series thus continued under a mathematical cloud even after its use in physics had begun to spread. Dirichlet finally provided the missing proof.

In 1837 the Berlin physicist Heinrich Wilhelm Dove inaugurated the annual review *Repertorium der Physik*. Although the review would deal primarily with experimental physics, as did Dove himself, he had invited Dirichlet to join the editorial consortium as the representative for mathematical physics. For the first volume, Dirichlet chose to present his most important results on Fourier series, since, as he put it:

> "The remarkable series, which in a determined interval represent functions which follow no law or in different parts of this interval follow entirely different laws, have found [...] numerous applications in the analytical treatment of physical problems" (Dirichlet 1837: 152).

Through the *Repertorium* Dirichlet reached a very broad audience of experimental as well as mathematical physicists with his message, which deserves reiteration in terms of curves.

He argued first of all, following Fourier and in agreement with formal definitions since Euler, that a mathematical function – more specifically, a single-valued piecewise continuous function – need conform to no law but could be defined geometrically, as any freely-drawn curve: "This definition prescribes no general law for the individual parts of the curve; one can think of them as composed of the most diverse parts or as drawn entirely without any law" (Dirichlet 1837: 153). For physicists, this meant that if any such curve could be obtained empirically, say as data from an experiment, it constituted a valid mathematical function (e.g., displacement of an elastic string, temperature distribution in a bar). Secondly, he proved with full rigor, that such arbitrary curves could be analyzed mathematically by

23 (Lützen 2003: 156-158). See also (Grattan-Guinness 1970).

representing them as Fourier series, as sums of simple waves with different wavelengths and amplitudes. Figure 16 gives an elementary example for a constant function (level straight line) between two points which drops discontinuously to zero at the endpoints. Thereby, the most non-lawlike looking curve could be analyzed into the simplest of harmonic laws, often taken to represent the underlying rhythms of nature.

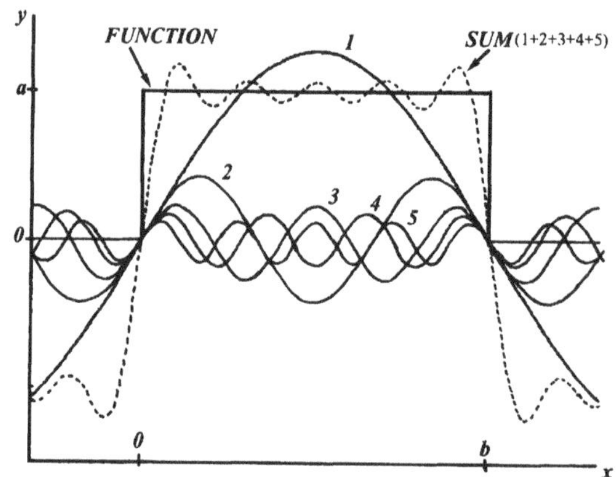

Figure 16: Sketch of the Fourier series for the function $y = a$ over the domain $(0, b)$ showing the sum (dashed line) of the first five sine wave contributions. The series converges slowly because of the discontinuities at 0 and b.

It is difficult to overemphasize the importance of the Fourier series and Dirichlet's proof for the history of physics at the time. In a most lucid and accessible way, it turned a whole range of purely experimental physics into mathematical physics through the curve, at least in principle. So enthusiastic was Bernhard Riemann, who studied with Dirichlet in Berlin from 1847 to 1849 and wrote his *Habilitationschrift* of 1854 on Fourier analysis, that he claimed Dirichlet's results covered

> "all cases in nature [...] for however great is our uncertainty about how the forces and conditions of matter change infinitesimally in position and time, we can nevertheless assume with certainty that the functions to which Dirichlet's investigations do not extend do not occur in nature."[24]

The second person who brought Fourier analysis to the attention of a broad audience in Berlin was Georg Simon Ohm, who had actually taught Dirichlet as a Gymnasium student in Cologne. Ohm moved to Berlin in 1826 to complete his now classic book on the electric circuit and then took up a part-time position at the War College for five years, where Dirichlet was also teaching. Ohm is known today largely for Ohm's law, which says that the current I through any section of a circuit is proportional to the electrical tension E (potential difference) across the section divided by its resistance R. Ohm himself expressed a more general time-dependent relation for the electric potential at any point, closely resembling Fourier's differential equation for the temperature in a bar conducting heat. Drawing on this

---

24 (Riemann 1990: 237); also quoted in (Koch 1998: 37).

analogy with heat conduction, Ohm solved the equation for electric potential with a Fourier series. Although Ohm himself supplied no illustrations of curves, his results showed once again the great power of the harmonic decomposition as an expression of physical processes (Ohm 1989: 170-176).

I will not pursue further either Fourier analysis or Ohm's work except to note that in the 1840s the members of the Berlin Physical Society would adopt Ohm, whose work had been only slowly recognized, as one of their heroes in the cause of rigorous experimental and mathematical science. His analysis of the physics of hearing, based on the assumption that the ear behaves essentially like a Fourier analyzer, so that we hear only the harmonic components of any complex sound wave, stimulated a long-standing debate over combination tones, including Helmholtz's well-known work from the mid-fifties on combination tones and on the sensations of tone.[25] Fourier analysis had by then become a pervasive tool of mathematical physics.

Return now to Dove, who had published Dirichlet's review article in his *Repertorium*. In the 1830s and 40s Dove was omnipresent in Berlin education. In addition to teaching physics at the University, where du Bois-Reymond attended his lectures, Dove taught at various times at the *Friedrich-Wilhelms-Institut* for army doctors, where he was among Helmholtz's teachers, the *Kriegschule* (where he lived with his family), the *Artillerie- und Ingenieur-Schule*, and at one or more *Gymnasia*, including the *Friedrich-Wilhelms-Gymnasium*, where his geometry course included exercises in geometrical drawing.[26] He was also one of the pioneers in Berlin of the use of curves to represent physical laws, especially laws that seemed to defy mathematical expression. This work connected his interests directly to Dirichlet's.

Already in his first publication as a newly habilitated *Privatdocent* in Berlin, "Meteorological Investigations of the Wind" (1827), Dove sought to show that the direction of the wind, which appeared to change so arbitrarily "that people had given up trying to discover anything lawlike in it", actually obeyed a regular law that could be revealed by barometric observations. On the basis of published measurements taken over 10 years in Paris, he succeeded in representing the average yearly barometric pressure (and temperature and humidity) as a periodic function of the direction of the wind, from 0° to 360° around the compass. For that purpose he borrowed an astronomical technique from the Königsberg astronomer Friedrich Wilhelm Bessel, remarking on how strange it was that the use of periodic functions in meteorology had languished for fifty years, ever since Lambert (Hummel's referent for geometrical projection) first employed them.[27]

Dove wrote the barometric pressure $b(m)$ around the compass as an arbitrary function, represented as an infinite sum of sine functions, much like a Fourier series, but evaluated only for n discrete directions of the wind, indexed by m,

$$b(m) = u + u'\sin(m\cdot 2\pi/n + U') + u''\sin(m\cdot 4\pi/n + U'') + \ldots,$$

25 (Turner 1977); (Jackson 2006: ch. 6). (Ohm 1843) and (Ohm 1844) both in (Lommel 1892: 587-633, 634-649); (Helmholtz 1856: 497-540) in (Helmholtz 1882: vol. I, 263-302).

26 (Neumann 1925: 13-14). Gymnasium courses in *Programmschrift, Königliches Friedrich-Wilhelms-Gymnasium* (Berlin 1838: 53-60).

27 (Dove 1827: 545, 550). Revised version in (Dove 1837: 97-120); (Bessel 1815: 9).

where the u's and U's are arbitrary amplitudes and phase angles. He then, using a least-squares procedure of Bessel's, fit the first three terms of the series to average barometric data obtained for winds directed toward sixteen compass points to obtain an empirical law of atmospheric pressure as a function of wind direction (and similarly for temperature and humidity). Since the third term turned out to be small, he could write an approximate law for a continuous function of the angle x in the simplified form,

$$b(x) = a - c \sin(x + \alpha).$$

It will be apparent why Dove would have been interested in publishing Dirichlet's proof of the validity of Fourier analysis of empirical curves. Although Dove did not publish the curves for his laws, he suggested that the reader should construct them from his tables to show the cycle of the barometer as the wind moved around the compass. Such a graphic representation, sketched in figure 17, would also show the symmetry between opposite sides of a line drawn from NNE to SSW, identifying what he called the "meteorological meridian," by analogy to geographical and magnetic meridians (Dove 1827: 585, 590).

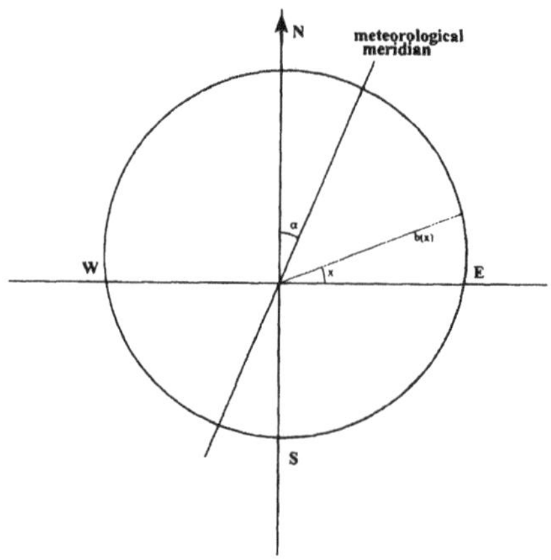

Figure 17: Sketch of Heinrich Wilhelm Dove's approximate law for atmospheric pressure as a function of angle around the compass, $b(x) = a - c\sin(x + \alpha)$.

Better known among Dove's analyses with curves is his "law of storms", according to which storms are rotating systems, rather than centripetal winds blowing toward a point of low pressure, and normally rotate counter-clockwise in the northern hemisphere and clockwise in the southern. Dove explained the law as resulting primarily from the effect that the rotation of the earth has on storms moving from lower to higher latitudes, with secondary effects on both rotation and progression resulting from prevailing winds. Figure 18 exemplifies the law of storms for the rotation and progression of a hurricane in the West Indies.[28] This is

28 (Dove 1841), using charts of storm rotation and progression from other meteorologists in the US and Britain. Several later editions and translations were published, e.g., *Ueber das Gesetz der Stürme* (Berlin: Reimer, 1857).

one of the more impressive of Dove's attempts to articulate meteorological laws through the use of curves.

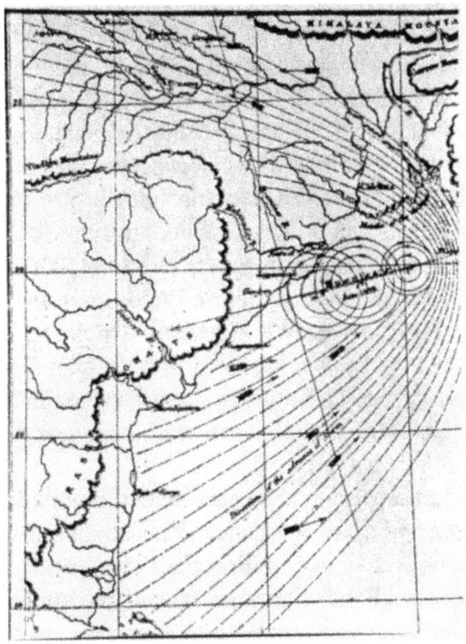

Figure 18: Heinrich Wilhelm Dove's use of flow lines to show the pattern of a storm spinning off a hurricane south of Calcutta. The hurricane rotates counterclockwise and moves to the west (Dove 1857).

Alexander von Humboldt: Patron of the Curve

If by 1840 Dove and Dirichlet represented the pinnacle of current practice in the physical and mathematical use of curves, there stood behind them, both intellectually and politically, a patron of enormous prestige. Alexander von Humboldt had come back to his native Berlin in 1827 after twenty years in Paris, where he published the grand volumes that contain the scientific account of his travels in South America and Mexico with Aimé Bonpland from 1799 to 1803. A favorite of King Friedrich Wilhelm III, Humboldt had officially held the title of Chamberlain since 1805 and returned to Berlin at the king's insistence to take up his duties, with an enhanced salary of 5000 Thaler. He returned like a modern Columbus. His lectures at the University and at the *Singakademie* in 1827-1828 – overlapping with the Dürerfest at the *Singakademie* – made him a sensation in Berlin society and laid the foundation for the most popular scientific book of the 19th century, his five-volume *Kosmos* (1845-1862).

Not so well known is that images of lush tropical landscapes from his *Vues des Cordillieres* (1810) and of mysterious peoples associated with them had already become familiar to Berlin opera-goers in the scenery that Schinkel designed for Mozart's *Magic Flute* (1816) and for a series of other operas: *Athalia* (1817), *Fernand Cortez* (1818), *Armida* (1820). Especially relevant for capturing the exotic and romantic aura of Humboldt's reception was Schinkel's scenery for the Spontini opera *Nurmahal,* which complemented Humboldt's popular lectures of 1828. It was an adaptation of a part of Thomas Moore's lyric romance *Lalla Rookh,* which in 1821, on the occasion of a visit by the king's daughter Charlotte and Archduke Nicholas, had been converted into an immensely popular musical. Performed before 3000 invited guests at the royal palace, it featured music by Spontini and "living pictures" by Schinkel, consisting of Humboldtian scenery animated by 168 living figures from the court, starring the honored guests. These scenes of tropical imagination were ultimately realized inside one of Schinkel's exquisite architectural productions, designed to house a large collection of palm trees, acquired in part with Humboldt's advice. The *Palmenhaus,* completed in 1832, would stand as the signature building on the *Pfaueninsel* [Peacock Island], a Royal estate in the Havel River near Potsdam, which had been developed as an English garden and menagerie from the beginning of the century. The enchanting *Palmenhaus* became the subject of well-known paintings by Karl Blechen in 1834, depicting an interior that immediately recalled *Lalla Rookh* and *Nurmahal,* complete with seductive Indian women reclining on carpets.[29]

Images of exotic lands and peoples thus surrounded Humboldt's popular persona as he went about establishing his scientific prescence. The famous lectures at the *Singakademie* were followed by an epochal meeting in Berlin of the *Verein Naturforscher und Ärzte,* which Humboldt organized with his longtime friend, the professor of zoology Heinrich Lichtenstein. It attracted participants from all over Germany, as well as foreign guests, and helped to instill a new sense of pride and confidence among German scientists. Staging was important, as Humboldt clearly understood, for with Schinkel's help, he employed some of the same sort of operatic scenery that already carried his popular image. For a celebratory evening session, held in one of Schinkel's greatest buildings, the *Schauspielhaus,* with Humboldt himself as the featured speaker and with King Friedrich Wilhelm III in attendance, Schinkel adapted his earlier Queen-of-the-Night scene from *The Magic Flute* for a backdrop. Like stars in the heavens above Zoroaster's temple, the names of Germanic scientists, from those of the first magnitude, like Copernicus, Kepler, Euler, Leibniz and Kant, to somewhat lesser lights like Stahl, Bernoulli, Lambert, Fraunhofer, and Klaproth, shined down on their earthly heirs. Music too enhanced the unifying spirit of the evening, including a choral piece composed at Humboldt's request by the precocious young Felix Mendelssohn-Bartholdy, one of the talented family of Abraham Mendelssohn-Bartholdy, at whose home Humboldt was a welcome guest.[30]

---

29 (Humboldt 1810: e.g, plates 31, 33, 41-42, 63), (Harten 2000: 132-135, 228, 233-35, 237, 266, 271, 274, 340). (Muthmann 1955: 91-102). Schinkel's twelve living pictures, as drawn by Wilhelm Hensel (husband of Fanny Mendelssohn-Bartholdy, Dirichlet's sister-in-law) were published as *Die Lebenden Bilder und pantomimischen Darstellungen bei dem Festspiel Lalla Rukh* (Hensel 1823). Full report with songs in *Neue Berliner Monatsschrift für Philosophie, Geschichte, Literatur und Kunst* 1 (1821): 315-343. (Seiler 1989) gives a rich account of the *Palmenhaus.* Humboldt's indirect presence at the *Pfaueninsel* is described in (Wise, Wise 2003).

30 (Humboldt, Lichtenstein 1829: 19) (schematic order of names for the backdrop). (Rave 1962: 363) (drawing of the hall with backdrop in the *Schauspielhaus*). See also (Jackson 2003), who gives a fascinating account of the role of music among the *Naturforscher*, especially at the Berlin meeting, and whom I thank for

As the cultural life of the city shaded seamlessly into the life of science, Humboldt went about promoting the first-class research structures that he envisaged for Berlin, drawing heavily on his personal relationship with the King, with his Minister of Culture, Altenstein, and with numerous friends throughout Berlin society.[31] He recruited young talent whereever he saw it, including the newly arrived Dirichlet and Dove. Almost immediately he set up a magnetic observatory to extend earlier work from 1806-1807, recording hourly and daily variations in the direction of the earth's magnetic field. For the purpose, Abraham Mendelssohn-Bartholdy offered the large garden of the family home, while the ubiquitous Schinkel contributed the design for a small iron-free observing house, built of brick with copper nails and hardware. Around this small observatory, linked through Humboldt's promotional activities into an international network of similar sites taking corresponding observations, the charismatic organizer attracted the young physicists and mathematicians of Berlin: Dirichlet, Dove, Magnus, Encke, and Poggendorf, among others. It was Dove who proudly took charge of publishing their results of 1829-1830, represented graphically in sixteen plates of curves of magnetic declination, comparing the Berlin observations with simultaneous readings from the string of observatories from South America to Russia.[32]

Humboldt's own use of curves was already well-established. Famous are his "physiognomic" projections of landscapes in South America and Mexico. In their simplest form they were vertical cuts, yielding a silhouette of rising and falling elevations over mountains, plateaus, and valleys, from which much could be read about the character of the landscape. These physiognomic silhouettes had deep significance for Humboldt. "The physiognomy of a country", he wrote,

> "the grouping of its mountains, the breadth of its plateaus, and their elevation, which determines their temperature, everything that belongs to the structure of the globe, stands in intimate connection with the advances of the population and with the well-being of the inhabitants".

It functioned much like Johann Caspar Lavater's physiognomic reading of human character. As a young man, in fact, Humboldt had taken drawing lessons from Chodowiecki, who did Lavater's illustrations.[33]

In the more elaborate form of the "physiognomy of plants" Humboldt inscribed characteristic species on his vertical projections as well as on the more usual surface projections, yielding changing zones of vegetation with changing elevation, latitude, and longitude. Through physiognomy, Humboldt sought not a botanist's taxonomic classification of the vegetation but rather "that through which its mass individualizes the total impression of a

---

very helpful discussions. We differ on whether the Schinkel backdrop was at the *Singakademie* or the *Schauspielhaus*.

31 Humboldt's promotional efforts are best captured in the collections of his letters published by the Akademie der Wissenschaften, e.g., (Biermann 1985).

32 (Humboldt 1829: 333); (Dove 1830), with a foreword by A. v. Humboldt; (Dove 1869: 22-23).

33 On Humboldt's physiognomic vision see (Dettelbach 1999). Alexander von Humboldt, *Essai politique sur le royaume de la Nouvelle-Espagne* (1808-1814), German in (Humboldt 1989-1997: vol. 4, *Mexico-Werk: Politische Ideen zu Mexico, Mexicanische Landeskunde* (1991), with plates from the *Atlas géographique et physique du royaume de la nouvelle-espagne* [*Mexico Atlas*], (1811): on 119f). (Lavater 1968 – 1969); (Beck 1959: vol. I: 11).

region," a mode of description suitable for a landscape painter in an analytic mood.[34] From this macroscopic analysis he identified eighteen main forms of plants characteristic of different climate zones, from the tropics to northern latitudes and from sea level to the tops of mountains.

Importantly for the present discussion, and as Michael Dettelbach has persuasively argued, the painterly character of Humboldt's physiognomy of plant zones cannot be split off as a romantic remnant from his equal emphasis on precision measurement of physical quantities: temperature, pressure, altitude, and magnetic parameters. In fact, for Humboldt, quantitative mapping was precisely what would reveal the qualitative landscape. This passage between quantitative and qualitative is particularly apparent in the curves with which he attempted rigorously to define the distribution of climatic zones over the surface of the earth. His pathbreaking "isothermal lines" of 1817 were curves of constant annual mean temperature mapped over the northern hemisphere for both surface position and mountain elevation (figure 18).[35] For Humboldt the isothermal lines continued to express the physiognomy of nature, a concept that included both art and science, somewhat like descriptive geometry.

By the time he arrived in Berlin, Humboldt envisaged a much broader program to incorporate variations over time into the curves of nature, as in the variations of magnetic lines, and to extend his method to a whole range of meteorological measurements. The time had come to discover whether "the pressure of the atmosphere, the quantities of rain falling from the air, the relative frequency of prevailing winds, and the direction of isothermal lines, like the distribution of magnetism over the earth, are subject to secular variations" (Humboldt 1829: 319). This is the program that Dove made into his life's work, adapting Humboldt's method of curves to reveal the laws of meteorology. Indeed, the curve of figure 17, relating the direction of the wind to average barometric pressure, was inspired in part by Humboldt's isothermal lines, which Dove cites, and Humboldt's extensive observations on climate. Dove had just completed his *Habilitationsschrift* on the distribution of heat over the earth, the distribution that Humboldt had depicted with his isothermal lines.[36]

The isothermal lines inevitably recall also Humboldt's personal acquaintance with Joseph Fourier in the Paris Academy of Sciences and with his *Analytical Theory of Heat* of 1822. On the basis of calculations dependent on the mathematical theory of the diffusion and radiation of heat, Fourier had become a leading proponent of the view that the earth was a cooling body, most likely having been formed originally as a molten mass. That view, which informed all of Humboldt's work in physical geography, had major implications for the isothermal lines at the surface of the earth as well as for the internal temperatures that Humboldt and others had measured deep in mines and in the water issuing from springs. He looked forward in 1817 to the "beautiful analytical work with which Fourier will soon enrich general physics" (Humboldt 1817: 94).

---

34 (Humboldt 1987: 184). Humboldt's physiognomic perspective attained its most extensive form in (Humboldt, Bonpland 1989: 43-161).

35 (Dettelbach 1999: 473-487); (Humboldt 1817); strangely, the chart appeared only in a separate publication (Paris: Perronneau, 1817); German in Beck, *Studienausgabe*, vol. 6, *Schriften zur physikalischen Geographie* (1989), 18-97, chart on 19.

36 (Dove 1827: 578). (Dove 1826a); (Dove 1826b).

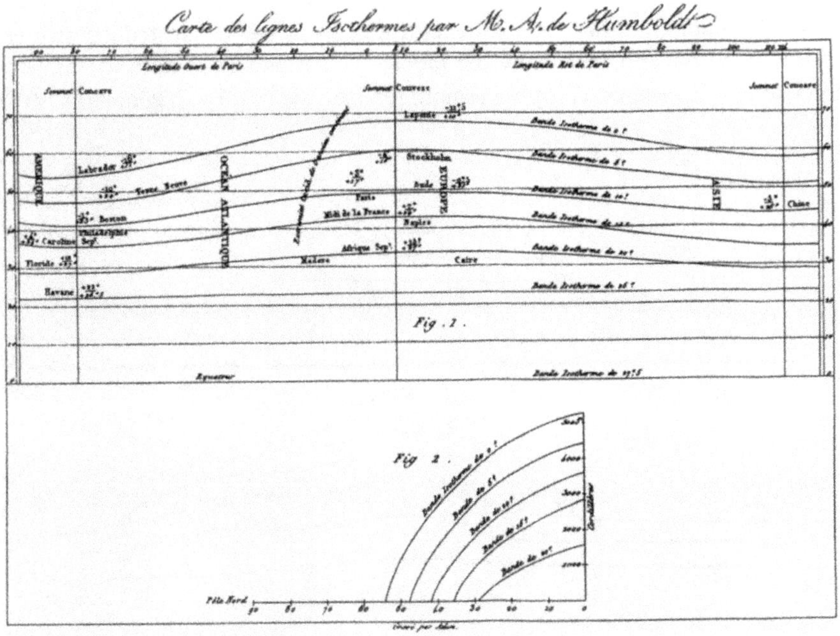

Figure 19: "Isothermal lines" (Humboldt 1817) from America to China (horizontally) and from the equator to the north pole (vertically). The lower figure shows how the isotherms vary with altitude (vertically) and latitude (horizontally).

Humboldt knew Fourier well by the time the young Dirichlet joined Fourier's circle in 1825. And it was at Fourier's instigation that Humboldt arranged for Dirichlet to obtain his appointments in Prussia, first at the University of Breslau in 1827 and then from 1828 successively at the *Kriegschule*, the University, and the Academy of Sciences in Berlin. Humboldt's loyal friendship smoothed Dirichlet's entire career, including even his marriage in 1832 to Rebecca Mendelssohn-Bartholdy (daughter of Abraham), following Humboldt's introduction to the family.[37]

1828 was a great year for curves. As though in a stellar conjunction, Dürer, Humboldt, Dirichlet and Dove arrived simultaneously in Berlin. Specifically, the careers of Dirichlet and Dove were hitched to Humboldt's star, in terms of both patronage and intellectual interests. Unlike Dirichlet, Dove never formed a close personal relationship with Humboldt, but for both of them their pursuit of curves as the expression of nature's laws had direct roots in Humboldt's isothermal lines. Those lines (as well as the patronage) would continue right into the next generation. In 1849, while writing his *Kosmos*, Humboldt turned to his much younger friend Emil du Bois-Reymond for an opinion on what, according to Fourier's theory, the isothermal lines would look like on a vertical projection through a mountain, assuming the earth cooled by radiation into space and neglecting both the sun's heat and the cooling air. Responding that Fourier's treatise contained nothing specifically on the temper-

37 (Biermann 1982), see the Introduction and early letters from 1825. (Biermann 1959: 12); (Kummer 1897: esp. 314-324).

ature inside mountains and that the mathematical analysis had exhausted his own capacities, du Bois-Reymond nevertheless knew the theory well enough to discount Humboldt's suggestion of parallel lines and to propose a more realistic distribution (figure 20 a, b).[38]

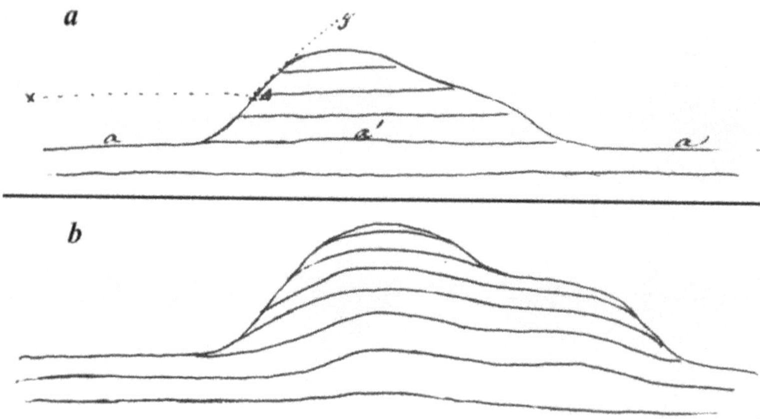

Figure 20: (a) Humboldt's suggested isothermal lines through a mountain. (b) du Bois-Reymond's proposed correction (Schwartz, Wenig 1997: 98-100).

Laws as Curves in the Berlin Physical Society

In practitioners like du Bois-Reymond and Helmholtz we see how the Humboldt-Dove-Dirichlet nexus of curve production became a part of the scientific literacy of a new generation that sought to make physical science the basis of all natural science, using curves to capture the lawlike character of apparently non-lawlike phenomena in nature. We see too how crucial was Dove's role both as a ubiquitous teacher in the Berlin educational network and as editor of the *Repertorium der Physik.* Dove taught physics to du Bois at the University and to Helmholtz at the Friedrich Wilhelm's Institute. And while he no doubt exposed them to the virtues of graphic representation, he also brought to their attention the latest works on electricity by Ohm and by Faraday, whose lines of magnetic force prefigured his own lines of storms. Du Bois carried both the electricity and the lines into physiology in the early 40s, where he began to study the electrical stimulation of nerves and muscles, culminating in his *Untersuchungen über thierische Electricität* of 1848-1849. His usual source of experimental material was the frog (figure 21), here with the nerve of the left leg connected to an electrical circuit through the sponge-like electrodes on the left. Du Bois-Reymond's drawing will recall his talents as a draftsman. Normally he operated on excised muscles and nerves and made every effort to represent his results graphically.

38 Du Bois-Reymond to Humboldt, shortly after 18 August 1849, in (Schwarz, Wenig 1997: 98-100).

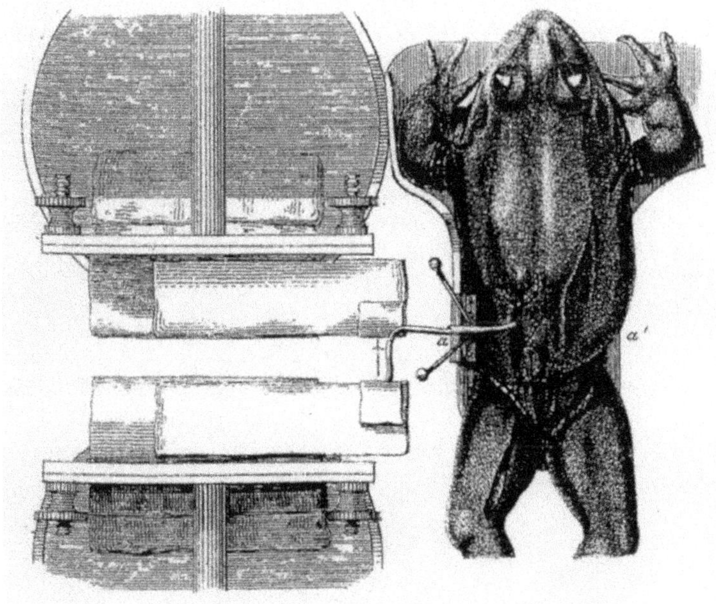

Figure 21: Prepared frog. (du Bois-Reymond 1848-1884: vol. II (1849), plate IV, Fig. 129).

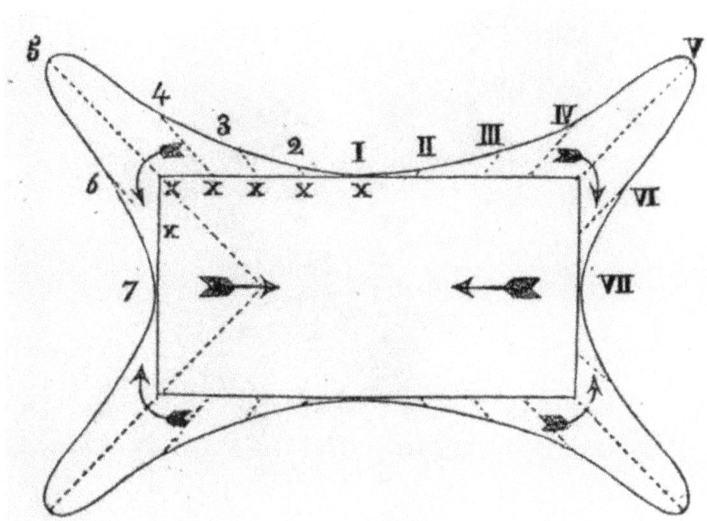

Figure 22: Diagram of du Bois-Reymond's law of the frog current. (Ludwig 1852: Fig. 81).

Figure 22 depicts his first major discovery, the law of the frog current. The diagram shows a rectangular section of freshly prepared muscle, with fibres running longitudinally, and a curve of current strength, which surrounds the rectangle. This curve of current (for the top left portion) results from placing two electrodes a short distance apart and moving them in steps between the x's along the longitudinal surface and down the cross-section. Du Bois-Reymond's sensitive galvanometer showed that a current will always flow between the electrodes in the direction of the arrow around the corners, with a strength increasing to the corner (5) and then decreasing to the mid-point on the cross section (7). The ordinates of the curve are the dashed lines parallel to the bisecting line of the corner. No current flows with the electrodes placed symmetrically on two sides of the mid-points at I and 7. This inventive if complicated representation, with the curve superposed on the muscle section and showing all of the symmetries of the phenomenon, suggests du Bois-Reymond's fascination with laws as curves.[39]

Frogs, however, were not his only experimental animal.

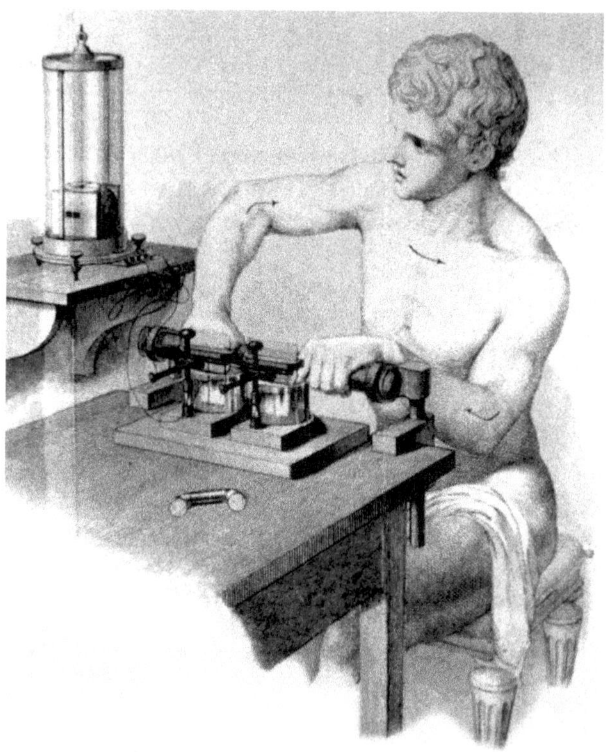

Figure 23: Self-depiction as experimenting Greek athlete.
(du Bois-Reymond 1848-1884: vol. II, part 2 (1884), plate V).

39 This description refers to the somewhat simplified picture in (Ludwig 1852: 316-317, Fig. 81). The original is in (du Bois-Reymond 1848-1884: vol. 1 (1848), 583-595 and plate V, Fig. 57).

In figure 23 he has drawn himself with the youthful beauty of a Greek athlete. He is measuring the current that passes over his body when his right bicep is strongly contracted while the left remains relaxed. Du Bois-Reymond's ability consistently to produce deflections of his galvanometer actually resulted from a highly skilled performance, mastered only after extensive practice in controlling his own body. As Sven Dierig has emphasized, this bodily control as experimenter mirrors du Bois's bodily control as a gymnast, attained through many years of exercise on the bar, beam, and horse. Some other experimenters, apparently lacking du Bois's coordination, had difficulty achieving his consistent results.[40]

In a preceding experiment, using the same setup, du Bois's two index fingers were immersed in baths of water at different temperatures, one at 0° centigrade and the other varying from 0° to 45°.

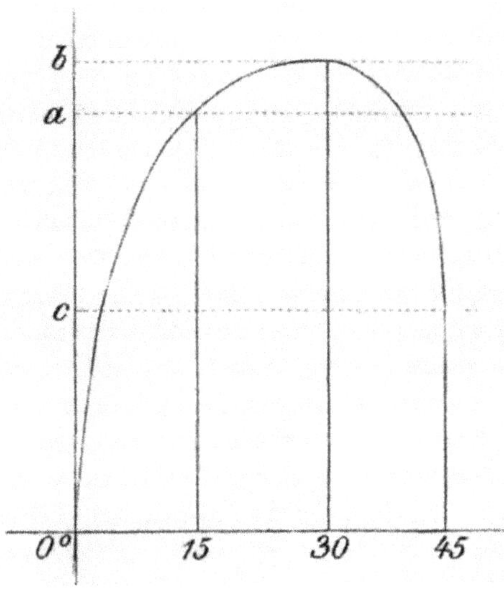

Figure 24: Curve of current between immersed fingers – in setup similar to figure 23 – with the temperature of one bath held constant at 0° and the other varying from 0° to 45°.

In this case the curve he ultimately obtained of current versus temperature difference (figure 24) was quite simple, increasing from zero to a maximum at 30° and then decreasing rapidly to zero again at 45°.[41]

Dierig suggests that in his classical self-representation, du Bois intended to portray himself as an Apollonian figure. Within the pervasive neo-humanist value structure of the *Bildungsbürgertum*, Apollo epitomized manly beauty and the virtues of athletic excercise, particularly gymnastics, as a component of *Bildung*. One of the original promoters of the gymnastics movement, J.F.C. Guts Muths, took the most famous sculpture of Apollo, the Apollo

40 (du Bois-Reymond 1848-1884: vol. 2, part 2 (1884), 287 and plate V, Fig. 147). Dierig, "Apollo's Tragedy," this volume; (Dierig 2006b: 122-134 and n.95). In a letter to Helmholtz on 3 August 1852 du Bois stated that the experiments and drawing were complete. (Kirsten 1986: 135-137).

41 (du Bois-Reymond 1848-1884: vol. 2, part 2 (1884), 208 and plate V, Fig. 146). Again, the experiment and figure date from 1852.

Belvedere, to epitomize the shaping of moral and physical manliness through gymnastics.[42] This interpretation of du Bois's self-image is thoroughly consistent also with the idealist conception of beauty that his closest collaborator in founding the Physical Society, Ernst Brücke, held throughout his life, reflecting the neo-classical aesthetics that he had learned from his artist father and uncles and that he recognized in the smoothly muscled bodies of trained gymnasts and acrobats (Brücke 1928: 139-146). Extending this reading along Dierig's lines, it was through similarly disciplined training with his electrical apparatus that du Bois attained his smooth curve of current as a function of temperature difference. His Apollonian stance thus represents an effacement of his own individuality and the attainment of a view from Olympus, a universally valid view in which his trained body behaves like the other mechanically and electrically stabilized parts of his apparatus, consistently producing a response curve which carries the verisimilitude of nature's laws.

Among the members of the Physical Society, such attempts to represent laws as curves ultimately aimed less at employing the disciplined self as a recording instrument than at developing the skills to use mechanical and electrical instruments that would draw the curves directly, thus "self-recording" or "self-registering" instruments. But as observed previously with respect to Hummel's "Origin of Drawing" and to descriptive geometry, the relation between the disciplined hand and the automatic machine remained quite subtle. It had not yet become the dichotomy of artist and machine that would demote Hummel's projective realism to photographic mechanics in the 1840s and that would put descriptive geometry more nearly in the category of mechanical drawing than of artistic perspective. For the experimenters of the Physical Society, in particular, self-recording instruments were aids in the effort to reveal the essences of nature in the form of curves. They sought ideal forms, not photographic realism and not the confused and contingent appearances of particular events.[43] One aspect of this perceived intimacy of values between the mechanic and the disciplined artist was surely the instability of the sensitive instruments they were attempting to perfect. Simply to calibrate du Bois's galvanometer, for example, required practiced skills analogous to those of his Apollonian gymnastics. The same is true of a variety of self-recording instruments that the members of the Physical Society would develop. But more generally, du Bois, Brücke, and Helmholtz all maintained the classicizing aethetics of their youth throughout their lives. I will indicate briefly how this simultaneous commitment to mechanics and aesthetics played out for Helmholtz.

Primary sources in mechanics for the proliferation of self-recording instruments in the 1830s and 40s were the dynamometers and indicators developed by engineers to record the work being done by any working machine, whether powered by muscle, wind, water, or steam. The indicator for a steam engine, because it responded to the pressure inside the cylinder, was often said to take its pulse, like a stethoscope: "In the hands of a skillful engineer, the Indicator is as the stethoscope of the physician, revealing the secret workings of the

---

42 Dierig, "Apollo's Tragedy," this volume. (Guts Muths 1804: 9), cited in (Mosse 1996: 41).

43 For this reason, the term "mechanical objectivity" employed for atlas makers of the 19th century by (Daston, Galison 1992); (Daston, Galison 2007: Chapter 3), is not appropriate to the aims of the Physical Society. More generally, while they make a very persuasive case for the atlases, which belong to the tradition of natural history, I am skeptical about its extension to natural philosophy, primarily because of the strongly idealizing practices of both mathematical and experimental physics.

inner system, and detecting minute derangements in parts obscurely situated" (Hopkinson 1857: title page) from (Main, Brown 1847: 5).

Figure 25a shows how an indicator from the 1840s was screwed into the top of the cylinder of a steam engine (Hopkinson 1857: title page). The string attached to the connecting mechanism of the engine's main beam produces one revolution of the recording cylinder of the indicator (figure 25b) for each cycle of the engine's piston, while a stylus rises and falls with the pressure in the cylinder. Although originally invented (in a much simpler form) by James Watt and his master mechanic John Southern in 1796, the indicator remained almost unknown until the 1820s and received rapid development only in the 1830s and 40s, when various versions of the rotating drum were introduced.[44]

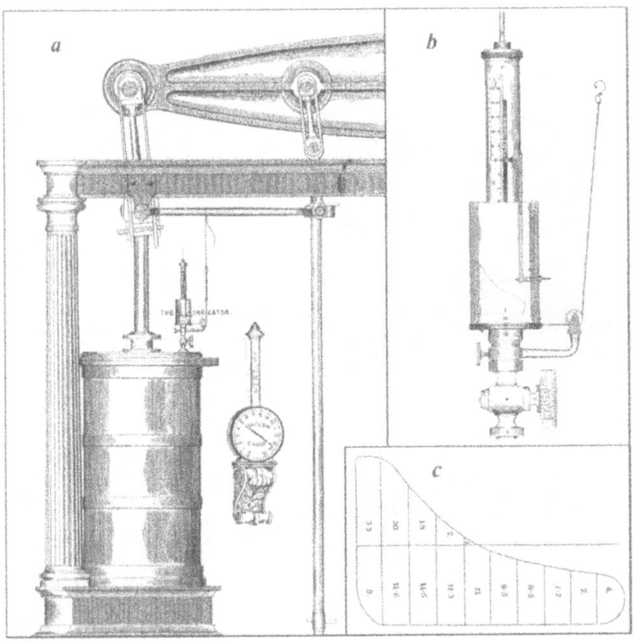

Figure 25: (a) Steam engine indicator mounted on top of engine cylinder.
(b) The indicator, with stylus drawing on rotating cylinder.
(c) An indicator diagram. All in (Hopkinson 1857: frontispiece, facing 9, 70).

Since the recording drum rotates with the motion of the piston, and thus in proportion to the volume of the cylinder, while the stylus records the rise and fall of pressure in the cylinder, the indicator effectively traces a curve of pressure vs. volume. Consequently, the area enclosed by the resulting "indicator diagram" (figure 25c) measures the work done by the en-

44 (Dickinson, Jenkins 1981: 228-233). (Hills, Pacey 1972) Indicators were described in Berlin for Beuth's in *Verhandlungen des Vereins zur Beförderung des Gewerbefleisses*: L. G. Treviranus, "Instrument, das Vacuum und den Druck des Dampfes in dem Cylinder einer nach dem Prinzip von Boulton & Watt gebauten Dampfmaschine zu messen und danach die Kraft der Maschine zu berechnen," (1830), 72; Haun, "Instrument zur Beobachtung der Kraft bei Dampfmaschinen" (1830), 228. For the Physical Society, Werner Siemens provided an obvious source.

gine in one cycle: i.e., the integral of pressure times change of volume over one revolution. The diagram provides both a measure of the capacity of the engine and simultaneously a visual control on adjustments intended to optimize its performance.

Helmholtz was familiar with the use of these diagrams for steam engines by mechanics and with their transformation into the idealized Carnot diagrams for generalized heat engines by mathematical engineers and physicists. He was equally familiar with the physiological instruments derived from them, especially the "kymograph" invented by his friend Carl Ludwig in 1847, which produced a graphical recording of blood pressure or respiration. The object of my present interest, however, is the frog-drawing-machine [*Froschzeichenmaschine*, later myograph] shown in figure 26, which Helmholtz developed between 1848 and 1852.[45]

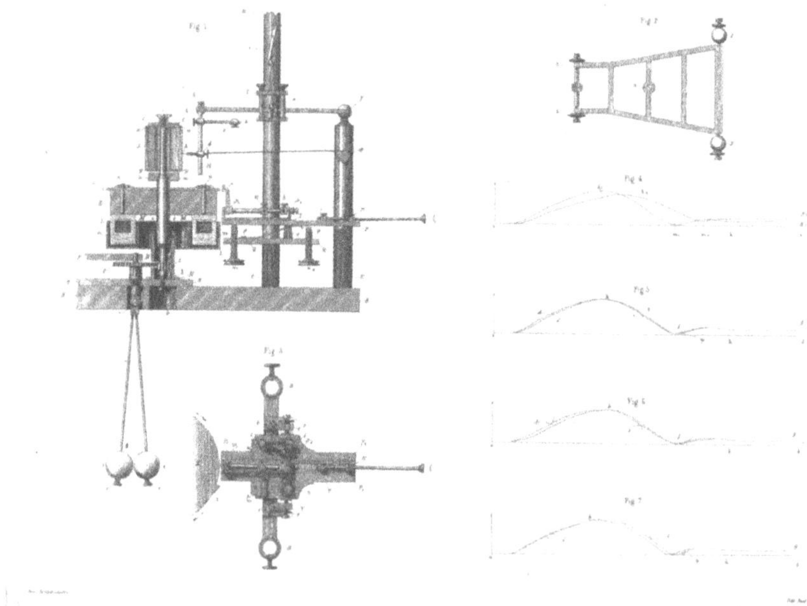

Figure 26: Froschzeichenmaschine (Helmholtz 1882-1895: vol. II (1883), plate II).

According to Ludwig, both his own and Helmholtz's instruments derived directly from the indicator. Helmholtz's device treats the contracting and relaxing frog muscle quite literally like an engine burning fuel to produce work. The frog muscle (not shown) pulls on the hook at top center in the main drawing. When contracting, it lifts a frame (top view on the right) which carries a stylus on its left end. The stylus draws a curve on the rotating drum to the left of one cycle of contraction and extension of the muscle, like an indicator curve, but here

45 I give a reinterpretation of this work, focusing on its relation to Helmholtz's conservation principle and his aesthetics, in *Bourgeois Berlin and Laboratory Science* (in preparation), ch. 8, "*Ein Schauspiel für Götter.*" This account builds on the papers of Olesko and Holmes cited below (Olesko, Holmes 1993). The main primary sources are (Helmholtz 1850a), (Helmholtz 1850b), (Helmholtz 1852), all in (Helmholtz 1882-1895).

representing at any point the net work that has been done by the muscle.[46] The conical pendulum on the bottom left, originally intended as a regulator of rotational speed, may have derived from a differential governor for steam engines developed by Siemens. Finally, the timing scheme for triggering the electrical stimulus to the frog's nerve at a definite point in the drum's rotation (detailed at bottom left), was adapted from a precision technique that Siemens had developed for measuring the muzzle velocity of cannon balls. Interestingly, at one of the first meetings of the Physical Society in 1845, Siemens and du Bois-Reymond had reported on a closely related technique and du Bois had suggested adapting it for muscle contractions.[47] In realizing this idea, Helmholtz obtained a fairly precise measure of the temporal process of muscle contraction following a stimulus.[48]

Helmholtz's results for what he called the curve of *Energie* of the frog muscle, appear on the right. The curves write out the muscle's action in the language of engines. His Fig.'s 5 and 6 show results for fresh and slightly tired muscles while Fig.'s 4 and 7 are controls, for tiredness and irritability. Figure 5 may be taken as the iconic result. Its two curves demonstrate two things in a perspicuous manner.[49] First, the muscle only develops its *Energie* over time (about 0.15 seconds to raise the weight to maximum height). This was a surprising result for physiologists when Helmholtz first announced it in 1850, although he fully expected it on the assumption that the *Energie* resulted from purely physical-chemical processes going on in the muscle. The result served to confirm the thesis of conservation of force that he had developed at length in his (not yet) classic paper of 1847, *Ueber die Erhaltung der Kraft*. Secondly, the observeable delay time suggested to Helmholtz that he might actually measure the time it took for a stimulus to reach the muscle when propagated along the nerve. The two curves of Fig. 5 depict Helmholtz's famous demonstration that the nerve impulse requires time for propogation. They are displaced relative to one another because they are stimulated from different points on the nerve. Dividing the distance between the stimulation points by the time represented by the displacement, shows that the nerve impulse travels at the relatively slow speed of 27 meters per second, less than 1/10th the speed of sound in air, again confirming the assumption of ordinary physical processes.

So far we have seen only Helmholtz's mechanics. But he had had five years of training in drawing at the Potsdam Gymnasium and when he began his work on the *Froschzeichenmaschine* in 1848 he had just taken up a position at the Academy of Art in Berlin teaching anatomy to art students, a position in which Brücke had preceded him and du Bois-Reymond would follow. The relation between these two activities can be symbolized by the fact that he made the drawing of the machine himself, entering *Gez. v. H. Helmholtz* (drawn by H. Helmholtz) on the lower left, in the manner of artists, with the lithographer on the bottom right. More deeply, his aesthetic values can be seen in his usage of *Form*, in both contexts. In his *Probevortrag* for the Academy of Art, his conception of *Form* appears in the adjectives he regularly associates with the term – *lebendig; ideal; harmonisch; geistig* – and

46 (Ludwig 1852: vol. 1, 333). (Brain, Wise 1994); reprinted in (Biagioli 1999).

47 Protocoll der Physikalische Gesellschaft zu Berlin: 1845, 4te Sitzung, 7 March 1845, reports 6-8. Archiv der Deutschen Physikalischen Gesellschaft.

48 (Siemens 1891 [1845]), (Siemens 1891 [1845]), (Siemens 1847). Protcoll der Physikalischen Gesellschaft zu Berlin: 1845, Archiv der Deutschen Physikalischen Gesellschaft, Nr. 10001, 7 March 1845.

49 (Helmholtz 1882-1895: (1852) plate II). A nice analysis focusing on "qualitative precision" is (Holmes, Olesko 1995).

with the artist's capacity to express it – *Anschauung der Form; künstlerischer Schönheitssinn; künstlerischer Geist.* Unlike the closely related term *Gestalt*, *Form* refers in Helmholtz's usage not to the particular shape that a muscle may have on a specific body but to its type and especially to its Idea. The artist must be able to perceive this *Form* in an immediate, intuitive, and lively Anschauung. Training in anatomy is crucial to recognizing the *Form* and its causes and to differentiating essential from non-essential features of a particular shape [*Gestalt*], although it can never replace the *künstlerische Geist.*

> "It is a means which facilitates for the artist his spiritual victory over the ever-changing manifoldness of his earthly object, the human *Form*, a means which should sharpen his view of the essential in the *Gestalt*, and which should equally make transparent to him the entire *Gestalt* [...] But art, I would like to say, begins only beyond anatomy. The artistic spirit reveals itself first in the wise application of the *Forms* whose interconnection and elementary features anatomy has taught; it reveals itself in the decisive characteristic of the *Gestalt*" (Helmholtz 1848: 102-103).

Thus it is through the realization of the *Form* that an artist produces the beauty of a particular *Gestalt.* And just because it is the *Form* and not the *Gestalt* that is of primary interest, the artist's task is not to copy nature but to capture ideal beauty, to awaken in the viewer "*das Gefühl harmonischer und lebendiger Schönheit.*"

> "The artist should never attempt to imitate in the truest possible way, because his model is always only a person grown up in earthly imperfection, never corresponding to the Ideal; rather he should modify the individual *Gestalt* until it is the perfected impression of its spiritual content" (Helmholtz 1848: 101).

To reveal the *Ideal* of the curve of *Energie* as opposed to any individual *Gestalt*, was precisely Helmholtz's aim in his four years of work on the frog-drawing-machine. His entire argument, in fact, rested on establishing this *Form* as a constant of the natural process of contraction. For example, the measurement of propagation speed from his Fig. 5, as he fully elaborated, depends on the two curves being congruent throughout their length, so that the displacement is uniform throughout and therefore dependent only on the propagation time, not on such contingencies as the state of tiredness or intensity of stimulation. These perfected curves actually make his accomplishment seem too easy, for he was able to produce them only two years after having laboriously established the *Form* of the rising portion of the curve ["*Die Form der Ansteigung der Energie*"] using a precise point by point method that depended on a different technique for measuring small time intervals, which he had adapted from another cannon-ball velocity measurement by Pouillet. Because this method required elaborate procedures for dealing with highly discrepant data and depended on a model of the muscle as an elastic band characterized by a modulus of elasticity, Helmholtz went to considerable length to argue that the general *Form* ["*allgemeine Form*"] of the smooth curve he drew to represent his data points was indeed the proper *Form* rather than a variable *Gestalt.*[50]

50 (Helmholtz 1850a) usages of *Gestalt* and *Form* appear at 768, 770, 791-794, 820. (Olesko, Holmes 1993) analyze especially Helmholtz's accounting procedures using the method of least squares.

The frog-drawing-machine represents the pinnacle of self-recording instruments in the early days of the Berlin Physical Society. But it was only one of a considerable number of curve-producing techniques and instruments that they developed in the 1840s, the conditions for which I have attempted to draw out of the artistic, scientific, and industrial life of the city.

Return now one last time to du Bois's certificate of membership, and particularly to the lush tropical tree (figure 27) on which the modernizing scientists perform their feats. It is apparently a tree of knowledge, to whose fruiting and flowering higher branches one of the heroes of enlightenment has tied the banner of the Physical Society. Unlike the pudgy cherubs or Putti familiar in many other such fanciful images at the time, the analogous figures here are athletic youths.

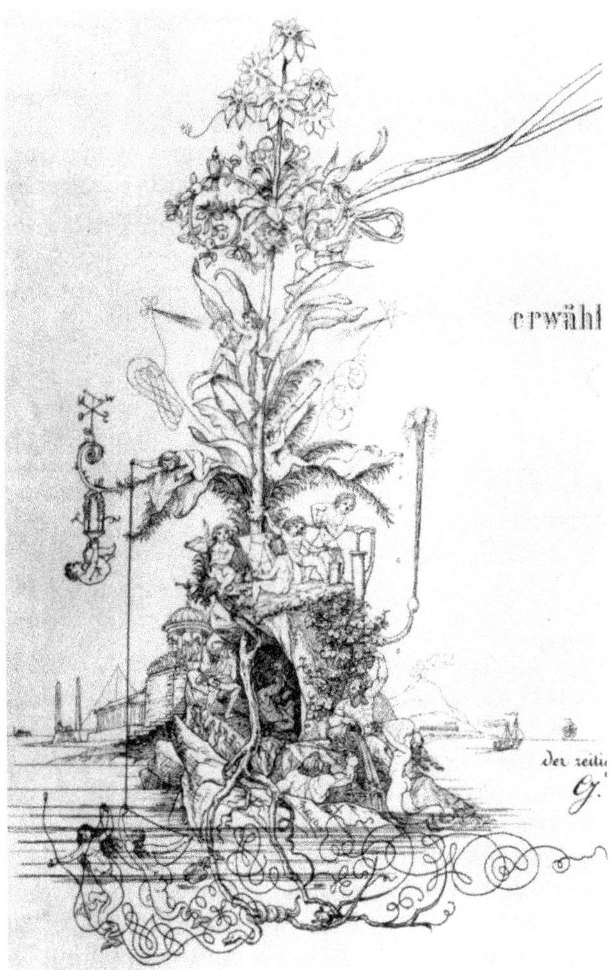

Figure 27: Detail of Figure 1.

They conspicuously employ physical instruments to carry out analytic experiments in the various "branches" of physics, which the group planned to review in its new journal, *Advances in Physics* (*Fortschritte der Physik*). From his perch in the higher branches, a young astronomer aims his telescope to reveal the line of epicyclic motions of a comet, now tied to the tree of knowledge, while a symbolic Newton with a large prism similarly analyzes the spectrum of sunlight, symbolized as an arabesque of harmonic loops, reminiscent of both Dürer and Fourier. A new Galileo on the right demonstrates the law of falling bodies, showing that the distance increases with the square of the time. The gymnast on the left, surely representing du Bois himself, performs his exercises on an electromagnet. His neighbor attempts to fathom the watery depths with a perfectly straight plumbline, unaware that the deceitful nymphs below are busy making a tangle of it. Of the performers at the base of the tree, one investigates the lines of Chladni figures (produced by bowing a metal plate covered with dust),[51] another carries out a geometrical analysis, a third draws electrical sparks from a Leyden jar, and a fourth tests the laws of hydrodynamics with a curiously phallic pump. The prominence in this scene of physical instruments and of curves is everywhere apparent.

Equally apparent, the curves are conceived as inscribed by the instruments, which read out the language of nature. The heroes themselves, in their classical beauty, seem to play much the same role as du Bois's Apollonian experimenter. If so, they should be seen as operating outside the realm of individual subjectivity, with a classicized universality obtained through disciplined performances on their instruments of analysis and revealing nature's true Forms. Thus du Bois's iconography states my central theme, that curves, especially harmonic or rhythmic curves, conceived as representing the essence of natural objects and processes, played an extraordinarily important role in the view of knowledge held by the members of the Physical Society, a view that carried across the boundaries of art, science, and technology. Brücke, in articulating the notion of beauty that he held in contrast to the decadent modern art of the late 19th century, focused on the line as the critical feature of "ideal art":

> "In order to be beautiful in our sense, [...] an image must present good lines in all postures whatever that occur in ideal art and from all perspectives, for the leading role of the lines is first and most important in every work of art that makes higher claims" (Brücke 1892: 3).

The same was true of the curves produced by the Physical Society, which aimed for ideal mathematical truth. As du Bois himself expressed it for physiology, in the introduction to his *Animal Electricity*, even though one could only rarely obtain knowledge of causal relations in a mathematically expressible form, it was always possible "to conceive of the magnitude of an observed effect [...] as an unknown [mathematical] function of all the circumstances which have an influence on it." Then changes in the effect could be observed in relation to changes in any one of the circumstances, while holding all others constant. In this way, "The dependence of the effect on each circumstance presents itself in the form of a curve [...] whose exact law remains [...] unknown, but whose general character one will in most cases be able to trace" (du Bois-Reymond 1848-1884: vol. I, 26-27). This is a view of

51 On Chladni's figures and their fundamental significance for music and science, see (Jackson 2006) especially chapter 2.

the relation between laws and curves that Hummel, Burg, Dirichlet, and Dove had all expressed in their own ways and that Helmholtz realized with his frog-drawing-machine. It carries Dürer's Platonism into the 19th century, but with the crucial addition of Fourier analysis, precision instruments, and graphical representation, which seemed finally to provide the tools to put Dürer's idealist vision of knowledge into a realist form.

Looking once again at du Bois's allegory, if we follow the vertical display downward, we leave the light of day and the scenes of rational analysis above ground and move under ground, where the roots of the tree of knowledge lie buried in the mythological past. This underworld recalls du Bois's well-known polemic, immediately following his extended discussion of mathematical-physical methods and the use of curves, launched against the dark and vitalistic notion of a *Lebenskraft*, which he ascribed to the speculative romanticism of contemporary physiologists and to their ignorance of physical methods (du Bois-Reymond 1848-1884: vol. I, xxxiv-xl). Here Mephistopheles steps out of the flames of hell to observe the searching figure in the cave, from Plato's *Republic*, who with book in hand is vainly attempting to decipher dim shadows on the wall, imprisoned in his own imaginings. Apparently he lacks the instruments of enlightenment that the new physicists regard as necessary for ascending the Platonic scale from the visible to the intelligible: from the realm of mere images or imagination in the cave, through actual things, to the abstracted mathematical diagrams that represent these things in conscious thought prior to attaining intuitive knowledge of pure forms. Outside, bearded giants with torch and urn evoke the powers of fire and water while plucking grapes to share with another of the voluptuous water nymphs.

Finally, as the roots of knowledge descend to their primordial source in the feminized world of the nymphs, the aesclepius, ancient symbol of medical art, entwines itself on a root. The serpent is ironically juxtaposed with the lowly frog, whose muscles and nerves were already providing the primary material for the new electro-physiology of du Bois and Helmholtz. This juxtaposition again suggests Dierig's interpretation of du Bois's self-image: he deployed his instruments "in combat with the frog" and with the dark powers of vitalism, much like Apollo directed his arrows against the serpent Python. In this watery domain, the rationalized curves of nature's authentic script – cometary trajectory, spectrum of sunlight, Chladni figures, geometrical diagram, and the pronounced black plumb line – take on an inaccessibly complex variability, apparently reminiscent of the seductive charms lurking in a dangerously feminine sea of romanticism. It is as though du Bois's lines quote from the classicist tradition of water metaphors, referring perhaps to Ovid's *Metamorphoses*, where the water nymph Salmacis became enraptured by the young Hermaphroditus, who had inherited from his father Hermes and his mother Aphrodite an incomparable combination of classical beauty, firmly muscled yet softly curved. Unable to seduce the virgin Hermaphroditus, Salmacis held him in a forced embrace in the waters of her spring while entreating the gods to unite them forever. This they did literally, producing a half-man, half-woman. Anguished by his fate, Hermaphroditus obtained from Hermes and Aphrodite a similar curse on any man who entered the spring. From the renaissance onward, and certainly for the generations of young males who read the *Metamorphoses* in their Gymnasia, the story implied danger to masculinity: "Ein Mann, der in der Quelle der Salmacis badet, steigt als Halbmann heraus. Durch die Berührung des Wassers wird sein Körper sofort verweichlicht." If du

Bois's athletic heroes feel the seductive attractions of the vital force, their machines of objectivity elevate them beyond its emasculating grasp.[52]

For interpreting the cultural location of du Bois's playful but intense promontory of science, its horizontally arrayed background is significant. For the lush tropical growth emerges out of scenes of both classical purity and industrial progress. From a galley nearly lost in the distance on the left, classicism proceeds through the Egyptian obelisks, sphinx, and pyramid, to the Parthenon of Athens, and up to an 18th century scene of academic learning, with a professor in wig and frock coat lecturing to the passively assembled students outside his temple. Only in the present of the mid-19th century do the students themselves, freed from temples and priests, take over the task of making knowledge through experimental research. From the right, their new mode of action has emerged from the era of sailing ships and stands before those newly domesticated powers of Neptune and Vulcan, the steamship and the railroad, which appear on the Bay of Naples before a gently smoking Vesuvius.

The background panorama thus carries forward to the viewer dual ideals of knowledge-making, or *Wissenschaft* – namely, classical learning and material progress – which had continually played off one another during the period when the members of the Berlin Physical Society were forming their identities. Du Bois's imagery, in brief, places the young heroes wielding their implements of progress at the juncture of a vertical history leading upwards from mythology to truth and a horizontal history projecting forwards both from classical education on the left and from industrial drive on the right, uniting those forces in the movement to the future. That movement is carried by the instruments that draw the lines of nature's script. Thus the growth of the tree of knowledge seems to express du Bois's hopes for his own work, penned as an epigram from Goethe's "Hoffnung" [Hope] at the front of a laboratory notebook begun on New Year's Day, 1850.

	[literally]
Schaff', das Tagwerk meiner Hände, Hohes Glück, daß ich's vollende, Laß, o laß mich nicht ermatten! Nein, es sind nicht leere Träume, Jetzt nur Stangen, diese Bäume Geben einst noch Frucht und Schatten.[53]	Fulfill the daily work of my hands, Wondrous luck that I complete it, Let me, oh let me not grow weary! No, they are not empty dreams: Now only sticks, these trees Will one day yet give fruit and shade.

52 Nicolaus Reusner, *Emblemata* (1581), cited by (Fend 2003: n. 59), from (Henkel, Schöne 1978: vol. II, 1628-1629).

53 (du Bois-Reymond 1924-1926); varying slightly from (Goethe [12]1981: 131). I thank Sven Dierig for this reference.

Sven Dierig[*]

Apollo's Tragedy: Laboratory Science between Classicism and Industrial Modernism

The 1825 painting by the Berlin architect Karl Friedrich Schinkel, *Blick in Griechenlands Blüte* (View of Greece's Glory) shows a temple in the process of being built. Looking down from an elevation onto a Mediterranean city landscape, the viewer of the picture could imagine himself in the midst of a construction site: with muscled men and stone masons at work, cable winches, an iron jack, a cog-driven hoisting apparatus, and artisans and builders beneath a shading tent. This scene is a stark contrast to the visual expectations linked to antiquity in the age of neoclassicism. While standard depictions of Greece usually show the monuments either in their finished state or as ruins, Schinkel here painted ancient architecture in the process of growing and being created [*Bildung*].[1] Central in Schinkel's *Bildungsbild* were the creative artist and the role of craftsmanship in man's production. "In all things, man should form himself [*sich bilden*] beautifully, so that every activity exuding from him will be thoroughly beautiful in both subject and execution": this was Schinkel's famous artistic imperative (Haus 2001: 249). Every human act should be an artistic task. This artistic idealism was applied to practical concerns and transferred to then current needs in the foreword to *Vorbilder für Fabrikanten und Handwerker* [Models for Manufacturers and Craftsmen], a collection of prototype design drawings for commercial artisans that Schinkel edited together with the Prussian industrial reformer Christian Peter Wilhelm Beuth. The drawings of lamps, vases, furniture and other everyday items, shown in the 1837 book, were intended to show craftsmen

> "how necessary and useful it is to give their products not only technical perfection, but also the greatest consummation of form. Only an execution that unites the two can bring the work of the craftsman closer to artwork, moulding it with the stamp of formation [*Bildung*] and giving it a more enduring value than the costliness of the material from which it was made" (*Vorbilder für Fabrikanten und Handwerker* 1863: V).

This aesthetic link between *Bildung* and craftsmanship intended by Beuth and Schinkel also proved well suited for understanding the making and use of scientific instruments in workshops and laboratories as a form-giving artistic task. In this essay I present, as an example of this link, the experimentation of the Berlin physiologist Emil du Bois-Reymond. In the sense

---

* This article is dedicated to Ernst Florey (1927-1997). Parts of it have been already published in a book length study Wissenschaft in der Maschinenstadt (Göttingen: Wallstein, 2006) and in two essays written for C. R. Biologies (2006) and Kultur im Experiment (Berlin: Kadmos, 2004). I would like to thank M. Norton Wise, Laura Otis, Michael Hagner, Henning Schmidgen and Peter Geimer for their criticism and their helpful discussion. Address for correspondence: Sven Dierig, Institut für Nanotechnologie, Forschungszentrum Karlsruhe GmbH, P. O. Box 3640, D-76021 Karlsruhe, Germany (sven.dierig@int.fzk.de)

1 I follow (Vogt 1985).

of Schinkel and Beuth, du Bois-Reymond's *Untersuchungen über thierische Elektrizität* [Investigations on Animal Electricity] in which he described the relation between nervous electricity and muscle mechanics by way of a long series of experiments, are the result of an art of craftsmanship directed at the mutual interpenetration of technical perfection and consummation of form.[2] If one replaces in the passage quoted above "the work of the craftsman" with "the work of the experimenter", it would read like a du Bois-Reymondian model [*Vorbild*] for how to work in laboratories and workshops. Scientific experimentation, according to du Bois-Reymond, was simultaneously an epistemic and an aesthetic practice. The motivation of his science was to produce both knowledge and aesthetic success. This "Aesthetic of Experimentation" [*Ästhetik des Versuchs*], as du Bois-Reymond called his goal, could not be reconciled with laboratory research practices three decades later. I conclude the article by looking at du Bois-Reymond's later years, when laboratories were transformed with the growth of technology and a factory-like approach to experimental research. This left du Bois-Reymond an isolated, nostalgic and somewhat tragic figure, who reflected the new ways of doing science in terms of cultural pessimism and its critique of civilization. As indicated by the title, 'Apollo's Tragedy,' Germany's idealistic dreams dissolved into the realities of the industrial age, not only in the arts, education and cultural life, but also in the world of scientific laboratories.

Growing and Becoming

Like others in his generation, du Bois-Reymond was fascinated by phenomena of growth and development, both in the direct biological sense and in terms of categories of individual formation [*Bildung*].[3] When it came to the formation of individuals, children surrounded by vegetation, "the children in the garden", was the dominant emblem of the age. In 1845, in the style of the romantic painter Philipp Otto Runge, du Bois-Reymond drew a kindergarten of the natural sciences, using it to illustrate the membership card of the *Physikalische Gesellschaft zu Berlin* [Berlin Physical Society], of which he was one of the co-founders. The drawing showed an exotic-looking imaginary plant. Between its stems and leaves there are boys using their research instruments to foster the growth of the various branches of the natural sciences.[4] The athletic young bodies suggest that the decisive prerequisite for the growth of natural scientific knowledge is the individual development of the researcher actively engaged with his laboratory instruments. He experiences "pleasure" because he sees how he is "progressing", du Bois-Reymond wrote at the beginning of his experiments to his

2 (du Bois-Reymond 1848-1884). For summaries and the incorporation of du Bois-Reymond's research work into nineteenth-century electrophysiology see (Rothschuh 1964), (Clarke, Jacyna 1987), (Lenoir 1986). The most comprehensive historical analysis of the Untersuchungen – including a long chapter on du Bois-Reymonds famous self-experiments which brought him in conflict with the Paris Academy – has been written by Gabriel W. Finkelstein in his biography on young du Bois-Reymond (Finkelstein 1996). See also (Finkelstein 2003).

3 The notion of Bildung as a social and cultural label that defined the group of the educated Bildungsbürger has been extensively adopted to young du Bois-Reymond by Finkelstein (1996). See also (Lenoir 1992). On the later du Bois-Reymond as a representative of the Bildungsbürgertum see (Engelhardt 1981), (Gradmann 1993). A most helpful book on the rise and fall of the German Bildungsidee is (Bollenbeck 1994).

4 For a deeper analysis of this 'Tree of Knowledge' see M. Norton Wise (this volume).

friend Eduard Hallman. "I am growing, we want to see where to."[5] In a public lecture ten years later, he used a young child as a model to illustrate the development of an experimental scientist:

> "Observe a child in the tender age of development as it begins to discover the external world with a fresh gaze and to place the causes of his sensations outside himself. He sits at a table: he has been given a spoon to play with. Accidentally, the spoon reaches the edge of the table and falls clamorously to the floor. His small face is transfigured as often as one repeatedly raises the spoon for the child, it repeats joyously the same attempt; but he still did not know that bodies are heavy, that an unsupported body rushes toward the earth, how should it? Only experience, some of it painful, will in the course of time impress this truth upon him so effectively greatly that he will think it self-evident" (du Bois-Reymond 1912a: 29).

The spoon in the nursery corresponded to du Bois-Reymond's major precision instrument on his laboratory table. The galvanometer he used consisted basically of two magnetic needles hung one over the other on a silk thread beneath a glass cylinder. The lower needle floated inside a copper wire spool, while the upper needle, visible from outside, hung over a scale divided into degrees. On the copper wire, electrical current directed the magnetic needles around the axis of the thread. The operation of the magnetic system, which was hard to calibrate and highly subject to disturbances, required a significant degree of hand-eye coordination. In other words: a successful experiment required the experimenter's having gone through a rigorous process of physical exercise and development. Self-perfection [*Selbstvervollkommung*] in using the instrument was rule number one in du Bois-Reymond's laboratory. In January 1848, he wrote to his friend Carl Ludwig:

> "Very soon I was also able to translate the pain that the burning of a frog's foot caused to the animal into electromagnetic motion, and with tireless exercise and perfection of the experimental technique [*bei unablässiger Übung und Vervollkommnung der Versuchsweisen*] I don't see why it should not ultimately also be possible to translate the change in the so-vital current of the *opticus* of a pike into a magnetic equivalent" (du Bois-Reymond 1927 [January 1841]: 5).

A researcher exercising the galvanometer is figured in the branches of the science plant drawn by du Bois-Reymond. One of the trained boys is shown doing chin ups on a magnet. Du Bois-Reymond here depicted himself twice: as gymnast and as experimenter. In so doing, the use of the galvanometer was linked to the physical experience in the gymnasium [*Turnanstalt*] that du Bois-Reymond regularly visited while working on the *Untersuchungen*.[6] The drawing of the boy exercising on the magnet suggested that gymnastic equipment and laboratory instruments had a similar kind of relationship to the body working on or with them. The practiced gymnast and the practiced experimenter were both the result of a physi-

5 (du Bois-Reymond 1918: 93); (Finkelstein 1996: 204).

6 Du Bois-Reymond as a physically trained physiologist is well known because of his committed promotion of parallel bars during the 'Barrenstreit' in the 1850s and 1860s. On du Bois-Reymond in that debate see (Buomann 1932: 40-54); On du Bois-Reymond as a gymnast see also (Lenoir 1992: 41); (Finkelstein 1996: 48-49).

cal self-perfection. In the appendix to his *Untersuchungen* du Bois-Reymond included a series of plates, drawn by himself. Plate V, drawn in 1852, shows a handsome, scantily-clad experimenter seated at a simple table, busily working his experimental set-up.[7] In this depiction, du Bois-Reymond gave himself the appearance of a figure from ancient art: An idealized image of beautiful manly youth. The classical symbol for physical perfection works at an experiment using a galvanometer. The experimenter was not a child [*Kind*], standing at the beginning of *Bildung*, but a fully developed young man [*Knabe*]. The beholder was thus to understand that experimentation in the laboratory is a form-giving act of increasing physical improvement. If the exercising experimenter is an artist on his own terms, the well exercised experimenter himself is a work of art. Just as the parallel bars perfected the gymnast, instruments shaped the experimenter. As in a gym, du Bois-Reymond converted himself in the lab from a crude, organic mass – the untrained nerves and muscles of his own experimenting body – into the ideal form of an ever youthful Apollo.[8] Wilhelm von Humboldt's educational ideal *Bildung durch Wissenschaft* meant *Bildung* trough *Geisteswissenschaften*, especially by incorporating Greek language, poetry and culture. Du Bois-Reymond's experimenting Apollo pictures Bildung through the natural sciences, through scientific experimentation. As in Schinkel's *Blick in Griechenlands Blüte*, du Bois-Reymond's experimenting Apollo can be interpreted as a *Bildungsbild*, and, as Schinkel's painting also intended, as a critique on Humboldt's onesidedness, which ignores gymnastics, the body, and the use of the hands as relevant instances for individual Bildung.

Collaborating

In du Bois-Reymond's years as a student, around 30,000 Berlin residents were craftsmen of some kind. Around 200 of these were professionally categorized as "mechanical artists" [*mechanische Künstler*], highly qualified jacks-of-all-trades in mechanics and optics. In the literature on du Bois-Reymond's *Untersuchungen*, the mechanical artists have only been given a marginal treatment. In this way, over and over again the image of an autonomous experimenter has been conjured up; an experimenter who, at his own whim, relying on his own ability and own intuitions, drove forward his scientific work. The opposite was the case. Du Bois-Reymond's experimental work on *Untersuchungen* was a shared undertaking, the result of a close collaboration [*Zusammenarbeit*] between experimenter and mechanics. The most important mechanical artist involved in *Untersuchungen* was Johann Georg Halske. Better known as the co-founder of telegraphy workshop *Siemens & Halske* (along with Werner Siemens) in 1847, du Bois-Reymond got to know him at the beginning of his experimental work when Halske worked as an apprentice in the workshop of the mechanic W. Hirschmann in 1841. Between his apprenticeship under Hirschmann and his later collaboration with Siemens, Halske established with F. A. Boetticher a workshop that was located in

7 See M. Norton Wise (this volume, fig. 23: 93).

8 During du Bois-Reymond's youth the Apollo Belvedere had been the most popular reference of the classic Bildungsideal and had been established among the Bildungsbürgertum as a common and essential Bildungsgut. The sculpture represented the heroic athlete and creator of ideal bodily form, harmonic order and mental clarity. "For the generation born around 1800," wrote Bonn archaeologist Reinhard Kekulé in an 1876 lecture, "the Apollo Belvedere was more than an artistic impression, it was a life experience" (Kekulé 1877: 3).

walking distance from du Bois-Reymond's laboratory. In his introduction to the *Untersuchungen*, du Bois-Reymond expressly emphasized this profitable relationship with Halske and Boetticher. Without the work of the two mechanics, *Untersuchungen* would have been "impossible" (du Bois-Reymond 1848-1884: vol. 1, LII). This was thus more than just a simple relationship between a customer and a manufacturer. Du Bois-Reymond did not just turn to mechanical artists in order to quickly have an order made, and then again leaving the workshop, returning later to pick up the completed apparatus. He remained in the mechanic's workshop as his instrument was being made, watching and participating in the process. It was in Boetticher's and Halske's workshop, where the galvanometer to be seen on the Apollo drawing was made. Halske and Boetticher took charge of making the instrument's mechanics, while du Bois-Reymond himself took on the task of winding the copper wire spool. With his detailed description of the making of the galvanometer's spool in the *Untersuchungen* du Bois-Reymond presented himself as a mechanical artist who was able to build scientific instruments. Being a scholar [*Gelehrter*] and a craftsman [*Handwerker*] in one person, this romantic self-concept was realized in du Bois-Reymond's *Untersuchungen.* Decades later it has been labeled by the hybrid term *Handwerksgelehrter,* meaning ironically a whole generation of experimenting reseachers and their outmoded style of doing science.[9] In du Bois-Reymond's own words:

> "We had to roll our own coils [of copper wire], solder our own elements [batteries], even glue our own rubber tubes together, for at that time you couldn't buy any rubber hoses. We sawed, planed, drilled, filed, turned, and sharpened. Our need for help and advice in mechanical matters drove us to the workshops, where we learned all kinds of maneuvers through exchanges with talented artists [*talentvolle Künstler*]. We became as familiar with the construction of instruments, down to the last screw, as we did with the anatomy of an animal" (du Bois-Reymond 1912b: 634).[10]

Step by Step

While Beuth and Schinkel were collecting *Vorbilder für Fabrikanten und Handwerker*, Ludwig Tieck was completing his novella *Der junge Tischlermeister* [The Young Master Carpenter] published by Reimer Verlag in 1836. Reimer was the publisher of literary romanticism in Berlin and his program in the 1840s included the writings of E.T.A. Hoffman, Jean Paul, Novalis, and Grimms' fairy tales.[11] It was also Reimer that published du Bois-Reymond's *Untersuchungen über thierische Elektrizität.* In Tieck's *Tischlermeister*, a tract against the beginning industrial age, the protagonist Wilhelm Leonhard is an independent autonomous craftsman who saw himself reflected in his own products, and was repelled by all imitation and the factory-like. Tieck's Wilhelm Leonhard had nothing in common with the design drawings that Beuth and Schinkel had suggested in the *Vorbilder.* A master of craftsmanship had to be his own draughtsman. The "relation of art, but without wanting to be art", Wilhelm Leonhard says in the novel, drew him to craftsmanship: "I thus

9 Heinz Otto Sibum found and introduced this highly useful term in his studies on the interrelationships between artisans and experimenters. (Sibum 2003).
10 See also (Lohff 1981: 117); (Rothschuh 1953: 140); (Finkelstein 1996: 207).
11 On Reimer Verlag and his authors see (Reimer 1999).

dedicated myself to drawing untiringly" (Tieck 1996: 57). Leonhard's tables and chairs emerged first as "shapes," as "things" that floated about in his "imagination" and were "turned back and forth," then to be drawn and finally built. From the idea to the drawing, and from the drawing to the final product: in the workshop of the mechanical artists, design and construction before building were at this time still the exception, and considered the latest innovation. Craftsmen working according to plan raised scientific instrument making to a new level. A report by the head of Berlin's observatory, Johann Franz Encke, on the precision mechanic Carl Otto Albrecht Martins describes this modern type of the constructing craftsman:

> "From the very beginning, Herr Martins made it his approach to make a detailed drawing the foundation of his work, and thus made it possible to form a rational judgement by improving each individual part. As unimportant as it might seem, I do believe it vital to place a great stress on this point, for I learned to treasure in Herr Martins a thoughtful artist who does not just try to discover something by trial and error, but gives his experimentation a sure foundation by making it completely clear to himself what he intends, and thus anticipating the problems that might hinder his intention" (Zaun 2002: 41).

Like Martins or the literary protagonist Wilhelm Leonhard, Halske was a craftsman who used draughtsmanship in order to explore the mechanics and the operation of the apparatuses that du Bois-Reymond used in his laboratory. Du Bois-Reymond later reported: "Halske was much more than just a talented worker". To a "rare degree", Halske possessed a "constructive talent" and a "sure intuition" for finding the "simplest and best way" to solve the task at hand:

> "It was a great pleasure that I often enjoyed watching him half the night, with a pencil in hand approaching step by step the complete perfection of an idea for an experimental set up or a device" (du Bois-Reymond 1890: 40).

The conceptual construction of the laboratory apparatuses with the pencil in hand, in order to anticipate how what was assembled in the workshop would later function in the laboratory, making clear how an instrument emerges and what is intended with a mechanical device or an experimental arrangement, drawing, tinkering, and assembling: Du Bois-Reymond's actual laboratory consisted of the triumvirate, drawing table, work bench, and experimental table. Like the workers, builders, and artists at Schinkel's construction site, du Bois-Reymond took part in a shared process of construction and step-by-step formation at Halske's workshop. In the mechanics atelier, du Bois-Reymond assisted in building and conceiving the instrument, got involved, and watched the process of planning, learning how technical things took on shape and form. Step by step over time the apparatuses he needed for the laboratory developed. The growth and development of technical things might well have enthused du Bois-Reymond just as much as nature's own processes of step-by-step development. The report on Halske's art of draftmanship thus sounded almost identical to a report on the drawing abilities of his former teacher and mentor Johannes Müller. As du Bois-Reymond remembered, Müller was a "master of drawing at the chalkboard": "It was a great pleasure to watch him gradually taking an animal form in the process of development through a series of intermediate steps to the final shape" (du Bois-Reymond 1912c: 272).

Mechanical Beauty

The craftsman should not be misguided "to compose himself", as Beuth and Schinkel warn in their *Vorbilder für Fabrikanten und Handwerker.* Instead, the craftsman should limit himself to internalizing the spirit and taste of historical models and imitating such models. Craftsmen should not seek to be artists. In his novel *Der junge Tischlermeister*, Tieck drew the same dividing line : It was a "relation to art, but without wanting to be art" that constituted the aesthetic autonomy of craftsmanship. In Tieck's text, Baron Friedrich Elsheim asks the carpenter the question, why he became a craftsman: "I have always been surprised, my friend, that you with your open mind and varied scope of knowledge, your pleasure on all developed things [*allem Gebildeten*], that you did not prefer to chose the status of the artist." Wilhelm Leonhard answers:[12]

> "Daß ich nicht zum Gelehrten paßte, sah ich früh ein, weil Sachen mich mehr als Gedanken, Worte und Formen interessirten. Zum Künstler fehlt mir ganz jener Enthusiasmus, jener strebende, fliegende Geist, der alles neben sich vernachlässigen und vergessen kann und darf, der in fremden Welten, aber nicht in der hiesigen einheimisch ist; mein Gemüth im Gegentheil ist beschränkt und wahrhaft bürgerlich mein Eifer für Arbeit, Nützlichkeit, meine Lust an Dingen, die brauchbar sind und fest stehen: alles dies überzeugte mich früh, daß ich zum Handwerker bestimmt sey, und zwar zu der Beschäftigung, welche ich erwählt habe." (Tieck 1996: 53).

Wilhelm Leonhard chose craftsmanship over the sciences and art. While scholar, artist, and craftsman are clearly distinguished in Tieck – the scholar is a pure intellectual, the artist is not of this world, and the craftsman creates useful things - in Halske's workshop, in contrast, craftsmanship, science, and art all mingled with one another. In du Bois-Reymond's view, Halske's creations were far more than just useful things: "Halske's fundamental attitude and goal was to make every piece as consummate an artwork as possible, up to the very last screw." In the same way that Beuth and Schinkel demanded this for the products of craftsmanship, Halske's instruments united both: technical perfection and a consummation of form. Du Bois-Reymond valued Halske's instruments as things of "mechanical beauty". In his 1851 lecture *Über thierische Bewegung* he saw in mechanical beauty "the same beauty that might strike one in a physical instrument or machine when it gives the unconscious impression of perfect functionality [*maximaler Zweckmäßigkeit*] with the greatest possible simplicity" (du Bois-Reymond 1912a: 32). Throughout his life, du Bois-Reymond celebrated the ideal of mechanical beauty, the unification of beauty [*Schönheit*] and functionality [*Zweckmäßigkeit*]. In an 1890 lecture, *Naturwissenschaft und bildende Kunst* [Natural Sciences and Fine Arts], he declared:

> "One case in which beauty can be particularly well dissected is that of the beauty one might call mechanical. This kind of beauty is seldom recognized, for its appreciation requires a special training of the eye. [...] Perhaps it fits the definition of unconscious conformity to reason, for here the pleasure is derived – and rightfully so – from the fact that we, with sufficient *Bildung*, unconsciously note how perfectly it meets a need. We see the way it combines strength with ease and the way it creates motion

12 As the following quotation is delicate to translate, the original German wording is given here.

with the most advantageous transfer of forces, without any useless expenditure of material."[13]

The mechanical beauty of Halske's art work was thus something quite different from the beauty of the use-objects in the *Vorbilder* or those imagined by Tieck. Beuth and Schinkel, just like Tieck, relied on the forms of the past, albeit in a different way. While Beuth and Schinkel related on the aesthetic models of antiquity, Tieck turned to the middle ages. Wilhelm Leonhard wants "to ornament hard straight lines and square corners with flowers and garlands or with light figures that border on the arabesque" – "the superfluous and unreasonable is what gives a work of craftsmanship beauty" (Tieck 1996: 60). Mechanical beauty as understood by Halske and du Bois-Reymond, was exactly the opposite: the beautiful is only what looks rational: The impression of beauty arises not by decorating the useful with classical or medieval forms, but by intentionally avoiding any superfluous decoration. There were no instruments placed on classical columns or featuring romantic ornamentation in du Bois-Reymond's laboratory. Even Schinkel, the architect of Berlin neoclassicism, had by now taken another route. Berlin's *Bauakademie*, built in the 1830s, was no Greek temple. The new building was a functional cube, a factory-like red brick building with iron window frames.

Aesthetics of Experiment

Du Bois-Reymond's experimenting Apollo links the ideas of neohumanism, classic aestheticism, technology and science and is thus emblematic for the cultural and social self-image of Prussians at that time. In Schinkel's *Blick in Griechenlands Blüte* muscled workers use an iron machine in building their temple, the young Greek in du Bois-Reymond's laboratory experimented with the newest tools and instruments. For Schinkel, as for du Bois-Reymond, the well-rounded, educated citizen, modern technology and antiquity were not opposite poles, but an invitation to expand one's intellectual horizons in both directions. In his lecture *Goethe und kein Ende* [Goethe without End], du Bois-Reymond spoke of a German tendency "to be equally drawn to the classical and modern cultural worlds" (du Bois-Reymond 1912e: 179). Since the 1820s, Berlin had been not just an "Athens on the Spree" but also a city of machines. Like the plaster casts of Greeks in the Royal Museum, machines for du Bois-Reymond were an educational resource. Railway excursions, trips to trade fairs, the Borsig factory in Berlin or the nearby Potsdam gardens and its engine houses permitted viewings and shaped the form of thought characteristic of du Bois-Reymond's mechanistic physiology: its notion of the "animal machine [*thierische Maschine*]," which could be understood only through "mechanical analysis [*mechanische Analyse*]."[14] The frog legs du Bois-Reymond investigated in his *Untersuchungen* were "physical instruments that could be studied like

13 (du Bois-Reymond 1912d: 397); my emphasis.

14 Particularly well-known was the Babelsberg Engine House display of a steam engine, effectively staged for the public. Here, from a gallery that extended all the way around the engine, one could observe all of its moving parts, including its six-meter flywheel and its crankshaft with connecting rods and control mechanism. See (Pieper 1994). Robert Brain and M. Norton Wise extensively investigated these local machineries as a major intellectual resource for the young generation of experimental scientists in Berlin and Potsdam (Brain, Wise 1994); (Wise 1999).

machines" (du Bois-Reymond 1848-1884: vol. 1, 458). Steam engines also offered blueprints for the construction of experimental set-ups in the lab that du Bois-Reymond or his friend Helmholtz called "machines". The most prominent one was described by Helmholtz in 1848 as a "frog writing machine," a device that recorded the contractions of a twitching frog leg onto a rotating metal drum covered with carbon paper.[15] It is "a spectacle for the gods to see the muscles at work, like the cylinder of a steam engine," du Bois-Reymond wrote in a letter to Helmholtz, describing one of his own experimental set-ups.[16]

In his *Theoretische Kinematik*, Berlin mechanical engineer Franz Reuleaux summarized the fundamental principle of engineering that had ruled machine designers since James Watt:

> "In the machine, the cosmic freedom of natural phenomena is transformed into law and order. [...] The elimination of interfering movements [*störende Bewegungen*] caused by latent forces is the first principle of the machine. [...] We regard any movement as interfering [*störend*] that is a deviation from the one intended" (Reuleaux 1875: 37-48).

Maximizing functionality by preventing interfering movements was also a central goal of du Bois-Reymond's training program when he served as a gymnastics squad leader.[17] Like a machine, the gymnast trained by du Bois-Reymond was treated as an aggregate of mutually interacting parts, and just as mechanics sought to optimize the internal play of movements, bodily training tried to order movements functionally. As in the construction of machines, one achieved the most efficient functional interplay of the body's limbs not just by practicing movements but by eliminating the "secondary movements" [*Mitbewegungen*] that accompanied them. Here again, the rules from the gym held true in the lab. Persistent exercise and perfection in the use of an experimental set-up gradually eliminated the "secondary effects" [*Nebendinge*], created by the components of the experimental system itself. Just what was meant by secondary effects can be best learned from Helmholtz:

> "What costs the most time and work is secondary effects, which in most cases are only remotely related to the goal of the investigation. Very quickly we have to start studying instrumental errors in order to eliminate them or, when they can't be eliminated, to avoid their disadvantageous effects. Or in other cases, we lose time and valuable opportunities trying to find an animal in the circumstances we need in order to investigate it. Then, during the investigation, we learn about new experimental errors that have damaged the results or maybe just created doubts about whether they've been damaged. We're then forced to resume our work from the beginning, until every shadow of a doubt has been eliminated [...]. Every one of you must know the disproportionate amount of time spent on preparations, secondary effects, and control of possible errors in a good experiment, as opposed to the relatively small amount of time needed to make the final observations or perform the procedure. You

---

15 See (Brain, Wise 1994).

16 Du Bois-Reymond to Helmholtz [Februar 1852], (Kirsten 1986); (Brain, Wise 1994: 140).

17 Du Bois-Reymond expresses his ideas on body motion in his two writings Über das Barrenturnen und die sogenannte rationelle Gymnastik (du Bois-Reymond 1862); Hr. Rothstein und der Barren: Eine Entgegnung (du Bois-Reymond 1863). See also his speech 'Über die Übung' (du Bois-Reymond 1912f).

> know how much more keenness and deliberation must be dedicated to mastering a stubborn piece of glass or brass than to planning the entire investigation."[18]

Just as the squad leader in the gym trained the young men on the bars to eliminate secondary movements, the experimenter in the lab smoothed troublesome corners and edges. Only when the play of parts in an experimental set-up was optimized could it make a well-timed intervention in the wheel works of the animal machine. In his lecture *Über die Übung* [On Exercise], du Bois-Reymond later declared that mastery [*Meisterschaft*] was created gradually, through exercise. Training leading to mastery resulted from an inborn artistic drive [*Kunsttrieb*] recognizable when "the smooth performance of a exercized series of movements" created a "certain pleasure [*Behagen*]." In other words: eliminating secondary effects in the gym or laboratory through exercise brings beauty to light:

> "One watches a boy pulling himself up a ladder for the first time. Although it doesn't do him any good, he doubles every movement of his arms with a kick of his legs. After a few weeks he presses his legs firmly together and keeps the joints of his hips, knees, and feet beautifully extended. In the repression of secondary movements, we unconsciously recognize the pleasant appearance of the trained soldier, the adept gymnast, indeed, of the cultivated man [*gebildete Mann*]" (du Bois-Reymond 1912f: 113).

At the experimental table, du Bois-Reymond acted as a squad leader who disciplined the "limbs" of his set-up until they created the same impression of beauty that Halske's instruments did. Just as a wriggling boy's body could acquire a perfect appearance with exercise and time so that "every part [assumed] the right dimensions, form, and position for performing its task," an experiment could be "mechanically beautiful" when it was fully controlled and optimally functional [*maximal zweckmäßig*], when it was free from all secondary effects. In his 1858 memorial lecture on his teacher and mentor Johannes Müller du Bois-Reymond used the phrase "aesthetics of experimentation" [*Ästhetik des Versuchs*] and he declared that it was "foreign" to Müller, whose experimentation was "rough in its secondary effects [*roh in den Nebendingen*]."[19] Like a neo-classic artist du Bois-Reymond based his aesthetic principles on the proportions, symmetry and harmony of ideal forms – that of a well-trained young body, a well-designed machine, a mechanically perfect scientific instrument or an optimally functional experimental set-up.[20] Experiencing the harmony of the optimally functional interplay of the parts of a set-up, along with the beautiful mind and body of the Apollinian experimenter shaped by his instruments leads to an impression of beauty: "There is an aesthetics of experimentation which drives one to give the set-up mechanical beauty [...] and no experimenter will ever regret his efforts to realize its demands" (du Bois-Reymond 1912d: 421-422).

---

18 (Helmholtz 1969: 370); on "Nebendinge" in du Bois-Reymond's view of scientific practice see also (Lohff 1981). On the epistemic value of "Nebendinge" (Hoffmann, Schickore 2001).

19 (du Bois-Reymond 1912c: 210). On Müller's style of experimentation and du Bois-Reymond's reflections on it see also (Lohff 1992).

20 On "classicistic taste" in science and art and its overlappings see Gary Hatfield's study on Helmholtz (Hatfield 1993: 556).

The Decline of Apollo

During the first years of experimentation du Bois-Reymond lived in his parents' house at Potsdamerstraße. In 1845 he moved into a converted two-room apartment at Karlstraße next to the workshop Boetticher & Halske. Hermann von Helmholtz later described du Bois-Reymond's live-in laboratory as follows:

> "I used to go there often. In the first of the two little rooms on Karlstrasse, each with its one window, was a small bookshelf and a sofa to sleep on, so that it gave the impression of a ship's cabin. In the second room, which offered an unobstructed view of the greenness outside [the garden of the Berlin veterinary school], the galvanometer occupied its place of honor. By the window were the dissection table for the frogs and some other scientific devices."[21]

In 1853 du Bois-Reymond set up a workroom in the main building of Berlin University. There, as part of the anatomical museum, Johannes Müller offered him a small room as a laboratory and later two others where a few students interested in experimenting had their own workspace. After taking Müller's chair when he died in 1858, du Bois-Reymond institutionalized the "Royal Physiological Laboratory" in Müller's rooms. In the same year he launched a persistent and long lasting campaign with the Prussian Ministry of Education for a new institute building, a project that was postponed again and again because Bismarck's "blood and iron" policy monopolized Prussian financial resources.[22] Only in the 1870s, when Prussian finances were reconstituted after winning the Franco-Prussian war, did du Bois-Reymond succeed in providing an up-to-date experimental work place for physiologists. The new Berlin Physiological Institute opened its doors in 1877. Meanwhile du Bois-Reymond had reached his sixtieth birthday. The title page of an ornate portrait album given to du Bois-Reymond by former students on this festive occasion reveals how drastically the space of physiological laboratories had been reorganized since du Bois-Reymond's youth. Below, in the idyllic, almost rural setting of the nearby Berlin Veterinary School garden, one can see the old house on Karlstraße where du Bois-Reymond experimented at the time of the *Untersuchungen*. Above, framed by the laurel crown of a triumphant Olympic champion, stands the new Physiological Institute, the biggest research institute of its kind at that time. In his speech at the institute's opening, du Bois-Reymond proclaimed that "for physiology, the queen of the natural sciences, a royal palace has been prepared" (du Bois-Reymond 1912b: 645). In no other place were "the research and teaching needs of physiology" fulfilled "so perfectly, wonderfully, and completely." He was certain that in his Physiological Institute, "nothing would be lacking that had not been thought of in advance."[23]

Between the 1840s and the 1870s, the surroundings of the physiological laboratory underwent the same metamorphosis as the laboratory itself. Biedermeyer Berlin – which could

---

21 From a news paper article honoring the fiftieth anniversary of du Bois-Reymond's doctoral degree, Neue Freie Presse, Vienna, 23 Sept. 1893. Nachlass Emil du Bois-Reymond: Staatsbibliothek zu Berlin – Preußischer Kulturbesitz, Sammlung Darmstaedter.

22 On du Bois-Reymond's role in the institutionalization of physiology in Berlin see (Genz 1976); (Lenoir 1982); (Lenoir 1988).

23 A contemporary description of the institute can be found in (Guttstadt 1886).

be taken in at a glance – became Germany's largest industrial city, "the Birmingham of the Mark," as one observer put it. This new Birmingham offered models for organizing scientific work: "The main principle, which he placed at the forefront and followed until the end with beneficial effect, was the division of labor", remarked Gustav Fritsch about du Bois-Reymond's management style at the institute. Like Johann Georg Halske and Werner von Siemens, who during the 1860s were intensively involved in the reorganization of their Berlin telegraph workshop, du Bois-Reymond divided work among departments [*Arbeitsabtheilungen*], in terms of both space and personnel, in which, under the leadership of the director, department supervisors [*Abtheilungsvorsteher*] were to lead subordinate workers. Considering the innumerable tasks to be performed, reasoned du Bois-Reymond, what could be "more desirable" than that

> "in the vast crowds of workers [*Arbeiter*] performing specific tasks with restless energy, they all assume their proper places? Why should the business of science not be run *like a factory*, where people at vices perform the most excellent work, even though they have no idea what will be done with the parts they are filing? Why should there not be foremen [*Werkführer*] who know how to insert these parts, but who do not know about the larger decisions affecting the whole product, and finally, better informed, far-seeing masters [*Meister*]?" (du Bois-Reymond 1912g: 148).[24]

The division of labor also applied to the making of scientific instruments. Prepared under factory-like conditions, laboratory instruments lost their image as mechanically beautiful pieces of art, developed through extended negotiations between experimenter and instrument maker. The new conditions under which scientific instruments were now prepared were described in this way:

> "As expected, a much more convenient system has now been developed than previously existed for scientists seeking the help of precision mechanics. One simply approaches a well-known mechanic with a detailed description of what one needs, and then orders it. One does this without worrying about the aids and methods, or even about the reliability, with which this work will be conducted, usually without any clear idea of how the thing is going to be made. [...] Even though some of these well-informed, highly talented men are aware of the latest developments in scientific knowledge, they still have to deal with the disadvantage of increasing specialization and the division of labor in scientific fields. Clearly, they can no longer concern themselves with the preparation of their research instruments to the degree that they did before."[25]

To be "familiar with the construction of instruments, down to the last screw" as du Bois-Reymond had been in his collaboration with the Halske – who left the Siemens & Halske Company in 1867 when it embraced mass-production – had become unthinkable for yet another reason: the increasing complexity of laboratory instruments and experimental set-

24 My emphasis. On another late nineteenth-century factory-like physiological institute, see Daniel P. Todes' extensive study of Ivan Pavlov's St. Petersburg institute. (Todes 2002). On technology and machinery aspects see (Dierig 2003, 2006).

25 Anonymus, 'Über die Förderung der Präzisionsmechanik durch den Staat,' Nationalzeitung, no date. Nachlass Emil du Bois-Reymond: Staatsbibliothek zu Berlin – Preußischer Kulturbesitz, Sammlung Darmstaedter.

ups. Du Bois-Reymond's own research field offers the best example. During the times of the *Untersuchungen*, the simple needle galvanometer was the chief instrument of electrophysiology, but by the late 1870s, it had been replaced by the far more sensitive capillary electrometer. Rather than simply observing the movements of a magnetic needle deflected by electrical current physiologists relied on complex optical set-ups, light sources, and photographic processes which led eventually to the development of photokymography. Faced with the problem of increasingly complicated laboratory instruments, Claude Bernard warned in 1865:

> "Answering specific scientific questions often requires complex, expensive instruments, so that one can say that the answer to a scientific question is now a question of money. In my view, we must make every effort to simplify these instruments, not just because of financial considerations, but for scientific reasons as well. One must bear in mind that the more complex an instrument is, the more sources of error it will introduce to the experiment. The experimenter does not become any more powerful through the number and variety of his instruments – quite the contrary. [...] The instrument should be an aid and a tool for the experimenter, not an extra source of error because of its complexity" (Bernard 1961: 214).

In a research factory experimentation was no longer what it used to be. In the eyes of du Bois-Reymond, the mechanically beautiful instrument and experimentation free from 'secondary effects,' had turned into something ugly. Despite these new relations, in du Bois-Reymond's lecture *Naturwissenschaft und bildende Kunst*, he still called for an "artistic sense [*Künstlerische Sinn*]" on which he believed experiments depended, as opposed to "an attempt to represent natural phenomena in a beautiful way" (du Bois-Reymond 1912d: 421-422). But his outlook was out of date. Johann Nepomuk Czermak's advertisement for a "young man" to do experimental work at his Leipzig institute in the 1870s epitomizes the new concept of the non-apollonian experimenter: "Manual dexterity and mechanical skill [required]. Scholarly *Bildung* not necessary."[26] Du Bois-Reymond wrote to Anton Dohrn:

> "Young people who can do anything besides the same monotonous jobs [*Dutzendarbeiten*] are extremely rare. [...] Those who don't do monotonous jobs have always worked, which shows the difference between our time and earlier periods, when people came to work who were incapable of doing monotonous jobs. I'm seeing exactly the same thing here at the institute that you're seeing in Italy. I have a whole host of beautiful, important research topics, but the young men prefer these intolerably dull studies of respiration, drawing curves you get as sick of studying as the houses on a street. And then they look at you disdainfully over their shoulders, as if you were working on something obscure."[27]

In his complaint about young scientists, du Bois-Reymond was echoing the pessimistic discourse of German imperial culture.[28] In the final third of the century, an entire generation of conservative intellectuals decried the decline in German intellectual life and made the dubi-

26 (Czermak 1879: 121); my emphasis.
27 Du Bois-Reymond to Dohrn [Januar 1882], (Groeben, Hierholzer 1985).
28 On du Bois-Reymond in that debate, see (Bollenbeck 1994: 225-229); (Gradmann 1993).

ous intellect of contemporary youth one of their leading arguments. "Just listen to a teacher of these young people, who has been watching our forthcoming generation develop," lamented historian Heinrich von Treitschke, speaking for many in a lecture at the German *Reichstag*. "It would shake the soul of any one of you, even the most optimistic, to see our young generation's growing hedonism, materialism, and rejection of life's intellectual treasures."[29] In identifying the "poison" that caused this wretchedness, du Bois-Reymond again relied on a contemporary stereotype, the emergence of "Americanism." By this he and others meant "the penetration and overrunning of European culture by realism and the rapidly growing preponderance of technology."[30] He warned that "in the struggle with realism, idealism is yielding, and the kingdom of material interest is upon us." "Neobarbarism" was threatening German culture just as it had once threatened the culture of Greece. Imported from America, "the home of utilitarianism," du Bois-Reymond saw a "rushing haste," a general "haste to produce," and even more, "the curse of everyday appearances" at work. This attitude had moved into the institute and poisoned the new generation of scientists:

> "In the intense battle for truth, who has time and desire to descend to an overgrown path and seek the font of eternal beauty? All too often, today's *Bildung* is an inorganic patchwork of ready-made results cut off from their roots, useful but dry facts, and gross opinions. Few people worry about how the truth was found, or about the connections between things that can be understood only by studying their development. No one heeds the fascination of the perfect form."[31]

In du Bois-Reymond's eyes, these rushing, Americanized "young men" had no interest in "inner and outer perfection" or even in *Bildung* itself. For him, a gift presented by former students for his seventieth birthday in 1888 must have illustrated this lost interest in "the fascination of the perfect form" quite perfectly: a failed sketch of a muscleman, offered along with a transcript of his lectures. This new Greek body, so distorted as to be almost grotesque, suggests the dissolution of cultural dreams of Greece. The least beautiful, least heroic, most hideous qualities of humankind, unintendingly depicted by the students, make the poor muscleman an almost anti-classical work of art. Du Bois-Reymond had no use for an artistic drive that reproduced life's realities. As the turn of the century approached, he clung ever more tightly to the classic artistic ideals. The artful creations of Greek fantasy – "perfect human forms" [*unentstellte Menschengestalten*] – emerging from the graceful material of a Greek landscape, constituted for him the highest norm in any kind of art (du Bois-Reymond 1912h: 572-573, 607-608). "The Apollinian Greeks fancied that Dionysus' works were barbaric," wrote the young Nietzsche in *Die Geburt der Tragödie* [The Birth of Tragedy] (Nietzsche 1988: 40). This was just du Bois-Reymond's attitude toward the widely condemned "neobarbarism." In his lecture, *Naturwissenschaft und bildende Kunst* – delivered at a time when Dionysus had achieved cult status in the Berlin artistic and literary scene[32] – du Bois-Reymond could call non-idealistic art only the "near-intolerable outrages"

---

29 Cited in (Steinhausen 1931: 148).

30 Quotations of this and the following paragraphs are from (du Bois-Reymond 1912h: 602-607); (du Bois-Reymond 1912g: 143, 151).

31 My emphasis.

32 On fin-de-sciècle science in Berlin see (Dierig 2001).

of an "underage crowd." Non-Apollinian art was non-art. Du Bois-Reymond proved himself a faithful follower of the artistic view that any art failing to preserve classical ideals was – in a well known phrase introduced by Wilhelm II – 'gutter art' [*Rinnsteinkunst*]. His assessment of the unfortunate muscleman's artistic qualities could hardly have been much higher.

One hears the physiologist's nostalgia in his praise for the 'vanished hero' at the 1883 unveiling of the Alexander von Humboldt monument in front of the Berlin University: "he must have been a wonderful youth. [...] the joy of the beautiful, self-begotten form was the good fortune of his age" (du Bois-Reymond 1912i: 259, 263). Du Bois-Reymond had a memorial dedicated to him while he was still alive. The portrait album presented to him on his sixtieth birthday was bound in a special decorative cover which again displayed the experimenting Greek youth in the *Untersuchungen*. Not the laboratory, but the monument became the place where the vanished hero worked, at peace with himself for the present and future. Following the stylistic lines of contemporary memorials, Menzel depicted du Bois-Reymond amidst the achievements of his youth: the improved galvanometer, the two volumes of the *Untersuchungen*, the slide inductor and the mercury key he invented.

In his essay, *Über die Klage, daß der deutsche Jugend der Idealismus fehle*' [On the Complaint that the German Youth Lacks Idealism], cultural critic Paul Anton de Lagarde offered a reasonable answer to intellectuals' laments. Though himself no stranger to idealist platitudes and the complaint of a 'New York tone' in Berlin, Lagarde still defended the young people of Germany: "those venerable old gray-beards" – in this category he included du Bois-Reymond – "who find that their sons and grandsons have not honored or have not sufficiently honored the ideals of the time in which they were raised should stop acting like wounded old bourgeois. [...] A German born in 1866 can and should not have to perceive the world like one who entered it in 1820 or 1830" (Lagarde 1903: 373). One of the institutions through which the "venerable old gray-beards" preserved "the joy of perfect forms," as in a holy temple, was the Prussian Academy of Sciences, of which du Bois-Reymond had been a member since the 1850s. At a "du Bois-Reymond banquet" in the 1880s, Berlin archaeologist Ernst Curtius, who had made his name by exhuming Olympus, called his fellow member "a sympathetic appreciator of classical beauty [...] a man who loves and values human beauty and greatness in all fields and in all ages."[33] In his lecture *Kulturgeschichte und Naturwissenschaft*, du Bois-Reymond regarded this appreciation of classical beauty as a cure for the Americanism threatening to overwhelm Germany's "lowbrow, flattened youth":

> "The human and godly forms of antiquity, surrounded by undying magic; the stories and legends of the Mediterranean peoples, in which almost all beauty and goodness are rooted; the mental exchanges of classical society, indispensable to the natural sciences and from which favored men have risen to almost unequaled heights: these are the things whose penetration of young people's minds offers the surest cure in the struggle against neobarbarism, whose iron arm now holds us loosely, but whose grip will almost certainly tighten. May Hellenism keep Americanism far from our borders!"[34]

---

33 In Neue Freie Presse, 25. Oktober 1883. Nachlass Emil du Bois-Reymond: Staatsbibliothek zu Berlin – Preußischer Kulturbesitz, Sammlung Darmstaedter.
34 (du Bois-Reymond 1912h: 607- 608), (Lenoir 1997: 173).

In the same lecture, du Bois-Reymond reveals the final stages of this dream of the past:

> "If we, who live in the cold magnificence of an imperial city, think back on our past in that poor, torn, powerless, petit-bourgeois Germany of our youth, seeing the homely, depressing gables of this central German village overgrown with grapevines and ivy – don't we sense something missing in the brilliant, stunning, surging present? Mustn't we sigh, as in a swallow's song, 'ah, how far away, what once was mine'? In our remodeling of Germany over the past generation, isn't it possible that we've thrown out the baby with the bath water? Haven't the Germans lost a lot of their enthusiasm for ideals, their unselfish quest for truth, their deep, silent inner life?" (du Bois-Reymond 1912h: 606).

In du Bois-Reymond's memory, 1840s Berlin, "overgrown with grapevines and ivy," seemed pleasant, despite the social and political realities of *Vormärz* Germany. In retrospect, the live-in laboratory on Karlstraße became quite cozy. All the troubles and the lack of money about which he had constantly complained somehow evaporated. If the "obstacles" were formidable, claimed du Bois-Reymond, they only "heightened" the "stimulus to do research." On this idealistic Olympus, experimenters elegized the role models of their youth. Helmholtz's characterization of Gustav Magnus, who performed the most "selfless tasks" in pursuit of an "ideal goal," offers an example of the way scientists came to look back on their beginnings:

> "That's the way he worked: not to increase his own wealth, but for science; not capriciously, like a dilettante, but tirelessly, in pursuit of a fixed goal; not vainly, chasing after any chance discovery that might quickly have brought him fame. On the contrary, he was a master of faithful, patient, modest work, a man who tested and retested his findings and never revealed anything until he was sure that there was nothing left to improve. But this is the kind of work that earns and wins the highest, most lasting fame, through the classical perfection of its methods [*die klassische Vollendung ihrer Methode*] and the exactness and reliability of its results" (Helmholtz 1903d: 37-38).

In his Physiological Institute, the wounded old bourgeois managed to build himself a fine little nest. In a side-wing, du Bois-Reymond set up his own private workroom, a self-appointed microcosm that recreated the intimate atmosphere of his former live-in laboratory. Gustav Fritsch wrote that:

> "In a big room back along the garden, E. du Bois-Reymond had set up his own little kingdom – his own private laboratory. There, cut off from the outside world and with his unique, inexhaustible endurance, he studied the most difficult problems of neurophysiology, often using the highly sensitive mirror galvanometer. Here a holy tranquility reigned which the undersigned was only occasionally permitted to disturb" (Fritsch 1910: 159).

Isolating the aging physiologist from the world outside, averting his gaze from the obscurities and contradictions of modernity, the walls of his 'little kingdom' blocked his view of the monotonous tasks [*Dutzendarbeiten*] the Americanized youth were performing. In an environment of such determined withdrawal from threatening, ubiquitous barbarism, the scientific virtues he perceived as particularly German – idealistic diligence and study independent of any pragmatic goal – could be cultivated and preserved.[35] The private laboratory was a refuge for perseverance and thoroughness, and a place where the experimenting Apollo kept himself alive. It is hardly surprising that du Bois-Reymond, preserved and hardened into a plaster cast in his own institute, lost contact with the present. Heinrich Boruttau remembers that du Bois-Reymond always offered the same demonstrations in his lectures and discussed almost exclusively the pioneering work of Helmholtz, Brücke, Ludwig, and Bernard in the classical age of physiology (Boruttau 1922: 57). His illustrations had not changed in years, and he always told the same jokes and stories, and said not to much about results that went beyond classical data. Gustav Fritsch remembered an inability to adapt himself to modern views – "when his subordinates approached him with problems, he took care never to refuse them directly, but he considered the issues in such a way that the matter ended right there" (Fritsch 1910: 159-162). As a result, two decades after its opening, the institute's technology was out of date. The institute was not even connected to the electrical power line. Perhaps the best representation of du Bois-Reymond's dissolution into modernity is the fate of his private library. After his death, his books were not passed on to the Berlin University Library; they were sold by a Leipzig dealer to the highest private bidder. On July 21, 1897, the *Berliner Börsen-Courier* noted briefly: "the purchaser of du Bois-Reymond's library is an American independent scholar."

---

35 "We [Germans]," Helmholtz believed, must "respect the truth with a passionate, unconditional, and selfless love which does not hesitate before any authority or appearance, which fears no end of work and sacrifice, and is extremely modest in its demands for outer success. For this very reason it continues to drive us to follow the greatest questions to the deepest tips of their roots and to worry less about matters with no connection to the deepest roots of things, or about their practical consequences and useful applications." (Helmholtz 1903d: 41). On another occasion, Helmholtz wrote that: "While the various European nations have made comparable progress in the study of inorganic nature, the new developments in physiology and medicine belong principally to Germany. [...] Currently, the central issue is the tireless diligence with which one must devote oneself to pure science, for ideal ends, regardless of practical applications. And we can certainly be proud that German scientists have distinguished themselves for just this kind of enthusiastic, self-sacrificing diligence, working for inner satisfaction rather than outer success" (Helmholtz 1903c: 397).

Simon Schaffer

Exact sciences and colonialism: southern India in 1900

> "He dated and cross-dated, pedigreed and triple-pedigreed, compared, noted, connoted, wove, strung, sorted, selected, inferred, calendared and counter-calendared for ten hours a day" (Rudyard Kipling, *Plain Tales from the Hills*, 1887).

> "If the forest is going to disgorge any more of its novelties upon me, I should prefer that it did so when I was not occupied in the observatory at night and alone" (H.G.Wells, *In the Avu Observatory*, 1895).

Astrophysics and anthropology may seem unlikely companions. Evident dissimilarities between observatory measures and participatory fieldwork make their juxtaposition jarring. But several historians of late nineteenth century sciences have found telling and intimate associations between studies of human and celestial populations. Imperial and anticolonial programmes have been peculiarly instructive cases. Relations between European and Indian cosmologies, for example, long provided a key theme in imperial and indigenous understanding of cultural conflict and contact. Both astrophysical and anthropological enquiries relied on socio-technical networks of imperial rule, and generated images, samples and numbers for processing in the imperial capitals. Both solar physics and survey ethnology were designed to master administratively challenging problems of social class and social survival in the name of the natural facts of high Victorian sciences. In the cases treated in this paper, both astrophysical observatories and anthropological surveys were pursued from the hill-stations of southern India, emblems and bastions of imperial rule and its vagaries. Yet the sense of fundamental contrast remains. In the studies just cited, it is judged surprising that such a lowly enterprise as field anthropology might have any similarities with lofty astrophysics. It is insinuated, for example, that whereas astrophysicists' photography was designed for reliable and precise scrutiny, ethnographic images were simplistic propaganda.[1] Whence the sense that the precision of observatory sciences and the impressionism of field sciences of the late nineteenth century were obvious contraries? Some reflexivity seems appropriate here, because social anthropological resources have so signally helped reorient historical understanding of exact sciences' pasts. Instead of chronicling grand temporal successions, such studies focus on places and their synchronic co-ordination. This reorientation inverts a familiar hierarchy. Historians of science traditionally took astronomy as exemplar, and for more than a century treated all other sciences as slowly advancing towards ideal exactitude and delocalisation. Accurate astronomy's seeming view from nowhere was supposedly a precondition of its triumph. But by attending to science's geographies, the pattern science becomes fieldwork. It is argued that all sciences are best understood as local, patchy,

1 (Bayly 1996: 261-264); (Pang 2002: 142); (Mahias 1998: 45-56); (Davis 2001: 216-230).

co-ordinated processes. This inversion gave some anthropologists' fieldwork in scientific workplaces an early rationale, and raises questions about an historical anthropology of astronomers' field sciences.[2]

There is an apparent contradiction between the need for delocalisation and the reality of locale. But the contradiction is only apparent. Work is done in specific sites to make them count as universally representative. Pastoral settings of contemporary labs and observatories are telling, since bucolic retreats with sure controls on access have long been taken as distinctively reliable places to make truth. For this reason, too, there are anthropological programmes conducted by the personnel of scientific sites in engaging the communities which surround them. This is not so much anthropology of science, rather anthropology for sciences. Observatories provide good examples. From the Vatican observatory on Mount Graham in Arizona to the KEK observatories in Hawaii, anthropological connexions between observers' cultures and the heavens' meanings were and are used to handle critical links between astronomers and hosts, and between hilltop ecologies and engineering. Sometimes, the linkages of these astronomical sites play complex roles in geopolitics. In late 2000 the Indian Institute of Astrophysics, originally based in the south at Kodaikanal in Tamil Nadu, opened a new observatory, now the highest in the world, far to the Himalayan north on Mount Saraswati in Ladakh. It is surrounded by managed wildlife sanctuaries and the threatening military tensions of Kashmir. Spokesmen made much of the political importance of the facility, especially because the observatory is run via satellite link from distant Bangalore, thousands of kilometers to the south. The electronic link is taken to embody the integration of northern Kashmir into the Indian state. It is also taken to solve the problem, much criticised, of the very remoteness of the expensive new observatory. The equipment there, a 2m optical telescope, was bought from an American firm, prompting denunciations of the threat to autonomous Indian technoscience.[3] A long political history of Indian astrophysics and an immediate scientific anthropology of Indian society matters to this installation, a geopolitical setting where the very naturalness of the field is secured partly by managing its sociability.

A complementary reorientation in anthropologically-inspired science studies is towards craft practices. Once again, the pattern science changes from a supposedly rule-governed calculus towards a somewhat unspecifiable family of embodied skills. The image switches from panoptic classroom to messy workshop. Here, too, an apparent dichotomy has emerged. On the one hand, we are to understand the sciences as metrological systems, networks which depend on the painstaking distribution of legally-enforced numerical standards. On the other, the sciences are to be seen as enterprises of tacit capacities, ineffably embodied in competent practitioners. The anthropologist Ward Goodenough urges in his account of Micronesian navigators that "there are relatively few ethnographic accounts of bodies of knowledge that tell us what we need to know to be acceptable practitioners of that knowledge". So the notion of a craft-like body of knowledge raises fascinating questions about the possibility of specifying necessary knowledge (Goodenough 1996: 41). A very ancient

2 (Agar, Smith 1998); (Galison, Thompson 1999); (Livingstone 2003).
3 http://www.space.com/scienceastronomy/astronomy/himlayan_scope_001017.html, "India's new scope is world's highest", 17 October 2000; http://www.dailyexcelsior.com/01june03/news.htm, "Chief Minister inaugurates Ladakh-Bangalore satellite communication link", 2 June 2003.

tradition reinforces the distinction between the imperium of exact standards and the pursuits of cunning intelligence, the *metis* of the Greeks. Cunning intelligence was akin to animal instinct, evident in the successful conjectures of ship's pilots, a set of ingenious semiotic skills proper to carpenters, politicians, physicians and midwives. Crucially for our understanding of the role of *metis* in the life of science, Greek commentators reckoned that the nature of things in themselves made cunning intelligence powerful. Because citizens, diseases and seaways were multiple and shifting, their mastery required multiply shifting skills. Those epistemologists who denied shifting ontologies denounced cunning methods. As Marcel Detienne and Jean-Pierre Vernant point out in their definitive account of *metis*, "the various forms of practical intelligence are sweepingly condemned once and for all in the name of the one and only Truth proclaimed by [Platonist] philosophy" (Detienne, Vernant 1978: 316). No doubt this condemnation has informed the sense that rationalist standardisation and cunning intelligence cannot easily be combined in the sciences.

Consider, for example, a vital term in the adoption of anthropological approaches by historians – 'microhistory' – the emphasis on the particular and the local, the suspicion of the abstract, general, and serial analysis of change, and, especially, the claim that the right method for the anthropological historian is an attention to the trace, the clue and the exemplar (Ginzburg, Poni 1991). The doyen of microhistory, Carlo Ginzburg, appeals to the contrast between standard measures and local ingenuity. He sees the classical scientist as denying the importance of individual traits, setting authorized knowledge on the Galilean path away from conjecture and ingenuity. According to Ginzburg, such a scientist must be "professionally deaf to sounds and insensitive to tastes and odours". This explained, he argued, why microhistory could never be like elite sciences, but relied on, and documented, an alternative paradigm, more interested in pursuing traces and clues than in establishing laws and standards. These are 'low sciences' which ground notions of probability and conjecture.[4] But a series of clues suggests a different reading of the relation between cunning intelligence and the sciences' life. According to Ginzburg, the first appearance of the word "microstoria" was at the very end of Primo Levi's *Il Sistema Periodico* in 1975. Levi closed his reflection on the crises of the often viciously modern projects in science and politics, a story of Auschwitz and of analytic chemistry, with a virtuoso evocation of the local particular act of tracing nature's capacities. "A chemist without a nose is in for trouble" (Levi 1986: 185). Levi's science was an enterprise both of signs and traces yet also of standard procedures and metrological obsessions.

> "You have to granulate tin so that afterward it can be easier to attack with hydrochloric acid. So now grub among poisons, lipsticks and chicken shit, granulate tin, pour hydrochloric acid, concentrate, decant and crystallize if you do not want to go hungry" (Levi 1986: 224).

Because Levi contrasted this lived experience of indicative experiment with the abstracted thus delusive deliverances of textbook versions of the sciences, he therefore decided to introduce the term "microhistory". "This is not a chemical treatise. It is in some fashion a history. It is – or would have liked to be – a *microhistory,* the history of a trade, and its defeats,

4 (Ginzburg 1990: 106-108). For a regretful remark on this passage see (Latour 1990: 63 n.14). For low sciences see (Hacking 1975).

victories and miseries" (Ginzburg 1993: 15). This is a moving version of an anthropological history of the sciences. The same text which first introduced the term "microhistory" did so to show that the sciences were always mixed from agonised standardisation and local, ingenious, conjectural skill. The received contrast between metrology and cunning is misleading and dispensable.

It is no coincidence that both microhistorians and anthropologists make the moment of imperialism and modernism, the decades around 1900, crucial in their stories. According to the microhistorians, in those years elite European scientists at last began to exploit humbler divination which produced reliable knowledge of individuals by pursuing inadvertent traces. Favoured examples include Sigmund Freud's analysis of parapraxes or Sherlock Holmes' infuriating ability to divine an individual's identity from marks on clothes, furniture or the body. In 1890 Holmes consulted ethnographic gazeteers of the Raj to help pursue an Andaman Islander; the following year, he published at least twice in the *Anthropological Journal* on these sophisticated divinatory techniques.[5] As scholars such as Nicholas Dirks, Paul Rabinow, and most recently Chandak Sengoopta have well explained, Holmes's exact contemporary, the statistician Francis Galton, appealed to fingerprint data gathered by British administrators in India.[6] Long in use among Bengalis, fingerprinting was now touted by British officials as a universal method of individuation. "It is worth taking pains to enquire whether each person may not carry visibly about his body undeniable evidence of his parentage and near kinships" (Galton 1888: 201-202). London anthropologists much debated these puzzles of standards and signs. In 1900 they were told that

> "measurements must either show racial characteristics and be of use to the ethnologist, or they must show individual peculiarities and therefore be of value to the prison expert. They certainly cannot do both at once" (Garson 1900: 194-198).

According to another story, this was a period of global measures, when large-scale networks of reliable standards were constructed by tying together local measures in systems of agreed and apparently universal values. The new teaching laboratories and statistical bureaux of late nineteenth century Europe were centres for standards systems. The careers of the new sciences of 1900 can scarcely be understood without these systems; indeed, recent historians of spectroscopy, relativity and radioactivity have shown that metrological puzzles often prompted the most radical innovations in the order of fin-de-siècle physical sciences.[7] The extension of these sciences, especially in colonial regimes, helped make the patchy imperial world-systems work. The capacity of imperial sciences was put on show, "captured and made miniature for all to behold", as Paul Tapsell, the Maori director of Auckland Museum, well puts it, alongside simulacra of colonised cultures at the international expositions, as evidence of the universality of those sciences' claims (Tapsell 2003).

---

5 (Ginzburg 1990: 96-123); (Eco, Sebeok 1983: 55-77).

6 (Dirks 2001: 186-187); (Rabinow 1996: 112-115); (Sengoopta 2003: 101-111).

7 For an overview see (O'Connell 1993). For spectrometry and light standards, see (Sweetnam 2000) and (Johnston 2001: 36-49, 94-117); for radiation see (Boudia 1997) and (Hessenbruch 2000); for special relativity see (Galison 2003: esp. 311-312).

So detective divination and metrological standardisation were both entangled with the production of typological universals.[8] Take the specialised case of solar physics, a telling combination of precision astronomy and experimental chemistry. The Europeans' scientific Sun was an exemplary measured, rational object. Its phenomena were seemingly mastered in observatories and almanacs. The use of almanacs was a proverbial sign of western power. Primitives gawped, so it was said, when threatened travelers predicted solar eclipses by consulting their tabulated handbooks. In 1898 an eminent British astronomer watching a solar eclipse in India was "somewhat nervous as to whether our native assistants, who were high caste Brahmins, might not fall down on their knees and begin to say their prayers".[9] Astronomers who utterly relied on these assistants nevertheless treated them as enslaved by solar fear. Solar theories of religion and magic were used to demarcate the culture of science from that of so-called primitive societies. The view of superstition as solar was common among Europeans in India. At Ootacamund in 1872 one European cameraman recorded that "it happened as the Brahmins foretold, a powerful good deity cut off with one blow the big serpent's head, and the Sun, instead of going down the serpent's throat, enmerged slowly in all his glory from the opposite side".[10]

But despite astrophysicists' arch anthropological condescension, the Sun was certainly an important cult inside their own sciences. Europeans had their own solar religions. The Oxford mythographer Max Müller helped himself to state-of-the-art solar physics "which teaches us how we live and move and have our being in the Sun" to explain "why so much of the old mythology, the daily talk of the Aryans, was solar – what else could it have been?" (Müller 1878: 207-208) At Cambridge, the eminent anthropologist J.G. Frazer's *Golden Bough* (1890) started its account of "magical control" with a list of cultures which held they could affect the Sun's course, and used solar physics' gloom at the poetic end of his work in 1913: "in the ages to come man may be able to predict, perhaps even to control, the wayward courses of the wind and clouds, but hardly will his puny hands have strength to rekindle the dying fire of the Sun".[11] If the Sun was dying, it was surely alive: "we are slowly pulling the Sun to pieces", European astrophysicists such as Norman Lockyer boasted. Sunspots and surface patterns were read as signs of organic behaviour. The Sun's fate taught social lessons. "The Sun is in the position of a man whose expenditure exceeds his income. He is living upon his capital and is destined to share the fate of all who act in a similar manner", they told bourgeois readers.[12] To know the Sun was to bring it under standard measures, yet identify it by tracing its vital motions. Looking at the Sun's surface, Lockyer was

> "reminded by the fleecy delicate cloud films of an English hedgerow with luxuriant elms; here of a densely intertwined tropical forest, the intimately interwoven branches threading in all directions, changing slowly, almost imperceptibly".

The pursuits of solar pastoralism seemed appropriate. Their advocates were detectives and metrologists. In solar astronomy, they argued, "the thing to hunt down is a cycle, and if

---

8 For a telling case of sign-reading and imperial standardisation at ths period, see (Brain 1998: 277-279).
9 (Maunder 1899: 65-80); (Christie, Turner 1898: appendix, 1-21).
10 (Boesinger 1872). For solar eclipses and orientalism, see (Pang 2002: 77-81).
11 (Frazer 1907-1915: vol. 11, 307). For Victorian solar religion see (Beer 1996: 219-241).
12 (Lockyer to Stokes, 1873?); (Stewart 1873: 152). See (Myers 1989: 322-326).

found, then above all things, and in whatever manner, lay hold of it, study it, record it, and see what it means".[13]

These astrophysical pursuits concentrated on the many dark lines in the solar spectrum. Were they due to absorption of light by elements in the solar atmosphere? Could their wavelengths be matched with spectra of terrestrial elements in specially-equipped laboratories? Could these lines then be used to analyse the chromosphere just above the Sun's surface? By the end of the nineteenth century, new institutions for solar physics had been set up in the industrial suburbs of European capitals. British solar physics surveys relied on networks of Indian colonial observatories. Astrophysicists guessed spectral lines from the Sun would act as signs of future weather – and got cash to support astrophysics as a branch of imperial climatology. The aim was to make atlases of the solar spectra with standard wavelength values matched with known terrestrial measures. Solar spectroscopy switched to tracing the individual character of these lines. Vast metrological systems, such as the prestigious International Solar Union launched in 1904, sought to marry agreed international standards for making spectral lines with debates about their indicative meanings.[14]

Reading signs and establishing standards were *both* key techniques for the global surveys of 1900, especially in imperial administration. So Alfred Cort Haddon, as 1903 president of the Anthropological Institute, celebrated "the intimate connection which is beginning to be felt to exist between anthropological studies in the wider sense and the *practical art* of government" (Haddon 1903: 8) in India.[15] This language of practical art reveals something of the process of translation and local agency in what has too often been seen as effortless subordination. Sengoopta's fine new book on fingerprinting reminds us that governance of the Raj long acted as a laboratory for political and scientific systems only later applied to the metropolis, where "it rapidly lost its imperial character" (Sengoopta 2003: 189-191). Consider two salient adjacent examples of science during the Raj: ethnographic surveys of the Nilgiri Hills and astronomical surveys from the Palani Hills. Both were run from local colonial headquarters at Madras, the Government Museum and the Government Observatory respectively. Both were launched in 1895-1896. Both were designed to generate metrological and conjectural knowledges, a mixture of careful study of individual marks and an equally strenuous attempt to collect standardised data. These two plain tales from the southern hills involve imperial attempts to survey recalcitrant phenomena with the aim of manufacturing tables, charts and atlases as aids to governance. Social puzzles, including police, land tenure, revenue and grain supply, were all to be solved by finding physical marks and measures, including the shapes of noses, the strength of hands, the breadth of spectral lines and the pressure of the sun's atmosphere. Local delegates were despatched from Madras, armed with standardised equipment and instructed in the routines of data collection. Neither survey ran smoothly, whether from the point of view of the imperial centre nor in locally troubled encounters.

Then, in 1902-1907, new representatives arrived in the hills – William Rivers, expert experimental psychologist from Cambridge, travelled to the Nilgiri Hills, while John Evershed, an amateur solar spectroscopist from Surrey, set himself up in the Palani Hills. In both

13 (Lockyer 1869: 355); (Lockyer 1873), cited in (Kutzbach 1987: 200).
14 (Hentschel 1993b); (Hufbauer 1991: 70-85).
15 See (Kuklic 1991: 194-199).

cases, these arriviste savants and their allies sought to generate rather novel forms of knowledge – ethnographic fieldwork could take over from precision ethnological measurement, while solar spectroscopy might displace more traditional versions of precision astronomy. These new sciences were neither universally welcomed nor compelling. But they were certainly important for the history of scientific practices and their cultures. They resulted in claims about the psychic unity of humanity, the tacit dimensions of social structure and the underlying structure of space-time. The success of these claims relied on the temporary collaboration of settler and indigenous experts, and depended on the places where collaborations developed. The same ambiguity applied to the observatories and the survey stations situated in colonial hill stations.[16] In many ways, they were central institutions. There and only there could reliable data be gathered by the application of unrivalled hardware to pristine phenomena in a state of nature. Yet they were highly marginal, isolated from the centres of their sciences and often deliberately cut off from support and intimacy. This ambiguous geography helps us understand how hill-stations like these worked as laboratories of modernity in 1900.

The wooded massif of the Nilgiri Hills rises to over 2500 metres on the borders of the Indian states of Kerala and Tamil Nadu. The original heavy forest cover, giving way at higher altitudes to grasslands and temperate woods (*shola*), the steep descents from the Nilgiris towards the Malabar coast to the west, and slopes towards the central plains to the north and east, all long acted as natural barriers. The hills have for centuries been the home of the Toda people, pastoralists of sacred buffalos which provide them with milk and the dairies which form the centres of their worship. According to the census of 1900-1901, the polyandrous Todas then numbered about 800 persons, organised in fourteen separate patriarchal clans. Impressed by the picturesque Toda villages (*mund*), their distinctive dress of thick white woven shawls (*puthikuzhi*) and their idiosyncratic language, social system and physiognomy, Europeans notoriously fascinated with cultural diffusion and isolation were much tantalized by the Todas. British power arrived in the Nilgiris in 1819 in the person of East India Company collector John Sullivan, accompanied by a French naturalist named Latour. By 1821-1822 Sullivan had begun establishing a settlement in the heart of the Toda lands at Ootacamund (now known by its Tamil name, Udhagamandalam). Within five years "Ooty" (its British sobriquet) had become the chief sanatorium and hill station for all the British personnel of the Madras Presidency (Walker 1986: 240-274). Soon large areas of forest were cleared for tea plantations, later for coffee and chinchona. "The European appearance of the landscape was greatly enhanced in the parks and gardens by the introduction of most of the scents of England; in various spots the illusion could be complete". The British residents at Anglicised Ooty started a local hunt, with jackals in place of their ritual foxes; the British members of the Ooty Club invented an appealing drawing-room game, snooker. The Todas' image was often defined for imperial consumption through highly standardised studio photography, one 1900 collection of *The Living Races of Mankind* displaying them in "a sort of tropical Switzerland". These Nilgiri institutions and showpieces were subjected to political rituals and metrological regimes of the Raj.[17]

16 For geography of hill stations, see (Spencer, Thomas 1948: 637-651); (Kennedy 1996). For politics of hill stations in the Madras presidency see (Kenny 1997).
17 (Hockings 1989: 338); (Pinney 1990: 279-280); (Walker 1997: 107).

Some 120 km southeast, the British of the Madras Presidency set up another hill station at Kodaikanal in the Palani Hills. Reached via the great pilgrimage shrine at Palani itself, easily cut off in the monsoon, Kodaikanal was founded in 1845 at a height of over 2000m to protect its European residents from heat and cholera. Not so much an enclave for Europeans as an exclave designed to bar all human and non-human threats, Kodai was never a serious rival in political and commercial importance to Ooty, not least because of the difficulty of access along a very steep bridleway, the so-called Coolie ghat. But its virtues were widely advertised. Unlike Ooty, the indigenous population in the Palanis were classified by Raj census-takers as farmers, not pastoralists, so not entitled to the large land grants sanctioned in the Nilgiris. By 1890 there was a European population of almost 2000. A Kodaikanal Club started in 1887; Afghan princes interned there as prisoners of the Raj brought a strange glamour. Physicians recommended the well-drained soil, the good food, the boating lake. Ooty and Kodai would then become centres for systematic surveys and the exercise of cunning intelligence by all their inhabitants.[18]

The most thorough government survey of the Nilgiris was launched at the century's end as part of an ethnographic census first tried in Bengal in 1884-1886 by the Oxford-trained director of ethnography in British India, Herbert Risley. Crucial in this mobilisation was the notion of demographic 'type'. Techniques dictated by such agencies as the International Statistical Institute in the 1880s were used to turn locally variable accounts of caste into defined things capable of weight and measure. Individuals were supposedly drilled as members of transparent classes visible to a governing power.[19] Risley authored standardised data entry systems for census-takers and enthused about "the great storehouse of social and physical data" accumulated in British India,

> "which only need be recorded in order to contribute to the solution of the problems which are being approached in Europe with the aid of material much of which is inferior in quality to the facts readily accessible in India" (Cohn 1990: 157).

Risley's racialist project involved precision correlations between the shape of the human nose, the census categories of caste and achieved social hierarchy. In India, he urged, "the process of fusion was long ago arrested [...] there is consequently no national type and no nation in the ordinary sense of the word" (Risley 1909: 288). Risley supposed that British imperial sciences would weld together a complex of races, types and castes otherwise incapable of self-determination. At the end of 1899, the head of the British Association for the Advancement of Science, the Cambridge biologist Michael Foster, lobbied the Indian Government to launch anthropometric surveys as part of its census operations, especially of "Dravidian tribes" and "jungle races – concerning which our information is very limited". The Indian Government, backed by Risley from Calcutta, agreed by spring 1901 that "anthropometry yields peculiarly good results in India by reason of the caste system" and saw an ethnographic survey as indispensable for "the purposes of legislation, of judicial procedure, of famine relief, of sanitation and dealings with epidemic disease". During 1901 the anthropometric project was launched, using state-of-the-art measuring equipment and "the only

18 (Sandwith 1907: 362); (Mitchell 1972: 102-129).

19 (Edwards 1990: 235-58), for type specimens; (Cohn 1990: 154-58, 246-247); and (Dirks 2001: 49-52, 218-224) for Risley; (Morrison: 1984: 141-169) for survey ethnographies.

method that is applicable to so vast a population – that of selecting characteristic tribes and measuring a sufficient number of specimens to determine the type".[20]

Scientific metrology continued politics by other means. Risley's nasal indices allegedly gave physical signs of caste and class. It sought to naturalise social facts. In the same way, British astrophysicists pursued celestial signs which would allow them to predict the greatest social crises, the catastrophic famines which followed on the failure of the summer monsoons and of the subsequent rice and grain harvests. Holocausts, so they judged, could be naturalised too. Successive government Famine Commissions recorded more than twenty major fatal crises of supply in the half-century after the War of Independence. Economic orthodoxy in the Indian Civil Service, inherited from the first East India Company professor of political economy, Thomas Malthus, made state intervention in the grain market useless or worse. Astrophysicists were convinced that "the same Indra and Váyu, the watery atmosphere and the wind, whom the Sanskrit race adored centuries before the commencement of our era, still decide each autumn the fate of the Indian people" (Lockyer, Hunter 1877: 594). By mastering solar cycles, Raj astronomers reckoned they could master what they judged an utterly natural phenomenon. Lockyer, veteran of several Indian eclipse expeditions, led a campaign to study sunspots and solar spectra to master crises of supply. "We are destitute," said British astronomers, "of the first conditions for a scientific study of the food supply until we reach the period of settled British rule and rain gauges" (Lockyer, Hunter 1877: 594, 597). None of them conceded that 'settled British rule' might in fact help cause these holocausts. The Parsi intellectual Dadabhai Naoroji, London professor and mathematician, angrily riposted that British administrators were responsible for the deaths of some 10 million Indians in the 1876-1879 famines: "why blame poor Nature when the fault lies at your own door?" As Mike Davis has demonstrated, free-market fatalism sat appallingly well with a search for famines' natural patterns.[21]

These patterns, it was argued, could best be surveyed from new scientific stations set above the sweltering plains. The very highest point in Kodaikanal lies to the west at Nadingapuram Hill, 2347m above sea level. There in 1893 the Government of India decided to build a dedicated observatory to survey the Sun. According to John Eliot, head of the Meteorological Survey of the Government of India, "the situation is almost unique and atmospheric conditions extremely favourable for these observations" (Eliot 1902). In this astrophysics project, the Cambridge-trained mathematician Eliot played the same role as that discharged in the ethnological census by the high-flying administrator Risley. Both sought natural marks for social realities. Eliot was convinced of a periodic relation between solar activity, monsoons and famines. Innovative hardware and the local expertise of Indian astronomers could analyse the light from sunspots and the whole of the solar surface, yielding signs of the pressure, temperature and movement of the gaseous chromosphere around the Sun's core. These signs, in turn, were supposed to unlock the mechanisms of terrestrial weather: "the cyclic character of the Madras rainfall cycle must henceforth enter into considerations connected with the food supply of the people". Just as Risley saw the Raj as a

20 For the establishment of the Ethnographic Survey see 'Resolution' (23 May 1901) in India Office MSS Eur E100 and Risley, 'Memorandum on the Ethnographic Survey of India' (23 October 1908) in India Office MSS Eur E.295 / 18. For Risley's caste ideology see (Raheja 1999: 128-131).

21 (Davis 2001: 216-224, 323-326); (Naoroji 1901: 216) is cited (Davis 2001: 58).

"storehouse" of ethnological material, so the astrophysicists held that "we have in India a monopoly of the raw material" for solar surveys. Eliot sought to extend political control over this databank. In the midst of a newly terrible famine which raged from 1898-1900, Eliot had himself named Director-General of Observatories for the entire Raj. Then, like Risley, he started recruiting local resources and personnel in the south to discharge these scientific programmes.[22]

Risley's southern Indian representative was Edgar Thurston, a old Etonian medical botanist appointed in 1885 as superintendent of the local "storehouse of ethnographic data", the Madras Government Museum. Thence he taught policemen the techniques of anthropometry and hoarded material and numbers from his many fieldtrips. Thurston's hunts have drawn important recent attention. Nicholas Dirks shows how Thurston's notable attention to criminological policing revealed "a displaced Victorian enthusiasm for the colonised body" (Dirks 2001: 188). Gyan Prakash rightly finds indigenous agency tacitly conceded yet publicly denied in Thurston's witless jokes and vicious metrology (Prakash 1999: 43). The many museum visitors could see the collection of well-standardised glass eyes Thurston used to classify racial types as illustrations of European ethnographic standards. The aim was governmental, to teach spectators how to see and to measure how they behaved. As part of these so-called 'gymnastics', they would be subjected to anthropometric scrutiny in the newly equipped museum laboratory he built in the 1890s.[23] Thurston first took his tent and his measuring equipment to the Nilgiris on an ethnographic survey in 1896. Ootacamund and its surrounding tea estates had already been a site of mixed photographic, anthropological and astrophysical work: twenty years earlier, for example, a French expedition there surveyed a solar eclipse and made photographs of Toda men and women. This solar data was widely reported in debates on the chromosphere. The Toda images were published in Europe as evidence of the physical parameters of seemingly primitive tribes (Mahias 1928). So Thurston had good precedents for his new hill-station surveys. He reckoned that

> "anthropological research among uneducated and superstitious people who believe in the efficacy of a thread in warding off the evil influence of devils, and are incapable of appreciating that one's motive is quite harmless, requires tact, bribery, coaxing and a large store of patience" (Thurston 1896b: 229).

But he also told the school groups who visited his museum how dangerous it was that the Todas were in the very process of being 'civilised': "I was lately shocked to see a Toda boy studying for the third standard examination in Tamil instead of tending the buffalos of his *mund*". This was the telling predicament of metrological pursuits. Designed allegedly to 'enlighten', their value depended on the *absence* of 'enlightenment' among the people they studied (Thurston 1896a: 20).

Thurston set up what he called "an improvised laboratory" in his hill camp and got Todas to visit. Using members of the Malabar police force as standards of physical capacity, he

---

22 (Lockyer, Lockyer 1928: 95-97, 337-343). See in particular (Lockyer 1877: on fol. 36). For the Meteorological Survey see (Kumar 1995: 86-88).

23 (Thurston 26 May 1898: 82-84) for anthropometry as gymnastics and the Museum anthropological laboratory; (Pinney 1997: 60), for standardised glass eyes. For the laboratory's establishment, see (Bourne, Thurston 1897: 4).

compared Toda strength determined with dynamometers and their skull shape and nasal size measured with Risley's expensive calipers. When he returned to Ooty in 1901, Thurston brought photographs of Todas taken on the first visit ("a source of never-ending interest"), and inadvertently started a competition among the Todas to win the dynamometer game. As in his Madras museum laboratory, so in the hills the fame of Thurston's machine

> "spread from *mund* to *mund* and was circulated at funerals. One Toda man of fine physique, not satisfied with his grip, went into training and fed himself up in a few days. Thus prepared, he returned to accomplish 103 pounds, the result of more skilful manipulation rather than of a liberal diet of buttermilk" (Thurston 1896b: 144, 229).

Skilful in their own preparations for dynamometric showmanship, the Todas were also extremely perceptive about the political motives behind the government's ethnographic survey. At least one of the Todas refused to let his height be measured since "he was afraid that the height-measuring standard was the gallows; nor would he permit me to take his photograph lest it should be used for the purpose of criminal identification" (Thurston 1901: 15). As Elizabeth Edwards and Chris Pinney have shown in their studies of photographic production in the India of this period, identification and typology were keys to these surveys. Thurston projected his huge account of *Tribes and Castes of South India*, with chapters on the Todas and their neighbours arranged alphabetically, as a putatively exhaustive tool for imperial government and control. Susan Bayly has persuasively argued that by turning caste into "theories of biologically determined race essences" (Bayly 1999: 127), Thurston and his colleagues "played a critical role in the intellectual history of India and the Empire at large".[24]

Thurston had his opposite number in the Presidency's astronomical establishment. The then head of the Madras Observatory was the forty-year-old Scotsman Charles Michie Smith, physics professor at Madras Christian College, director of the local observatory from 1891. Michie Smith had a thorough grounding in late Victorian metrology – he had already worked on Kelvin's submarine telegraphy projects. Just as Madras citizens visited Thurston's museum and his public library, so they used Michie Smith's observatory for their own time standards, weather forecasts and almanacs. Tamil almanacs were widely published and much used by indigenous experts in celestial prediction. In 1867 Chintamani Ragoonathachary at Madras Observatory made a decisive observation of a variable star. Michie Smith was the sole European on the observatory staff – four Brahmin assistants, led from 1894 by the brilliant K.V. Sivarama Aiyar, worked at Kodai, along with a mechanic, "two menials, four peons and two lascars" (Smith 1900).[25] They were all to be provided with hilltop accommodation and appropriate salaries. Michie Smith adapted himself to Eliot's astrophysical programme with efficiency, but no enthusiasm. "The daily presence of abundant raw material in the shape of sunshine must be the first point to be considered in an Indian solar observatory," Lockyer told the Raj government: "I would recommend that a competent observer be sent from Britain to carry on the work". Eliot agreed with his friend Lockyer that Michie Smith needed a permanent European colleague at Kodai. He contacted his old college friends, who recommended John Evershed, a competent amateur astronomer with con-

24 (Edwards 2001: 134-135); (Pinney 1997). For Risley's praise of Thurston's progress, see (Risley 1908).
25 (Kochhar 1991: 97-99); (Henderson 1923: 253-254); (Smith, May 1900).

siderable experience of eclipse photography in India: "there is no doubt that Evershed handles a spectroscope with great skill", even if he were "not exactly the man wanted on the theoretical side". By early 1911 Michie Smith had retired, and Evershed took over the entire project.[26]

As a young chemist Evershed had been inspired by Lockyer to build his own spectroscopes, then directed the solar spectroscopy section of the British Astronomical Association, a new organisation for amateurs. Once at Kodaikanal, he completely overhauled its spectroscopic equipment and work patterns. It was quickly noticed that there were many line shifts, typically towards the red end of the spectrum, between spectral lines observed from the Sun and the equivalent spectra from elements on Earth. Without a reliable standard for spectroscopic measures, the entire project linking solar and monsoon activity would be compromised.[27] Between 1896 and 1910 a range of explanations of solar redshifts were available to astrophysicists – Doppler effects due to motion in line of sight of gases in the solar chromosphere; shifts proportional to the pressure of the luminous gases; and anomalous dispersion, which would affect the apparent wavelength of light reaching Earth through the higher layers of the Sun's atmosphere. Evershed put his staff to work in teasing these effects apart. He recruited the young Lancashire physicist Thomas Royds from Ernest Rutherford's Manchester radiation laboratory to help with experiments on the electric arc generating comparison spectra in his observatory. In late 1912, German spectroscopists worked out that wavelengths depended on the length and arrangement of the electric arcs used to make terrestrial standards. Already trained by Rutherford to make photos of helium spectra in alpha-particle experiments, Royds took the first photograph of this 'pole effect' in 1914. The work helped rapid international standardisation of the precise choice of metal, arc conditions and dimensions. Meanwhile Evershed got the Cambridge Scientific Instrument Company to build a new spectroheliograph to take better photographs of spectra from the Sun. The image of the Sun was formed on a spectroscope slit, then shone through a second slit to isolate a very narrow wavelength sample. The sun's image would cross the first slit uniformly, while a photographic plate behind the second slit would be displaced at the same rate. This would give a remarkable monochromatic image of the whole Sun, allowing extremely precise temporal monitoring of sunspot spectra. In Evershed's scheme, the slits were shifted hydraulically, keeping the solar image and the plate both fixed. He soon then commissioned a high-dispersion spectroheliograph on site at Kodaikanal built by his technicians from Madras, the first state-of-the-art astrophysical instrument entirely constructed in India. Then his Brahmin assistants could read off solar features from drawings of projections onto large charts used for recording sunspots and prominences.[28]

The Kodaikanal astronomers decided there were no pressure effects in the chromosphere. Using Doppler shifts in spectral lines visible from Kodaikanal in conditions cleared by monsoon rain, they claimed in 1909 to see a radial acceleration outwards from the centre of sunspots at deeper levels and an inward motion at their higher levels, a phenomenon soon

26 (Lockyer 1898: fol.44); on Evershed see (Larmor 1902). For the difficulties of building Kodikanal, see (Smith 1895, 1898); for staff and training see (Smith 1904: 333-334).
27 (Stratton 1957: 40-51); (Evershed 1955: 33-40). For Evershed's Indian eclipse work see (Maunder 1899: 65-80). For troubles with solar spectroscopic standards, see (Hentschel 1999: 208); (Hentschel 1997: 206).
28 (Forbes 1961); (Hentschel 1997: 228-236); (Evershed 1911); (Evershed 1912).

baptised the 'Evershed effect'. Any motion of this kind would, presumably, vanish at the Sun's edge. But Evershed's team saw large redshifts at the solar limb too. Work on the solar limb was aided by a periodic lull in sunspot activity.[29] By the outbreak of World War I, when surveys at Kodai were temporarily compromised, Evershed even started to contemplate a possible repulsion exercised by the Earth on incoming sunlight. Were there such a strange repulsion, then light from the far side of the Sun reaching the Earth after being reflected by Venus should not show redshifts. This odd proposal seemed confirmed. Evershed confessed the Earth effect was "very astonishing and rather unexpected"; it was entirely "contrary to expectation" (Evershed 1920: 155). Then in the midst of this astonishment Evershed read a new paper by the Cambridge astronomer Arthur Eddington, who in 1917 gave his British colleagues a summary of Albert Einstein's novel theory of gravitation, with its apparent implications that starlight should bend under gravitation, and that there should be a detectable redshift of about 8 mA in the wavelengths of sunlight reaching Earth. "It is probable that the [redshift] effect, if it exists, is masked by other uncertainties of the same order of magnitude" (Eddington 1917: 380), Eddington explained to his fellow astronomers. Kodaikanal set out to make itself a key centre in the pursuit of evidence for the general theory of relativity.[30]

Anthropological surveys in the Nilgiris were also reoriented in these years. Thurston had close links with other biologists in Britain, notably Foster's student Haddon, who gave Thurston's anthropometry positive reviews and greeted the Madras scientist when Thurston visited Britain on leave in autumn 1895. Thurston and Haddon stayed close friends throughout the following decade.[31] Then in spring 1902 Thurston was contacted by another keen British scientist, Haddon's ally the physician and experimental psychologist William Halse Rivers. In his late thirties, Rivers was already a veteran of leading European medical and psychological laboratory programmes, and, like Evershed, a customer of Cambridge's scientific instrument workshops. In 1898 Rivers travelled with Haddon and his colleagues to the Torres Straits, taking light tests and clocks, colour and eye tests, dynamographs and electromagnetic machines. Rivers designed new methods of 'field work', a term Haddon now coined explicitly to describe such enterprises as his Torres Strait expedition. In particular, Rivers recognised the collection of kinship data, his so-called "genealogical method", as a key to unlock social structure in conversations with native informants. Rivers contacted Thurston because he wished to try these strategies in southern India, and got funds from the Royal Society and the British Association for the Advancement of Science to pay for his fieldwork. While firmly committed to the value of psychometric tests in the field, indeed better equipped than any previous field worker with high precision laboratory apparatus for field tests of sensory acuity, Rivers was already sceptical of the anthropometric programme which Thurston and his superiors advocated. Instead of ethnology, Rivers sought to launch what he called ethnography, a detailed and highly descriptive literary account of the entire social system of a defined group.[32]

---

29 (Evershed 1909); (Evershed 1913a); (Evershed 1913b).
30 (Evershed 1918). See (Forbes 1961: 138-142); (Earman, Glymour 1980).
31 (Haddon 1896); (Thurston 1904).
32 (Slobodin 1997: 19-30); (Langham 1981: 69-73, 86-93); (Herle, Rouse 1998: 137-146, 158-178); Kuklick 1991: 133-149).

The processes through which the local, particular performance of experiments, instruments and technologies is compared with other locales are called "calibration". Without calibration scientific technique doesn't easily work anywhere else. Calibration succeeds when the proper community agrees that two different phenomena are in all relevant respects identical. The techniques devised in European laboratories, field stations and observatories in the late nineteenth century relied on such calibrations. Reliable machines might make humans reliable delegates, while sensuous experience got entangled with precision measurement. Francis Galton, elitist guru of psychometric testing and social statistics, pulled off the trick of making his sense of smell into an efficient computer:

> "I taught myself to associate two whiffs of peppermint with one whiff of camphor, three of peppermint with one of carbolic acid, and so on [...]. I convinced myself of the possibility of doing sums in simple addition with considerable speed and accuracy by means of imaginary scents" (Galton 1908: 284).

Rivers and his colleagues liked this trick. They quoted it with approval in their reports from the Torres Straits as they tried poking glass tubes, baptised rather grandly 'olfactometers', into their own and the islanders' noses. The Englishmen put up with this, but the islanders didn't. "The attention of the subjects immediately flagged as soon as they disliked or lost interest in the experiment" (Rivers 1908: 88). Self-discipline made scientists into self-aware instruments sent out to measure, observe, detect. The implication of this strategy for the sociology of the scientific community was spelt out by Rivers.

> "Up till recently ethnology has been an amateur science. The facts on which the science has been based have been collected by people who usually had no scientific training [...] By means of the genealogical method it is possible to demonstrate the facts of social organization so that they carry conviction to the reader with as much definiteness as is possible in any biological science".

The idea was to use the routines of training to make field experience count elsewhere.[33]

The small size of the seemingly insular Toda community and their status as anthropological exemplar, if not tourist showpiece, made them apt hosts for Rivers' work. Just as Haddon's group in the Torres Straits had relied on local colonial infrastructure of police and administration, so in the Nilgiris Rivers made use of Raj government. Haddon's network of field scientists waited for information on Rivers' trip: "I should like to hear how he has got on and if he has found anything new", a Ugandan-based researcher wrote to Haddon at the end of 1902. Between June and December 1902, with Thurston in tow, and accompanied as interpreter first by a forest ranger, Albert Urrilla, then by one of the Anglican mission's Tamil catechists, P. Samuel, Rivers worked with the Todas throughout the Nilgiri Hills. Rivers also collaborated with a Toda herdsman, Kodmer, who brought others for psychological testing and acted as guide through the hills. "My guide at the end of the day would sometimes go for a distance of eight or ten miles and back to arrange for my supply of men for

33 (*Reports of the Cambridge Anthropological Expedition to Torres Straits. vol. 2: Physiology and Psychology*: 181, 170 (report by C S Myers)).

the next day's work". He recorded the names of every Toda, their genealogies and customs, their dairy cult and their patrilineal clans.[34]

The anthropometrist Thurston *carefully looked at Todas as though they were anonymous things*. The anthropologist Rivers instead worked out how *carefully specifically named Todas looked at things*. He measured his hosts' colour-blindness and their sense of smell, their response to pain and to visual illusions. He commissioned from local studios quite new photographs of Toda dwellings, landscape and customs, he corresponded enthusiastically with botanical and philological experts, he supplied friends among the Todas with medical remedies, spectacles and cash. The Todas themselves, through Samuel, communicated to Rivers their enthusiastic expectation of his return to the Nilgiris, their demands for his photos, and their interest in the possible foreign markets for their fine white cloths. Just before leaving Madras, Rivers gave a passionate interview to the local newspaper, summarising the point of genealogical enquiry and setting out his scientific aims.[35] Soon after Rivers left the Nilgiris, his former interpreter, Samuel, wrote to him about arrangements for the Ootacamund photography studio, Wiele and Klein, to make pictures of individual Todas named in the genealogical tables. Samuel rejected one individual because he "went to a Hindu temple and shaved his head and beard"; he reported delightedly that a Chief had just died, so "I hope a good many funeral scenes can be taken by the photographer" (Samuel 1903). It has been urged that Rivers innovatively pursued virtual witnessing of Toda ritual by integrating photography with transcription of indigenous language and culture. He helped make images into evidence in Britain and gift objects in the Nilgiris (Hockings 1992).

By tracing the kinship structure of Toda society, and mapping all but a few of the Todas on his charts, Rivers could explain such puzzles as why in a society devoted to marriage between cousins it was to be expected that the same term would be used for father-in-law and for maternal uncle; the method also let him follow some of the complex rituals of Toda life. He rejected picturesque discourse which saw in the Nilgiris a kind of pristine paradise – there were good sociological reasons, Rivers alleged, for Toda behaviour and their choice of such homelands. On his return to his Cambridge college in early 1903, Rivers began a seventy-page paper on "the senses of the Todas", published in the first number of his own new *British Journal of Psychology* in late 1905, and a seven-hundred page ethnography of Toda society, published the following year.[36] *The Todas*, Rivers explained, was "an attempt to apply rigorous methods in the investigation of sociology and religion" (Rivers 1906: 7). One of the most important implications of Rivers' work was an effective demonstration of the psychic unity of mankind, the claim that differences in psychometric performance were not strongly related to what evolutionists judged the relative cultural development of the subjects on trial. "Pure sense-acuity is much the same in all races" (Rivers 1905: 391, 393). Socialization dominated psychological performance. Psychometric testing in the field could plausibly reveal more adequately than laboratory trials the basic psychological parameters of humanity (Mandelbaum 1980: 298).

---

34 (Rivers 1906: 8-10, 14, 19); (Rooksby 1971: 109-121); (Mandelbaum 1980). For requests for information about Rivers, see (Roscoe 1902).

35 (Thurston 1905) and (Andrew 1903). Rivers' interview is in *Madras Mail* (16 December 1902).

36 (Rivers 1906: 466); (Rivers 1903), reprinted in (Slobodin 1997: 188-193); (Rivers 1905).

Standardisation matters partly because it helps agreement about the virtue of a trial independently of the result of that trial. The puzzle faced both by Rivers and by Evershed was to establish the repute of their collaborator's data when the correct outcome of observations was itself in question. In spectroscopic astronomy the very quality of metrological standards was itself under threat. Hence the difficulties of the Kodaikanal programme. Between 1914 and 1923 Evershed and his collaborators had to work hard to make Kodaikanal a reputable observatory of shifts in solar spectral lines. The Indian Observatory and Meteorology systems offered weighty support. The imperial famine project's status could warrant the quality of the solar spectroscopy work. Administrators were "impressed, more than by the routine work, by the keenness of the staff, Indian as well as English, and the number of ideas which were being investigated or were awaiting examination".[37] But Evershed's team faced a major puzzle: neither Einsteinian gravitation nor terrestrial repulsion seemed acceptable accounts of the redshifts they found. During 1918, he got his new chief assistant Narayana Aiyar to measure redshifts from the Sun's centre out to its limb. At the solar centre, the Brahmin found shifts of 4.3 mA, at the limbs as much as 6 or 8 mA. So in summer 1918 Evershed explained his dilemma, a choice between the 'intolerable idea' that the Earth repelled sunlight and the novel proposal from Einstein. "If we exclude relativity, we are up against the Earth effect, and even if we assign the shift to a combination of relativity and motion, we cannot by that means escape the Earth effect".[38]

Kodaikanal's comparatively large shifts could not be marketed uncontroversially to the wider astronomical community. At Mount Wilson, the newly prestigious California observatory which had become the metrological centre of astrophysics, Charles St John reported no visible redshifts at the solar centre, and barely 2 mA at the limb. Since he could scarcely assail Mount Wilson's equipment, Evershed criticised the choice of data: "because of their faintness, most of the weak lines measured by St John are very difficult and unsatisfactory lines to measure especially in the limb spectra". St John replied in kind. He alleged that Indian observations were made with inferior equipment and insufficiently well-calibrated against the standards for making comparison spectra with electric arcs.

> "Further elucidation of the discrepancy between Kodaikanal and Mount Wilson will be awaited with much interest. It seems clear that the absence of the effect would either be fatal to Einstein's theory or at least involve a rather fundamental reconstruction of it".

Evershed conceded the point: "we seem able even now to state, with St John, that our results are distinctly unfavourable to relativity".[39] But for the Kodaikanal astronomers, like many others, matters shifted decisively after dramatic 1919 reports of light-bending measured during a solar eclipse in the south Atlantic, an announcement stage-managed by Eddington in London at the Royal Astronomical Society. This startling eclipse report in the imperial capital helped make Einstein's general theory of relativity a discovery of cosmological truth. Evershed reorganised observatory work in line with new metropolitan orthodoxy and what

37 (Walker 1908); (Walker 1911).

38 (Ayyar 1915); (Evershed 1918: 371, 374-375).

39 (Evershed 1918: 373); (Hentschel 1993a: 151). For the status of Mount Wilson, see (Hufbauer 1991: 73-79).

he called its 'brilliant confirmation' of general relativity. What had once seemed strange reports of large redshifts from a remote Indian hilltop turned into confirmations of relativistic cosmology. Narayana Aiyar's brilliant observations of large limb shifts, "taken by themselves, appear favourable to Einstein's theory".[40]

Rivers faced very similar problems of replication, repute and reliability. Fieldwork culture had to be transferred elsewhere. This was hard work, not easily to be achieved simply by the publication of his results. Rivers tried to enlist the Raj census system for his own purposes. In March 1910, when preparations were being made in Calcutta for the new Indian census, Rivers sent Risley's team schedules of terms of relationship to drill census officers in the genealogical method, while officers ordered several copies of Rivers' work on cousin-marriage throughout the census network (Gait 1910). Such mobilisation did not on its own give demonstrative authority to Rivers' fieldwork. It revealed the effort required to make this work count. Alongside these crises in the regime of ethnological measurement, Rivers also offered, in his lengthy ethnography, what became a talismanic text for later fieldwork. It was "intended to be a sample of scientific method as applied to the collection and recording of ethnographical facts". Backed with newly standardised methods for anthropological enquirers, authored by Rivers then published by the British Association, *The Todas* was taken as manifesto. "Details which may seem insignificant or tivial are often of great importance in the comparative study of custom and belief". Frazer found Rivers' imagery familiar material for his own solar mythography. He told Rivers in 1906 that "the book is a model of anthropological research" and that "every Cambridge anthropologist will feel proud". The programme functioned as resource for fieldwork, notably that soon pursued in India by Rivers's followers such as Arthur Radcliffe-Brown (in the Andaman islands in 1905) and Charles Seligmann (among the Veddahs in Ceylon in 1906). Seligmann wrote to his London School of Economics colleague Bronislaw Malinowski, most eminent of Rivers' successors, that "field research in anthropology is what the blood of martyrs is to the Church". For Malinowski, in 1910, "the book on the Todas stands as a model".[41] In India itself, students of Rivers and Haddon such as G.S. Ghurye at Bombay, and L.K. Ananthakrishna Iyer and Sarat Chandra Roy at Calcutta, all pursued this kind of work. Such field study, notably that of Sarat Chandra Roy in Chota Nagpur, was often pursued in deliberate contrast to the surveys of Risley's census enumerators. So in his review of caste in Indian society, Ghurye insisted that projects such as Risley's census had in fact generated the very 'caste-spirit' they allegedly only reported. Rivers himself never returned to India, but just before his premature death in 1922 he was planning lectures at the new social anthropology department at Calcutta University. Though eventually excised from the official memory of British social anthropology, for which it was Malinowski who would henceforth act as founding father, Rivers' experiences at Ootacamund and the surrounding hills sought to turn the Nilgiris into an exemplary site for the human sciences and their fortunes elsewhere.[42]

It was hard to make such sites exemplars. The problem of observational independence was crucial in all Indian sciences. One leading British astronomy journal, now proclaiming

40 (Evershed 1920). For the theatrics of Eddington's eclipse expedition, see (Earman, Glymour 1980a); (Sponsel 2002); (Stanley 2003).

41 (Rivers 1914: vol. 1, vi); (Urry 1993: 27-31); (Frazer 1906); (Mandelbaum 1980: 294).

42 (Langham 1981: 179-180); (Mandelbaum 1980: 288-289); (Srinivasan 1998: 72); (Ghurye 1932: 158).

Einstein as "the leader of the revolution in science", nevertheless pointed out that unlike Eddington's apparently splendid (and metropolitan) light-bending observations, in the case of redshifts it was the 'weaker' data, that from India, which seemed to support Einstein. Evershed protested. His instruments and the local staff were equally splendid, the discovery of new solar physics showed his observatory's virtue, and "the difficulty of deciding for or against the Einstein shift in the Sun lies in the conflicting nature of the evidence itself. It is not a question of weighing the relative accuracy of the measures made [elsewhere] and at Kodaikanal".[43] These protests were vain. In summer 1923 Evershed resigned from the observatory, leaving Royds as his successor. When in autumn of that year St John at last agreed that all redshifts could be explained by a combination of relativistic gravitation, Doppler effects, and anomalous dispersion, his so-called 'conversion' was lauded as decisive evidence for Einstein's doctrine. Kodaikanal's track record was discounted. Evershed wrote to the Oxford astronomer Herbert Turner, who made much of St John's conversion in the pages of the London newspapers. "It is a great satisfaction to me to learn that Mt Wilson measures of spectrum lines have come out so closely in accordance with our Kodai results", Evershed ironised from his new suburban home.

> "Although I had to construct the instruments at Kodai, and did most of the work of exposing the plates and measuring them myself, I managed to get a body of evidence which ought to have been convincing, but does not seem to have made very much impression",

he complained. He recalled his criticism of St John's 1918 paper, which had then been used against Einstein; he boasted of the decade's 'extended study' his team had conducted of solar redshifts; "yet it seemed that anything from Mount Wilson must necessarily be superior to Kodaikanal". Turner agreed that "you have been hardly treated", but he counselled Evershed that "it is rather the result than the man who obtained it that will interest readers". The former Kodaikanal astronomer was never mollified. In 1930 Evershed restated to an American astronomer that his team had shown that the Einstein effect was responsible for 'part of the shift', but "there is a residual shift to be explained".[44] The Kodaikanal programme, reinvigorated after decolonisation and now the basis of work at the Bangalore IIA, continues to provide precedent for contemporary all-India astrophysics. Concern with post-colonial hierarchy stays crucial for these sciences. Whether in anthropology, astrophysics or meteorology, the dramaturgy of sudden proof at ambitious world-centres can, perversely, swamp painstaking surveys compiled at allegedly more remote out-stations.

Imperial apologists often used stories of empire to describe the role of science in changing the scope of power. Such systems prompt the modern Indian anthropologist Amrit Srinivasan to observe that

> "when field-work was colonial, it was understandable that we, as servants of the Raj, acted as modest assistants to the British, sometimes even taking the scholarly place, but only in the data-collecting half of the positivist praxis, leaving science to our betters" (Srinivasan 1998: 73).

---

43 (Evershed 1921a) and (Everhed 1921b); (Hentschel 1993a: 148-152); (Earman, Glymour 1980b: 198).
44 (Hentschel 1993a: 164-165); (Evershed 1923), (Turner 1923), (Evershed 1930).

Astrophysical and ethnological surveys manifested this kind of hierarchised politics, because they involved the difficulties of local co-ordination and standardisation for the welfare of imperial government. This is why conflicts about status, authority and credibility counted so much. Instead of asking how in 1900 scientific universality was produced in general, better to ask about the conditions which then picked out some places and persons as more reliable, secure and indicative. What exactly distinguished, say, Eddington's astronomical trips from Evershed's Kodaikanal, or Rivers' Nilgiris from Malinowski's Trobriands? And, even more importantly, what precisely was it which picked out any surveyors' work in the hills from among the customs of those who lived there?

The hill-station is a peculiarly indicative site at which to pose these questions. It seemed that by constructing their own asylum, governing Europeans could escape the otherwise inevitable infantilisation visited upon all others, thus know and rule them. The highly nuanced categories of hill-station life, with its European elites in hill-top compounds, native bazaars both essential and invisible to their survival, dramatised the gradations of colonial rule. "At any great station in India, society is just as exclusive as in any small country town in England". Edgar Thurston could joke in his *Nature* report on Madras anthropology that his move to the hills in every hot season since 1894 had "brought down on myself the unofficial criticism, that anthropological research at high altitudes is eminently indicated when the thermometer registers 100 degrees in Madras" (Thurston 1898: 84). Epicurean civil servants explained that in Kodaikanal the native bazaar and cemetery could not be seen from the European compound, which "commands a view over the plains which is comforting to those who agree with Lucretius that it is sweet to watch, from a safe spot, one's neighbours in distress".[45] When Eliot first proposed making Kodaikanal the centre of his astrophysical famine survey, he mixed the hierarchy of scientific labour with that of hill-stations' health. Such observatories needed three kinds of workers: observers/computers, technical managers and directors/publishers. The first task was for Indians, the second for local British officials, the last for men like himself.

> "It would evidently be advisable to place such officers in a climate most suited to hard and effective brain-work and in immediate connection with a large observatory and a scientific library &c. and this could most satisfactorily be obtained by locating them at Kodaikanal".

Others, especially Indian astronomers, protested. The eminent boss of Bombay observatory, Nanabhoy Ardeshir Framji Moos, riposted to Eliot that "it may be true, as Mr. Eliot presses, that brain work can be done better at Kodaikanal than at Bombay. But the records of this Observatory, to speak of nothing else, show that good brain work can be done in Bombay also. A survey with an pretensions to accuracy must be conducted personally by a man of recognised ability", and there was no reason to suppose that native scientists were devoid of this ability nor their cities necessarily lax and unreliable.[46]

---

45 (Mitchell 1972: 18-20); (Tilt 1875: 72); (Francis 1906: 247). For health and dress in the hills, see (Cohn 1996: 149-160).

46 (Eliot 1896) and (Moos 1896). For the work of the Bombay observatory, see (Lockyer 1898: 6-9, 18 fols. 26-28, 32). For support for hill-station hierarchy by the Astronomer Royal, see (Christie 1898: fol.57).

The work of ethnographic surveyors, astronomers and census-takers neatly dramatised perceived problems of indigenous morality, belief and technical expertise. The cult of precision instrumentation was often the focus of attempts to perform authority. Surveyors tried to use hardware to dramatise power. In many ways, European accounts of indigenous incapacity were designed rather to mask the immense reliance of the surveyors on the co-operation and the skills of all others. In these surveys, it has been very clearly demonstrated, Indian technicians and experts were essential actors.[47] The same kinds of skilful forms of collaboration and co-operation gradually impressed Rivers in his own fieldwork. Both astrophysicists and anthropologists began to realise the importance of working at sites already, if unevenly, engrossed by global networks. The predicament of the field sciences around 1900 involved systematic shifts and reconciliations between the values and geography of the transient tented camp, the more permanently established observing base, and the systems of communication and governance on which they utterly relied. In his influential accounts of method, Rivers urged from 1907 the importance of "intensive work [...] in which the worker lives for a year or more among a community of perhaps four or five hundred people and studies every detail of their life and culture" (Rivers 1913: 7). For his patron and colleague Haddon, the "intensive study of restricted areas" (Haddon 1906) had to displace survey ethnography. In astrophysics, general surveys which characterised the enterprise of spectral atlases were displaced by the intensive study of specific groups of carefully chosen emission and absorption lines pursued at Kodaikanal and at its rival and associated observatories. The construction of these highly specialised spaces, based on cooler hilltops and staffed by supposedly loyal and disciplined personnel in turn demanded the temporary alliance of indigenous and European experts. From the hill-stations new kinds of knowledge emerged which could never quite be dictated by, yet entirely depended on, the rules of imperial hierarchy.

These have been stories of salient linkages between empire, standardisation and locale. Such tales should trouble the claim of Paolo Palladino and Michel Worboys that "there is no evidence that physics or astronomy played any significant role in British imperial policy or colonial rule" (Palladino, Worboys 1993: 101). They should help challenge Lewis Pyenson's view that "those who achieved distinction overseas in the exact sciences did so by staying beyond the reach of the imperialist octopus that moved men across oceans to satisfy metropolitan appetites" (Pyenson 1990: 77). These accounts have also included sketches, literally and figuratively, of knowledges high and low. Return once more to Primo Levi. He recalled a disturbing conversation between a lofty astrophysicist and himself, a humble chemist, in the wake of a characteristically messy laboratory catastrophe. Here was the moral drawn by the astrophysicist: "Better not to do than to do, better to meditate than to act, better his astrophysics, the threshold of the Unknowable, than my chemistry, a mess compounded of stenches, explosions and small futile mysteries". But Levi "thought of another moral, more down to earth and concrete: differences can be small, but they can lead to radically different consequences. The chemist's trade consists in good part in being aware of these differences, knowing them close up and foreseeing their effects. And not only the chemist's trade" (Levi 1986: 60). This moral applies to anthropology and other knowledges too. We can illuminate the subtle differences which so mattered in science, empire and history by helping pursue high sciences' stations in the messy troubles of lowly life.

47 (Raj 1997); (Pang 2002: 137-139).

Hans-Jörg Rheinberger

The Art of Exploring the Unknown: Views on Contemporary Research in the Life Sciences

In a recently published essay collection aimed at a diagnosis of contemporary reason, the American anthropologist of science Paul Rabinow made use of the Deleuzian concept of "assemblage" to describe the particular constellations within which new knowledge arises. Assemblages are to be seen as conglomerations of actors, things, and institutions in which, according to Rabinow, something eventually "emerges out of a lot of small decisions; decisions that, for sure, are all conditioned, but not completely predetermined."[1] "From time to time," then, we read at another place in his *Anthropology of Reason*,

> "new forms unfold that have something peculiar around them; something that elevates existing actors, things, and institutions into a new mode of existence, engages them in a new assemblage; an assemblage that lets things"– not only appear in a different light, but – "happen in a different way" (Rabinow 2004: 115).

Experimental Systems

Translated into my terminology, *experimental systems* – a kind of assemblage specifically tied to the production of knowledge – are at the core of the dynamics of our contemporary empirical sciences. Within as well as among such systems, from time to time, and usually due initially to the concurrence of incremental decisions rather than of major revisions, *conjunctures* happen that subsequently let things appear in a different light and let things happen in a different way in the future. Experimental systems did not always exist, certainly not in early modern science. They are the products of more recent developments, structures that have proven useful for practicing the art of empirically exploring the unknown. And as far as I can see, they are also the moving forces for last century's shifting of disciplinary boundaries and perhaps even for the future dissolution of the boundaries of our classical disciplines. Disciplines appear to have been the structures for practicing the art of exploring the unknown that were developed and that dominated during the nineteenth century. In the twentieth century, experimental systems took over.

As an example, I would like to discuss here a particular experimental system. It came to play a decisive role in the history of molecular biology, and it stands as an exemplar for the kind of structures that I regard as the moving elements of contemporary research. It is the in vitro system developed for exploring the mechanism of protein synthesis.[2] Protein synthesis came to occupy the central place in the elucidation of the mechanism for translating genetic

---

1 (Rabinow 2004: 63). Since the papers in the book have been revised and the interviews not previously published, I quote from the German collection.
2 See (Rheinberger 1997).

information in the context of molecular biology. The starting point of this research endeavor, however, was very remote from the outcome. The first steps took place in the context of a specific conjuncture between sub-cellular cytology and cancer research shortly after World War II. Then, in the course of its early development, protein synthesis grew into a system that allowed biochemistry to be done in an unprecedented way and helped to establish a biochemical working culture that went way beyond classical enzyme chemistry as known from the first part of the twentieth century. Finally, it was inserted into the research core of molecular genetics, leading to one of the major feats that characterized the golden age of molecular biology, namely the deciphering of the genetic code. Before characterizing a few of the elements of protein synthesis that will lead us to a more general discussion of the role of model organisms, of test tube research, and of research instruments, let me make a few introductory remarks on experimental systems.

Describing his laboratory work at the Pasteur Institute in Paris, the French molecular biologist François Jacob has stressed that modern biology is impossible without what he called the "choice of a 'system'" (Jacob 1987: 261). Such systems enable the research process in that they permit the manipulation of certain parameters and thus allow for the stable representation of phenomena that otherwise would remain intangible. But they also confine the researcher's gaze and range of action. In the biological research literature, the notion of "experimental system" has been in regular use since the early twentieth century, in particular in connection with the establishment of a vigorous in vitro biology,[3] and with the coming into use of a variety of model organisms – as, for instance, the bacterium *Escherichia coli* – two essential features of experimental systems to which I will return.

A historical comparison may help to set the stage. Two centuries ago, toward the end of the eighteenth century, when natural historians and nascent biologists talked about "systems", they meant systems of speculation such as the "system of the earth", the "system of nature" or, with respect to the then current theories of generation, the "system of the eggs" and the "system of the sperms". Into these systems, sporadically, experiments or observations were inserted as supporting arguments. They were, however, not the driving forces for the establishment and change of these systems. Two hundred years later, it is just the other way around: Experimental systems determine the research context, and theorems eventually come to be inserted in them.

Despite its widespread generic use, the concept of experimental system was not seen for a long time as a historiographical and epistemological tool for the description of the modern research process. We can find first hints of such a use in the writings of Ludwik Fleck, although he did not make a systematic terminological use of it (Fleck 1979). But Fleck stressed the fact that the research process is based on a stream of experiments that constitute historical, materially continuous trajectories, and not on isolated experimental acts. Karl Popper, as we know, argued for the latter, and with that, in contrast to Fleck we could claim, delivered an epistemology for eighteenth century science rather than for our present times. It was only at the beginning of the 1990s and in the context of an ongoing replacement of theory-dominated perspectives of scientific change with practice-driven views on research that the concept of experimental system found entrance into the historical and philosophical literature of science (Rheinberger 1992). Within the same context, but in slightly different

---

3 See, e.g. (Gale, Folkes 1954: 1224).

perspectives, the terms "manipulable systems" (Turnbull, Stokes 1990), "production systems" (Kohler 1991), and "experimental model systems" (Amann 1994), have been used.

To pick up an actors' category, take it out of its laboratory context, and elevate it to a central epistemological category for the characterization of the dynamics of contemporary research must be justified. The term "system" is used thereby in a rather loose fashion. To begin with, it points rather simply to a kind of weak coherence, or "coupling" – again a term of Fleck's – both synchronically with respect to the epistemic, the technical, and the social elements that enter into an experimental system, and diachronically with respect to its persistence over time, its historical trajectory. The advantage of the concept lies in its capacity to tie together – and to think together – essential aspects of scientific research processes, such as instruments and measuring devices, scientific objects of various sorts, the skills and constellations of experience necessary to make use of them, and the spaces in which they are located. The experimental system approach refuses to describe science as a mere system of concepts. Rather, it characterizes *research* as a material process of the coming into being of scientific knowledge or, to speak with Bruno Latour, of science in action (Latour 1987).

How can experimental systems be described with respect to their more general characteristics? They exhibit epistemic and technical as well as social and institutional aspects. The social and institutional aspects lie in the fact that they display locally situated research connections that grant coherence to the activities of a group of researchers. At the same time, such systems differ sufficiently from other comparable units, and with that, they convey identity and individuality to the work of a particular group of researchers.

Let me say a few more words on the epistemic and technical aspect of experimental systems before I come to my example. Four basic features have to be distinguished. First, as I have argued extensively elsewhere (Rheinberger 1997), such systems are the smallest integral working units of contemporary research. Within them, scientific objects – "epistemic things" – and technical objects – the technical conditions of their production – are inextricably linked. The first kind of entity, the scientific object, is that badly defined something to be addressed as the target of the whole experimental endeavor. Paradoxically speaking, it embodies in an experimentally manipulable manner what one does not yet exactly know. The scientific object is therefore necessarily underdetermined; it is under-defined by definition. The technical objects, in contrast, are characteristically determined. They are the instruments, apparatus, and devices bounding and confining the assessment of the epistemic things. Their rigidity is necessary in order to keep the vagueness of the scientific objects in a hypo-critical condition. Within a particular research process, epistemic things can eventually be specified and turned into technical things and become part of the technical conditions of the system. Parts of the technical system, however, can also acquire epistemic status and thus turn again into research objects. The dialectics between epistemicity and technicity is at the core of an experimental system; it is its driving force. Thus, an experimental system is a kind of dynamic research body that conveys material shape to the scientific objects formed within it, and at the same time, determines the boundaries of their conceptual apprehension. Second, experimental systems must be able to undergo what I have called differential reproduction if they are to remain arrangements in which new knowledge is generated that lies beyond what one has been able to imagine and anticipate. Only in this way can they act, as Mahlon Hoagland once put it, as "research generators" (Hoagland 1990: xvii). Difference and reproduction are the two inseparable faces of a coin. Their interplay determines the delays as well as the precipitations in the course of a research process. In order to remain productive, experimental systems must be organized in such a way that the generation of differences becomes the reproductive driving force of the whole machinery. Differential reproduction conveys a special kind of historicity to experimental systems. Accordingly, experimentation acquires, to speak with Ian Hacking, "a life of its own" (Hacking 1983: 150). Experimental systems are thus units extending in time: emerging, taking shape, and eventually also disappearing again.

Third, experimental systems are the units within which the material signifying units of knowledge are produced, the items that are usually called data, but should be rather called *facta* in the sense of those primary things that are *made* during the research process. Their meaning is tied to the spaces of representation in which the traces and inscriptions become recorded, articulated, dislocated, reinforced, marginalized, and substituted. Researchers 'think' within the confines of such spaces of representation, within the hybrid context of the representational machinery at hand that makes up the technical conditions of a given experimental system.

Finally, ramifications and fusions of experimental systems lead to ensembles of such units. They can be addressed as experimental cultures. Fusions and ramifications themselves are, as a rule, the result of unprecedented events within experimental systems, events that are often connected to the introduction of new technologies of representation. In the last instance, such experimental cultures determine the emergence as well as the historical obsolescence of scientific disciplines. The concept of experimental culture as an articulated ensemble of experimental systems should allow us to write the history of research domains free from the burden of disciplinary history. This is, however, not only a matter of historiography. Today's experimental science itself derives its dynamics less and less from rigid disciplinary boundaries and their social solidification and more and more from the digressions and transgressions of smaller units of research in which knowledge is not yet labeled and classified, and in which new knowledge can take shape at any time.

An Example: Protein Synthesis

I will not enter here into the details of the development of the protein synthesis system. Nevertheless, I would like to give you an impression of the research trajectory it created (Fig. 1).[4] If the chart conveys the sense of a maze, it also gives an idea of the options and choices that come with the creation of such a system and from working within it. This particular experimental system is peculiar in several ways. It is characterized by binding together a number of heteroclite elements. I will identify a few of them, and then add some general remarks concerning contemporary experimentation to each of them.

First, the model organism of early protein synthesis research: rats, and more narrowly, a particular organ of the rat, the liver. People spoke about the rat liver protein synthesis system. This was the legacy of the working context out of which the research grew. It was a clinical context at the Massachusetts General Hospital in Boston, a place more specifically

4 Redrawn after a sketch of Douglas Allchin with his kind permission.

concerned with cancer research, and a few other medical schools in the United States and in Europe right after World War II. Rats had been used as clinical models for malignant growth for decades, and indeed, the establishment of the in vitro system of protein synthesis derived from the purpose of the first experiments: to find out whether the process of protein synthesis differed, in speed or in mechanism, between malignant and normal livers of the laboratory animals.

Model Organisms

This leads me to a few more remarks on model organisms and their use in biological experimental systems. In earlier centuries, it was the *differences* between organisms that attracted the interest of researchers who devoted themselves to the task of giving a full picture of the overwhelming diversity of life forms. This was the perspective of natural history. Under the epistemic regime of the early twentieth century, biological differences between research organisms were transformed into tools that could be exploited for the characterization of the most general features of living beings. In this perspective, the peculiarities of particular organisms are no longer interesting in themselves. They are only interesting insofar as they enable the search for generalizable features. A model organism is no longer analyzed in its own right, but for the sake of something that lies beyond it.

Under these conditions, our present notion of what constitutes a biological model organism came into being. One could even say that in order to leave the state of pure comparison and to become an experimental science itself, general biology as envisaged around 1900 *had* to create model organisms to work with. Biology around 1800 had asked what is intrinsic to living beings in contrast to non-living entities; biology around 1900 asked what pertained to all living beings as compared to each other. A model organism in this sense is defined as a living being no matter whether from the kingdom of plants, animals, or bacteria, whose manipulation might lead to insights into the structure, function, development or evolution common to a whole class of organisms, if not ubiquitous for the whole world of living systems.

It is in this context that model organisms started to play an increasing role, from the beginning of the twentieth century, in research on heredity, cytology, and embryology. First, they represent a particular phenomenon in an easily accessible fashion, and second, they can be handled in a productive operational way in the process of setting up an experimental system. This last point is particularly important: In order to function as model organisms, they need to be embedded in experimental systems, where they can play out their dynamics and function as exemplars. But their entrenchment in experimental systems, as a rule, has not only material prerequisites, but also material consequences. In the course of the work, the model organism under investigation is modified. The creation of pure lines or of particular gene combinations in genetic model organisms is an example. *Model* organisms are thus, as a rule, also organisms *modified* for particular research purposes.

Their entrenchment in experimental systems may also make them, at least to a certain degree, resistant to being replaced by potential competitors in a particular historical window of time. Model organisms embedded in experimental systems can, to speak with Gaston Bachelard, turn into "epistemological obstacles" (Bachelard 1969: 13). In a certain sense,

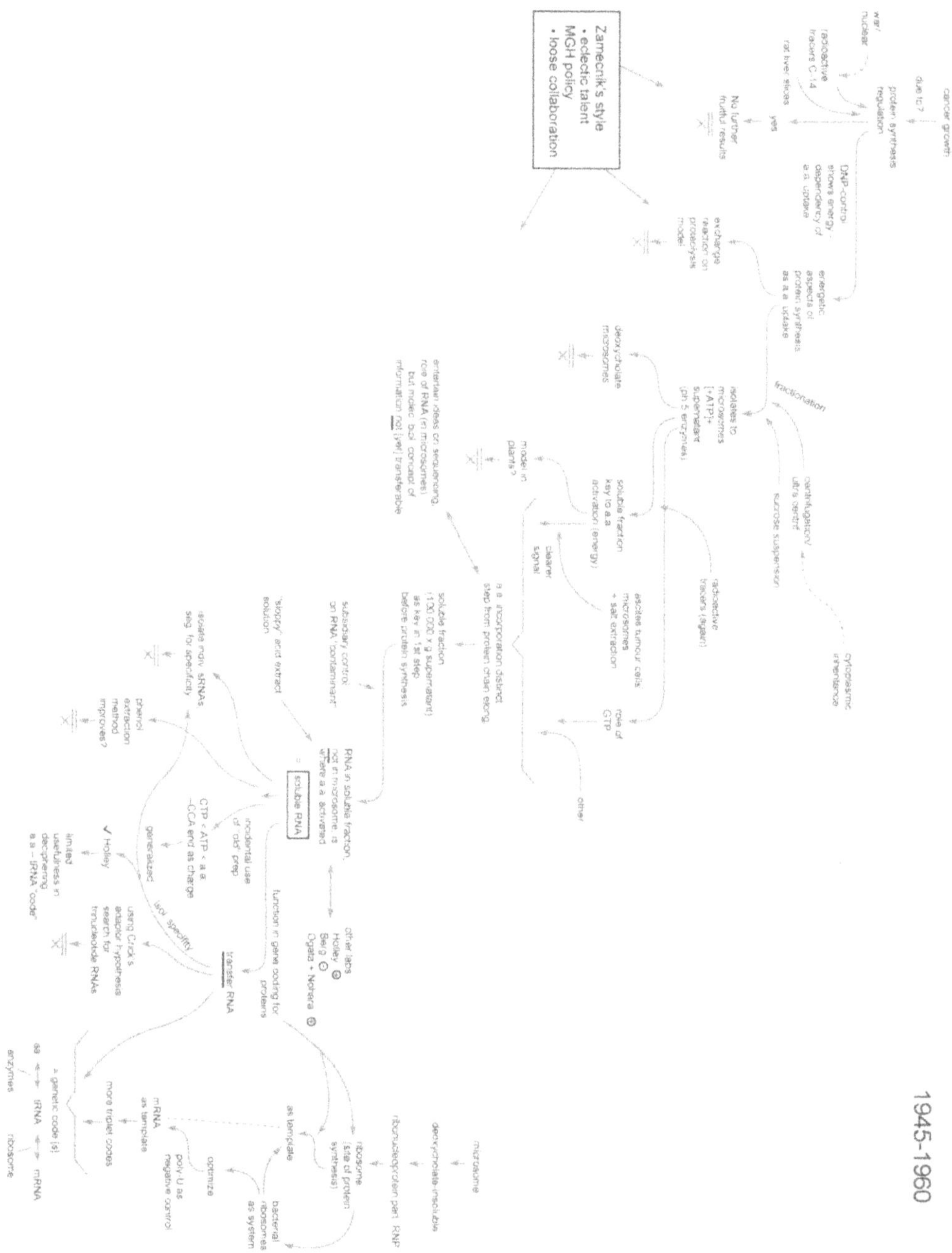

Figure 1: Protein synthesis research 1945-1960.

this happened with the rat in the protein synthesis system. It was only at the beginning of the 1960s that rat liver was replaced by the bacterium *Escherichia coli* in protein synthesis research, which turned out to be simpler and much easier to dissect into its components and therefore, to be handled analytically. It was clearly the legacy of the clinical context out of which the research grew and the way it was funded, namely, predominantly by cancer research money, that kept the rat for a whole decade as a model. Switching to bacteria at this time would simultaneously have switched off the connection to cancer research and its funding sources.

In Vitro

The second aspect of protein synthesis research is its test tube character. It originally also derived from cancer research in the medical context of the Rockefeller Institute in New York. There, the Belgian physician Albert Claude had set out to investigate the particulate character of Rous sarcoma virus in the late 1920s, and from this starting point eventually was led to the characterization of sub-cellular particles by means of high speed centrifugation of homogenized tissue. The technique of extracting cells and separating their contents by centrifugation was taken up by protein synthesis researchers. This appropriation inserted their research into a larger trend of physiological research into which I would like to make a short digression here because it characterizes the life sciences of a major part of the twentieth century.

The differentiation between in vitro and in vivo systems became established around the beginning of the twentieth century, after it had been demonstrated that not only enzymes secreted by glands but also cellular ferments were able to exert their action outside the cells or tissues or organs or the intact organism, in the test tube, if supplemented with appropriate ingredients and subjected to specified buffer conditions. To be sure, working on dead bodies and preparing specimens was a very old practice in the life sciences. But the in vitro systems of the first half of the twentieth century claimed to be able to generate artificial environments in which actions that normally went on in the living body took place outside the body and the cell. As such, they marked the transition from an organismic and cellular to a subcellular, and finally to a molecular biological knowledge regime. In vitro systems are usually reduced systems. They enhance certain features of a complex network by eliminating and purifying away others. They are prone to the production of artifacts, which is inherent in the approach. Therefore, a constant qualification is necessary when relating in vitro systems back to the in vivo situation. Much of the history of twentieth century biology is due to and derivers from this very specific game of checking and of rectification.

Radioactive Tracing

The third element of the protein synthesis system I would like to mention is the technology of radioactive tracing. It became a key technology of biochemistry within a decade after World War II, and it makes clear at the same time that the development of molecular biology, through its research technologies, is inseparable from the age of atomic physics and with that, from the broader context of the uses – and abuses – of atomic energy.

The production of radioactive sulfur, hydrogen, and finally carbon – key elements in biological molecules – was realized in principle with the advent of cyclotrons during the 1930s. But only in the middle of the 1940s did radioactive elements become available in larger quantities as an offshoot of reactor technology and the first atomic piles in the United States.

As a consequence, more complex, radioactively labeled molecules could be produced by means of biochemical manipulation and introduced into biological assays. It became possible to mark molecules such as amino acids, the building blocks of proteins, or nucelotides, the building blocks of nucleic acids, with radioactive labels. Until their decay, these molecules behaved chemically and physiologically just as their unlabeled counterparts. Consequently, what physiologists had in their hands was a kind of probe that could be introduced into the metabolism, and there it participated in the respective reactions. When it decayed, it released a signal, that is, it left a trace at the time and place of its breakdown.

The availability of these biologically relevant, weakly radiating isotopes in turn had an impact on measurement technology. In the small amounts optimal for biological assays, the isotopes could no longer be reliably measured with the traditional Geiger-Müller tubes. I cannot go into detail here on the history of "liquid scintillation counting," a technology that made the wide use of the new isotopes possible (Rheinberger 2001). Here it suffices to say that from a biological perspective, this measurement technology brought not only weak signals into the realm of the measurable, it also provided a 'wet' intersection for test tube experiments involving organic material. The liquid scintillation counter could accommodate fluid samples containing even water. Under these conditions, most biological samples could be made ready for radioactive measurement. The liquid scintillation counter technology was developed in close interaction with the revolutionary transformation of biological chemistry by radioactive tracing in the course of the 1950s.

The introduction of radioactivity into the laboratory cultures of biology also had broader consequences for laboratory architecture and laboratory life as a whole. The possibility of measuring radioactive traces of minimal strength in biological samples depended decisively on the precondition that the environment of the experiment remain uncontaminated by the radioactive probe. That condition not only resulted in a completely new laboratory regime; it also had massive effects on the very design and form of the experiments.

These few remarks may suffice to indicate that the technology of radioactive tracing cannot be reduced to either an instrument or a substance. Rather it forms a kind of capillary structure, which, with its components, penetrated and permeated a whole experimental culture. First, it introduced an indicator principle into the analysis of metabolic processes, and with that, it oriented biological chemistry as a whole in the direction of an in vitro experimental regime that could not have been developed without tracer technology. For the production of radioactive traces in the test tube also meant the possibility of bypassing a long-held principle of chemical measurement. Classically, chemical substances, in order to be measured at all, had to be rendered in pure form and in sufficient amounts for analytical micro-determination. In contrast, radioactive measurement could be performed without removing the 'impure' background of a mixture of all sorts of cellular components. In addition, sensitivity was enhanced by about a half-dozen orders of magnitude. Second, as already mentioned, radioactive tracing was the driving force for the development of new measurement technologies, of a whole research technology industry in fact, whose integration into the experimental systems of molecular biology not only altered their size, but also their structure and disposition. Finally, radioactive tracing became the material point of mediation between the know-how of biologists, chemists, physicists, and engineers. It was thus a technology that in its material structure – the chemistry of liquid scintillation, the physics of photomultiplication, engineering electronics, biological sample preparation – displayed the new interdisciplinarity of molecular biology in a paradigmatic form.

With the case of the in vitro protein synthesis system, we have thus come across, in an exemplary fashion, a few elements that are characteristic of the experimental structure of twentieth century life sciences. The comparative investigation of this complex network of experimental systems should help us also to understand how new, unprecedented knowledge arises in the process of research. Viewed structurally, novelty appears always to be the result of singular events in space and time. In the realm of our knowledge spaces, experimental systems are precisely those structures that allow for the generation of singularities. They allow, to put it paradoxically, the creation of new knowledge effects in a regulated manner and yet a manner that transcends our capacities of anticipation.

Shifts of Assemblage

It appears to me that the history of the molecular life sciences of the twentieth century is characterized by two decisive shifts of assemblage, and for that matter, of experimental systems. They were essentially unplanned and unprecedented in the form they took. The first happened in the two decades between 1940 and 1960, and the protein synthesis system I have been using as an example is a good case. This period had its crystallization point at the beginning of the 1950s with the characterization of the structure of the DNA double helix, and its apotheosis with the deciphering of the genetic code between 1960 and 1965. It happened, as I have already mentioned, in a bacterial version of the in vitro protein synthesis system. Generally speaking, this transition was characterized, first, by a series of new analytical techniques or, to describe it as the sociologist of science Terry Shinn does (Shinn 2002), of research technologies, such as ultracentrifugation, electron microscopy, X-ray structure analysis, radioactive tracing, chromatography, and electrophoresis. All these technologies had their origin in very different research contexts, which were often, at least in the beginning, rather remote from biology. Second, the transition was characterized by the shift from the classical biological model organisms, such as the fruit fly or corn, to new models such as lower fungi, bacteria and viruses. Third, it involved cooperations that extended over several disciplines. The molecularization of the life sciences happened thus – such is my hypothesis – by no means in the form of a linear continuation of classical genetics, but on the contrary, it came to form an active *Rabinow assemblage* of its own. On the methodical level, it was characterized by the deliberate import of analytical procedures from biophysics and biochemistry into biology. The landscape of these technologies did not take shape under the influence of a new theoretical paradigm, but in the end, it engendered one. What stood in the center of this resulting conceptual shift was a new notion of biological specificity that found its expression in the idea of genetic information and genetic program.

The second shift of assemblage, to which I can only briefly allude here, happened during the 1970s and can be connected with the beginnings of gene technology and genetic engineering. It is essentially characterized by the introduction of *molecular* technologies in the proper sense of the word, that is, technologies in which the function of biological macromolecules themselves, in particular enzymes and nucleic acids, plays a central role. Plasmids and other vectors, restriction enzymes and the polymerase chain reaction, in which a DNA polymerase is a central piece, are examples of such molecular biological tools. I have characterized the decisive step of this phase as the transition from the "extracellular" representation of intracellular structures and processes to the "intracellular" representation of an extracellular project (Rheinberger 2000). Classical molecular biology was biology operated and driven by the methods of biophysics and biochemistry; gene technological molecular biology is biology operated and driven by molecular tools that are active in the space of the living cell itself. At this point, the in vitro culture of earlier twentieth century biology began to be replaced by a new in vivo culture – the manipulations being shifted back from the test tube into the cell. It was the starting point of a thoroughly constructive and synthetic biology.

In summary, we could state with Stephen Toulmin:

> "A nutshell definition of science – as of anything else – inevitably floats around on the surface. An investigation of any depth forces us to recognize that the truth is much more complex. To understand the ways in which meritorious scientific ideas differ, in any age, from their less deserving rivals calls for a painstaking and laborious study: only in this way shall we bring to light the manifold functions that science has performed, performs now, and *might* perform in the future within our whole intellectual economy" (Toulmin 1963: 15).

At the beginning of my talk, I made a few remarks regarding disciplines. It would certainly be rewarding to think more intensely about the dynamics that came into the disciplinary landscape of the life sciences during the course of the twentieth century, and to think about whether, with these dynamics, the boundaries not only between biology, physics, and chemistry, but also between biology and medicine are in the process of becoming profoundly reconfigured, or even dissolved. Moreover, the admittedly strong hypothesis would have to be considered of whether the classical disciplines, as shaped during the nineteenth century, have altogether entered into a process of dissolution, a process of which the molecularization of biology is only a particularly prominent example. This of course could not remain without consequences for the possibility – or the growing impossibility – of understanding the dynamics of the contemporary sciences in the frame of disciplinary histories. In this context, Paul Forman has talked about the recent "devaluation of disciplines," and he has seen this trend relying not only on the growing problem orientation, but also the growing market orientation of the contemporary sciences:

> "This reorientation toward the market [...] together with the increasing orientation toward the particular problem, works powerfully to dissolve the scientist's attachment to his discipline, indeed to dissolve the disciplines themselves and their disciplinary authority" (Forman 1997: 185, 189).

Let me end with a final remark about model organisms in medicine. I have talked about the role of rats for cancer research, and the peculiar way this organism made its way into the in vitro system of protein synthesis of the 1950s. Now, with the recent achievements of in vitro human tissue cultivation, of cell proliferation in Petri dishes, of test tube fertilization, of cellular cloning and of stem cell manipulation in human reproductive and developmental biology, it appears that we are entering in an epoch of the eclipse of *animal models* in medicine and human research. The thorough going molecularization and the subsequent – as we could say – re-cellularization of research has created the perspective of an investigation of human specificity without intervening animal models. The potentials of this new research mode, however, in particular with respect to reproduction, embryonic development, and differentiation have also created the necessity of discussing new boundaries for human experimentation. I said that model organisms are always modified organisms. With man becoming, in a sense, a model organism of his own, "modeling" inevitably takes on the meaning and the form of *human modification*. There is thus an urgent need to discuss how far-reaching such modification – in particular genetic modification – can and should go in the name of exploring the unknown.

Evelyn Fox Keller

Contenders for Life: Approaches from Physics, Biology, and Engineering

Collaborations between physicists and biologists are flourishing today as never before. But relations between the physical and the biological sciences have a long and troubled history. And despite the heterogeneity of both sciences, and the vicissitudes of this history, a particularly prominent source of tension – one that has surfaced in virtually every encounter between physics and biology in the 20th century – is to be seen in the often dramatic differences in the significance attributed to particularity. Despite the manifest differences both among physicists and among biologists, one question repeatedly arises, and it takes various forms: Is it useful to think of the cow as a sphere? Of the cell as a bag of enzymes? Or of the organism as an ensemble of quasi-random networks? Early on, D'Arcy Thompson, often regarded as the father of mathematical biology, railed against colleagues for what he described as the

> "ingrained and deep-seated belief that even when we seem to discern a regular mathematical figure in an organism, [...] [it is] the details in which the figure differs from its mathematical prototype [that] are more important and more interesting than the features in which it agrees" (Thompson 1942: 1028-1029).

But I will not try to review this history here; rather, I want to focus on the ways in which these historic themes can be seen to play out to this very day, in particular, in the various perspectives currently being brought to discussions of the role and meaning of self-organization in biological systems.

What *is* 'self-organization'? What does the term mean? In its current usage, it has been subject to such vaguery as to lead a number of commentators to regard it as nothing more than an empty phrase. But the history of its meanings and associations is, I think, of interest – indeed, I suggest it provides a useful window onto the changing relations between biology, physics, and engineering. Of particular interest are the changing meanings of 'self-organization' in relation to organisms on the one hand, to machines on another, and to spontaneously emerging dynamical phenomena like eddies, vortices, and thunderstorms on the third.

Perhaps the most conspicuous feature of the history of the term 'self-organization' is its inseparability from the history of biology – that is, from the history of the science concerned with the problem of life that was first explicitly demarcated from other sciences only in 1800. That term is originally due to Kant, and it was introduced ten years earlier in an effort to address the fundamental question of the life sciences, namely: What is an organism? What is the special property, or feature, that distinguishes a living system from a collection of inanimate matter?

Of course, this was also the question that first defined *Biology* as a separate and distinctive science, and, indeed, the two words, *organism* and *organization* acquired their contemporary usage more or less contemporaneously. For what led to the common grouping of plants and animals in the first place – i. e., what makes "the two genres of organized beings" (as Buffon referred to them) *organisms* – was a new focus on just that feature, i.e., on their conspicuous property not simply of being organized, but of being organized in a particular way. As François Jacob observed, by the end of the 18th century, it was

> "[by] its organization [that] the living could be distinguished from the non-living. Organization assembled the parts of the organism into a whole, enabled it to cope with the demands of life and imposed forms throughout the living world" (Jacob 1976: 74).

Of course, the idea that organization was key to an understanding of living beings was not exactly new. Aristotle had argued along similar lines, and in more recent times, Locke had laid great emphasis on the importance of organization in establishing the identity of plants and animals. Of plants, he wrote:

> "For this organization, being at any one instant in any one collection of matter, is in that particular concrete distinguished from all other, and is that individual life, which existing constantly from that moment both forwards and backwards, in the same continuity of insensibly succeeding parts united to the living body of the plant, it has that identity which makes the same plant, and all the parts of it, parts of the same plant, during all the time that they exist united in that continued organization, which is fit to convey that common life to all the parts so united" (Locke 1841: Part II, Book 27).

Ditto for animals: what establishes an animal's identity, and keeps it the same over time, is its organization. But Locke's concern lay primarily with the identity of the individual plant and animal, whereas the end of the 18th century brought a particular concern with what distinguished plants and animals, now grouped together under the category of the living from that altogether distinct category of the non-living. And here the emphasis shifts to the particular ways in which living beings were organized. But what made it possible to distinguish an organism from its Greek root, *organon*, or tool, was that special arrangement and interaction of parts that brings the well-springs of form and behavior of an organism *inside* itself. A tool, of necessity, requires a tool-user, whereas an organism is a system of organs (or tools) that is self-steering and self-governing: it behaves as if it had a mind of its own.

And indeed, it was this property that was of particular importance to Kant. In his definition of an *organism* – given not as a definition *per se*, but rather as a principle or "maxim" which "serves to define what is meant as an organism", he wrote:

> "an organized natural product is one in which every part is reciprocally both end and means. In such a product nothing is in vain, without an end, or to be ascribed to a blind mechanism of nature" (Kant 1993b: 558).

Organisms, he wrote, are the beings that

> "first afford objective reality to the conception of an *end* that is an end of *nature* and not a practical end. They supply natural science with the basis for a teleology [...] that would otherwise be absolutely unjustifiable to introduce into that science – seeing that we are quite unable to perceive *a priori* the possibility of such a kind of causality" (Kant 1993b: 558, 566).

Elaborating on this kind of causality, he writes:

> "In such a natural product as this every part is thought as *owing* its presence to the agency of all the remaining parts, and also as existing *for the sake of the others* and of the whole, that is as an instrument, or organ. [...] [T]he part must be an organ *producing* the other parts – each, consequently, reciprocally producing the others. [...] Only under these conditions and upon these terms can such a product be an *organized* and *self-organized being*, and, as such, be called a *physical end*" (Kant 1993b: §65, 557).

Here is where the term "self-organized" first makes its appearance. An organism is not merely self-steering, self-governing, and self-maintaining: it is also self-organizing. More, it is self-generating. As Kant puts it, an organism is both "cause and effect of itself" Kant 1993b: §64, 555). The conspicuous organization of living beings and their consequent appearance of design are not due to any external agent, but only to their own nature. Organisms are products of nature. No external force, no divine architect, is responsible for this peculiar entity, only the internal dynamics of the being itself.

The beginnings of *Biology* thus prescribed not only the subject and primary question of the new science, but also the form of answer to be sought. To say what an organism is would be to describe and delineate the particular character of the organization that defined its inner purposiveness, that gave it a mind of its own, that enabled it to organize itself. What is an organism? It is a bounded body capable not only of self-regulation, self-steering, but also, and perhaps most important, of self-formation and self-generation. An organism, Kant writes, must be both "an organized and a self-organizing being" (Kant 1993b: §65, 557). It is a body which, by virtue of its *peculiar and particular organization*, is constituted as a self which, even though not hermetically sealed,[1] achieves both autonomy and the capacity for self-generation.

At the close of the 18th century and the dawn of the 19th, – to Kant, as to his contemporaries - that neither blind chance nor mere mechanism, and certainly no machine that was then available, could suffice to account for this special kind of organization. "A machine," Kant writes, "has only *motive* force, whereas an organism possesses *formative* power" (Kant 1993b: §65, 557). "Strictly speaking", he elaborates, "the organization of nature has nothing analogous to any causality known to us" (Kant 1993b: §65, 557). The science of such an entity would thus have to be a new kind of science, one which, not surprisingly, Kant assumed to be irreducible to physics and chemistry. As he famously wrote, there "will never be a Newton for a blade of grass". Furthermore, it is "contrary to reason" that "raw material could have originally formed itself according to mechanical laws, that life could have originated from the nature of the lifeless, and that matter could have arranged itself in the form of a self-maintaining purposiveness" (Kant 1913: §81, 424).[2] While mechanical devices might be built to simulate the behavior of a living system, no combination of mechanical and theoretical ingenuity could actually produce a physical system capable of self-organization. Machines and organisms belonged to radically distinct categories.

---

1 Or, perhaps it is more accurate to say, *because* it is not hermetically sealed, for modern studies of self-organization suggest that it is precisely the relation between a system and its environment upon which such properties depend.

2 Quoted in (Ginsborg 2001: 243).

Kant's confidence in the irreconcilability of machines and organisms was widely shared by his contemporaries, and it persisted pretty well until the middle of the 20th century, when the arrival of a new kind of machine, and with it, the vision of a new kind of science, turned that confidence on its head. In the 1950s and early 60s, self-organization resurfaced as the organizing theme for an ambitious series of projects aiming to probe the basic logic of living organisms and to implement it in the construction of man-made devices.

But finally, it was not the efforts of these early cyberneticists that put self-organization on the map of contemporary discussions of the nature of biological organisms. If we look at citation measures, we can see that it was not until the early 1980s that use of the term self-organization in scientific publications really begins to take off. Here, e.g., is a graph of the frequency of term as it appears in the cumulative citation index of INSPEC and Compendex:

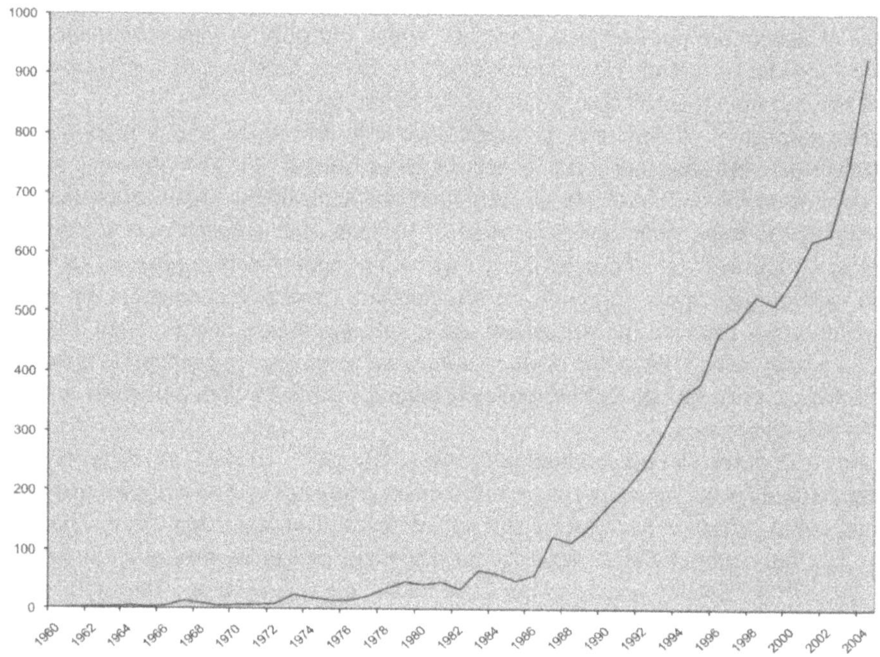

Figure 1: Self-Organization Citations. In INSPEC and Compendex, 1960-2004.

This turning point coincides not with an engineering vision, but with the entrance of physics and physicists into the field, and it is on this episode of the history that I want here to focus.

According to the most familiar account, the turning point comes with the Russian born physicist, Ilya Prigogine. In 1967, after over two decades of work on the thermodynamics of irreversible (but close to equilbrium) processes, Prigogine and Nicolis published a paper on "On Symmetry-Breaking Instabilities in Dissipative Systems", the main point of which was to argue for the "existence of instabilities in purely dissipative systems" – i.e., in systems far from equilibrium. The principle sources of his insight were the two best studied examples at that time of non-linear dynamics giving rise to the spontaneous emergence of stable spatial patterns in uniform systems: the first was the well known Bénard instability from fluid dynamics (in which a horizontal fluid layer heated from below becomes unstable at a critical rate of heating), and the second, rather less well known, a set of chemical reactions introduced by Alan Turing 15 years earlier to illustrate a possible mechanism for morphogenesis (Turing 1952). The term "self-organization" appears on the second page, almost as an aside, where, after introducing his main argument, Prigogine and Nicolis write, "We also have the opportunity to indicate briefly the implications of such 'self-organizing' open systems for biogenetic processes" (Prigogine, Nicolis 1967: 3543). Here, the term refers to the emergence of "dissipative structures"**3** in systems far from equilibrium, and low in entropy, and what made these patterns so interesting (unlike, e.g., the formation of oil drops or snow flakes) was precisely their dependence on a flow of energy (and, in some case, of matter). Indeed, they argued that given a sufficiently large flux of free energy in such systems, one can expect to see the spontaneous emergence of all sorts of novel spatial and temporal structures - like those seen in eddies, vortices, or Bénard rolls and cells. One might even say that the primary contribution of this paper (apart from introducing this term into a new context) was to link these early examples of non-linear dynamics to physics, and he does this, first, by redescribing both the Bénard and the Turing phenomena in thermodynamic terms, and second, by suggesting a clear analogy between hydrodynamic instabilities and nonequilibrium phase transitions. Prigogine followed up this them over the next decade in a number of papers, and in 1977, together with Nicolis, he published his very popular book on *Self Organization in Non-Equilibrium Systems.*

Hermann Haken, a theoretical physicist known for his work in laser physics and nonlinear optics, had been present at the Versailles conferences on Theoretical Physics and Biology in 1967 and 1971 that Prigogine had helped organize. In 1972, he organized his own workshop on what he called "Synergetics" – i.e., on "cooperative phenomena in multi-component systems."**4** "Synergetics" is Haken's term for the discipline he helped to found linking non-linear dynamical systems theory with statistical physics. And in 1977, the same year of Nicolis and Prigogine's book, he published his book on *Synergetics, An Introduction: Nonequilibrium Phase-Transitions and Self-Organization in Physics, Chemistry and Biology.* Except for the discussion of Chaos in Haken's book, and of Thermodynamics in Nicolis and Prigogine's, the two books covered much the same material, and they became part of a package. After their publication, things took off pretty quickly. In 1983, two international conferences on self-organization were convened – one held in Austin, Texas (Nicolis, Baras 1984), and the other in the Soviet Union (Krinsky 1984) – and in the published proceedings of the two conferences, both of these books are cited extensively. From then on, the literature on self-organization simply explodes, and it is this explosion that invites the framing of these events as signifying a "scientific revolution" and the emergence of a "new paradigm."

The dominant reference of *self-organization* in this new paradigm is to the production of stable patterns observed in physical, and sometimes in biological, systems governed by non-linear dynamics, and the dominant methods of analysis, those of non-linear differential equ-

---

3 This term he also introduces for the first time in 1967, and he does so at the first International Conference on Theoretical Physics and Biology, held in Versailles (Marois 1969).

4 Haken seems to have first introduced the term synergetics in 1971 to describe the cooperative behavior (*Zusammenwirken*) of individual atoms in a monochromatic light beam (Haken, Graham 1971).

ations, cellular automata, and Ising-type models from statistical mechanics. Enthusiasm for such analyses has run high, generating some extraordinary expectations. As Paul Davies wrote:

> "Mathematically we can now see how nonlinearity in far-from-equilibrium systems can induce matter [here quoting Charles Bennett (1986)] to transcend the clod-like nature it would manifest at equilibrium, and behave instead in dramatic and unforeseen ways, molding itself for example into thunderstorms, people and umbrellas" (Davies 1989: 111).

A noteworthy (and indeed, much noted) addition to these ideas comes in the late 1980's with Per Bak's introduction of the notion of "self-organized criticality" (SOC) (Bak, Wiesenfeld 1987).[5] Stuart Kauffman's 1993 book, *The Origin of Order* (a book Bak referred to as "the first serious attempt to model a complete biology"), sought to apply Bak's ideas to evolution, but his vision of evolution operating on "interacting dancing landscapes" spontaneously approaching criticality, turned out to be premature. Almost immediately after, however, Bak and Sneppen (Bak, Sneppen 1993) did succeed in developing a model for evolutionary dynamics that exhibited all the desired characteristics. Here was a model of an open and dissipative system that organizes itself into a critical state simply by virtue of its intrinsic dynamics, independently of any control parameter. Drawing on an analogy with the physics of phase transitions, the existence of a critical state is said to be signaled by a power-law distribution in some variable – to physicists familiar with the behavior of systems at thermodynamic critical points, a seemingly clear indication that short range interactions have induced long range correlations, and a kind of global organization in which details of the particular system get obliterated. The familiar image of SOC is that of sand being dripped onto a sand pile. Once the sand pile has attained a critical slope, it retains its conical shape as more sand is added; it manages this by setting off small avalanches. The timing and size of individual avalanches is unpredictable, but the *distribution* of avalanches (in both size and timing) displays the prototypical regularity of the power law.

Self-organized criticality soon became a new buzz word, the latest key to an understanding of emerging structure in complex systems. According to Bak and his colleagues,

> "The dynamics of complex systems in nature often occurs in terms of punctuations, or avalanches, rather than following a smooth gradual path. Earthquakes, stock market crashes, and mass extinctions in biology are examples of this turbulent behavior. It will be argued that this is a consequence of self-organized criticality, the process by which large dynamical systems evolve to a state with no characteristic time or length scale" (Paczuski et al. 1996: 414).

And just a few years later, Laszlo Barabasi and his colleagues extend Bak's idea of self-organized criticality to the world of network topology, providing yet another boost to expectations about what the physics of phase transitions might do for biology.

One of the great appeals of the analogy between biological organization and phase transitions is the hope that looking at organisms in this way might provide emancipation from

---

5 In 1996, Bak reported that, since the coining of the phrase SOC in 1987, "more than 2,000 papers have been written on the subject, making" this initial paper "the most cited in physics[...]" (Bak 1996).

biology's traditional dependence on and commitment to particularity. A major triumph of modern statistical physics was that it produced a theory identifying the universal aspects of critical phenomena – a theory in which the macroscopic (thermodynamic) properties of a system near a phase transition are insensitive to the particularities of the system, its underlying microscopic properties. Perhaps, some hoped, a similar approach would finally make it possible to identify those aspects of biological organization that were universal, and biologists could finally stop worrying about all that messy detail.

This literature on self-organization begins in the world of physics, written by physicists and published in physics journals, but it rapidly spreads to other fields, and has now been taken up by a flourishing industry of science popularization. Indeed, reference to a new "law of nature" and a "universal architecture" linking complex systems across the natural, social, and engineering sciences have become commonplace. It is worth noting however that the term self-organization, as used here, is severed both from its original biological (or Kantian) meaning and also from its later engineering sense. It is stripped of all resonances of design, in either the biological or engineering sense of that term, and instead, appropriated for the categorization of complex phenomena arising out of random ensembles of simple, generic, and (for the most part) uniformly distributed physical entities. Not only eddies, whirlpools, and Bénard cells are to be understood as arising from homogeneous gases, fluids, and lattices, but also more dramatic eruptions like thunderstorms, earthquakes, and living organisms. Indeed, this is a literature claiming that the emergence of life itself can be seen as a Self-Organized Critical Phenomenon.

Lee Smolin, one of the more thoughtful writers on this subject, argues that viewing the universe as a whole as a non-equilibrium, self-organizing system has many advantages: in particular, it allows for a world in which

> "a variety of improbably structures – and indeed life itself – exist permanently, without need of pilot or other external agent, [and] offers the possibility of constructing a scientific cosmology that is finally liberated from the crippling duality that lies behind Plato's myth" (Smolin 1997: 160).

But my concern is with the costs of such assimilation.

Despite Smolin's caution, and for all his hopes, I claim there is a critical elision here. Analyses of non-linear dynamical systems clearly demonstrate the ease with which complexity can be generated, but such arguments fail to recognize a crucial distinction – one that Warren Weaver described way back in 1948 as a distinction between disorganized and organized complexity,[6] and that we might characterize as a difference between the generation (or emergence) of complexity and its organization. In the words of John Mattick: "The problem is not how to generate complexity – that is easy – but rather how to control it to specify ordered trajectories that lead to highly organized and complex organisms" (Mattick 2004: 317-318). This is an absolutely crucial point for biology, and central to all questions about the origin of those attributes that continue to divide biology from physics, namely function, purpose, and agency.

6 "Organized complexity" was also a term that was sometimes used to refer to the theories of selforganization that were being developed in the late 1950's (see, e.g. (Rapaport, Horvath 1959: 90)).

Much has been claimed for capacity of non-linear dynamical systems to mold themselves, as Davies put it, "into thunderstorms, people and umbrellas" (Davis 1989: 111). Or as Stuart Kauffman wrote a decade ago, "metabolic networks need not be built one component at a time; they can spring full-grown from a primordial soup. Order for free, I call it" (Kauffmann 1995: 45). Kauffman is correct. Many complex structures – including networks – can and do arise spontaneously. Indeed, we can find examples of order-for-free all around us. The difficulty is that such structures do not yet have the properties we designate by function, agency, or purpose. They are not yet alive. Statistical mechanical analyses of nonlinear systems have been immensely productive, but for all the fecundity of the models that have been put forth, none of these models has yet succeeded in offering an account of how inherently biological properties (like function, purpose, and agency) might emerge from the interaction among simple elements of effectively homogeneous systems, however complex the dynamics of interaction might be. Rather, the appearance of such properties seems to require an order of complexity going beyond whatever might emerge spontaneously out of such systems – a form of complexity, based on specificity and heterogeneity that a number of authors, starting with Warren Weaver, and now rearticulated by John Mattick and John Doyle, have referred to as "organized complexity". As Bak and Kauffman have used the term, *self-organization* may be necessary to the emergence of organized complexity, but as I've already said, and as Kauffman now acknowledges,[7] it is not sufficient. Something else, something engineers have been struggling to characterize ever since the 1940s, is also required. And it is beginning to appear that the answer to what that something else is might be found precisely in the domain of particularity that physics, and especially statistically physics, has traditionally disavowed.

For Warren Weaver, the domain of problems characterized by organized complexity – problems like "What makes an evening primrose open when it does?" – lay in sharp contrast to the problems of statistical mechanics that made up the domain of disorganized complexity, and it may well be that the reliance on random ensembles, maximum likelihood, and generic components that is so fundamental to the methods of statistical mechanics preclude consideration of some of the most critically biological features of living systems. Following Weaver, the work of cyberneticians on feedback offered us important clues about how the design of a system with function; equally important, the work of Herbert Simon taught us of crucial links between organized complexity and hierarchy.

But recently, Weaver's contrast between the phenomena of statistical physics and organized complexity has been taken up by a number of engineers, and their discussion suggests that engineering may again offer a useful corrective to many of the assumptions physicists have traditionally brought to biology.

For example, Alderson and Willinger (Alderson, Willinger 2005) offer a perspective to self-organization perhaps more in keeping with the attachment to particularity against which

7 In his more recent work, Kauffman concludes that "We have no theory of organization", and asks: "What must a physical system be to constitute an autonomous agent?" (Kauffmann 2003: 1089). His answer is that it must couple a motor with self-reproduction/auto-catalysis in a non-equilibrium energy flow, and offers as a primitive example a DNA hexamer that can catalyze the linking of two trimers into a hexamer that is a copy of the initial hexamer, provided it is coupled with a PP (pyrophosphate) motor, in a system, far from equilibrium, with an excess of trimers and photons.

D'Arcy Thompson railed – a perspective in which system-specific details is the main driver, not randomness. As they write,

> "in contrast to the physics-based approach [...], where complexity is explained by simple underlying forces, the engineering approach seeks to create simplicity in function or performance through the use of underlying complexity. This approach requires incorporating knowledge of the systems fundamental objectives, the details of its component parts, and the specifics of its operating environment to yield descriptions that explain the observed structure or behavior" (Alderson, Willinger 2005: 96).

Such a perspective draws important grounding from the work of John Doyle and his colleagues who compare a number of different internet topologies, based on the statistical and the engineering approach. What they find is that those topologies which are designed to optimize performance under existing constraints are effective but improbable, where networks described as "scale free" are probable, but extremely ineffective. Indeed, they conclude that the "likely" topologies "have such bad performance as to make it completely unrealistic that they could reasonably represent a highly engineered system" (Li et al. 2004).

They suggest, and I concur, that there is a message here of great importance to biology – namely that biologists, like engineers have good reason to value specificity. As well as hierarchy. Indeed, I suggest that specificity and hierarchy, are the critical components of organized complexity, and that, together, they hold the key to the emergence of that feature that joins biological systems with engineering systems, and that separates them from the systems with which physics is concerned, i.e., function.

Elsewhere (e.g., Keller 2002; 2005) I have argued that, if the new rapprochement between physics and biology is to be successful, it will require rethinking the meanings of words like deep and fundamental, and I would like to reiterate this point here. What is fundamental in biology, I suggested, is far more likely to be found in the accidental particularities of biological structure arising early in evolution than in any abstract or general laws. Such particularities, probably arose as rare events, can be built upon over time to generate ever more complex structures. They are fundamental in the sense of having been built in on the ground floor, and hence having become most deeply entrenched, and not in the sense most familiar to physicists.

But perhaps engineers too have something to learn from biology, even now, for there remains one crucial fact that continues to separate organisms from machines, namely, the fact that the organized complexity of the former, unlike the latter, arose and evolved spontaneously. Machines may be self-steering, but even after all these years of mechanical ingenuity, it remains the case that only organisms can be said to be self-organizing in Kant's sense of the term.

The properties that continue, even now, to distinguish organisms from machines – the internal generation of agency, function, of Kant's reciprocality of ends and means – these properties too must have come from evolution, from the iterative processes of self-assembly and self-organization that occur in heterogeneous systems *in* time, and that can cumulatively build upon themselves *over* time. Often over a great deal of time. And here again, the power of randomness has been over-estimated, for it is becoming evident that the most dramatic powers of evolution come not simply from cumulative selection acting on random variations, but from natural selection operating on what Kirschner and Gerhart call "facilitated variation" (Gerhart, Kirschner 2007). Selection is not simply cumulative but also recursive, or hierarchical, always making use of the resources already at hand. And among these resources, early on in history, it now seems clear that a new kind of evolution evolved, something we might call adaptive evolution, or the capacity for evolvability, under which term I would include the remarkable capacity to generate biological novelties when they are most needed and not merely when random chance generates them.

This may be the next task for engineers, the design of evolvable systems that can not only learn from their environment, but that can learn to evolve in response to a changing environment. Maybe they're already doing so, I don't know. If they are, or if not now, then probably they will soon be doing so, and then maybe, just maybe, we will have breached the last conceptual gap dividing organisms from machines. As I say, I don't know, but I would not lay odds against it.

Dominique Pestre

What about Participation, Governance and Politics? Remarks on Contemporary Techno-science and the Field of STS

My main preoccupation in this text is to consider some of the ways of framing questions in the field of science and technology studies today, notably around what I would call the democratic management of techno-science, and of techno-scientific products, in and by society.[1] A first way to express what I have in mind is to say that I am in favour of returning to the core of our work, the question of power and of economic and political macro-regulations – because these questions matter, and because I feel we are not working enough at that level of analysis. I propose to advocate more down-to-earth, anti-euphemistic and realpolitik-oriented ways of describing what organizes the world today.[2] Since many approaches in STS have also, *de facto*, become normative – an attitude which I approve of – I would like to consider the political philosophy on which many texts in STS now rely, to open a broader discussion of the moral and social values STS practitioners mobilize when considering the question of the political. To put it another way – and having notably in mind my friends in the École des Mines in Paris – my suggestion would be that we should clearly acknowledge that the strategy of merely following actors has a cost and a politics which do not help to grasp some of the major changes in the world today.[3] As Haraway put it when considering gender relations nearly two decades ago, we cannot but face personality disorder if we go on only with symmetrical anthropology and/or agnostic deconstructionist moods at the very moment we consider it important to understand the forms of domination which redefine our everyday experience (Haraway 1991).

1. Recent changes in the regime of science in society

By way of introduction, I would like to briefly evoke some of the significant changes our society has undergone over the last 30 years. Of course, I can in no way justify now why I believe that such majors changes have happened, at which rhythm, according to which logic

---

1 The first version of this paper was written in the Wissenschaftskolleg zu Berlin in early 2004. It would have been impossible to write it without the many discussions I had with Amnon Raz-Krakotzkin, Gil Anidjar, Peter Hall, Zhiyuan Cui, Susan James, Beate Rössler, Heike Paul, Pascal Grosse, and Helga Nowotny and Eberhard Schmidt-Assman. This text is an extended version of an article published under the title "Challenges for the Democratic Management of Technoscience: Governance, Participation and the Political T'oday". *Science as Culture* 17(2), June 2008: 101-120. Thanks to its editor, Les Levidow, for allowing the publication of this version.

2 On this point and the (despised) status of "la critique" today among French sociologists and political scientists, see (Garnier 2001).

3 For details, see chapter 3 and the conclusion of (Pestre 2006). On the politics of SSK, see the special issue of *Social Studies of Science* 26(2), May 1996; add (Halffman 2003), (Fuller 2000) and (Pestre 2004).

and following which actors.[4] I would only summarize a huge volume of literature and my convictions by saying that we have passed from a group of dominant *cités de justice* that regulated social relations to a quite different arrangement. Freely borrowing from Boltanski and Thévenot,[5] I would characterize the 'common worlds of social justice' of mid-20th Century – at least in Europe – in the following terms: on the one hand, there were three major common worlds comprising 'good society': (1) the *civic world of the nation-state*, exemplified by the welfare state or the labor contracts signed between workers' unions and managers in France, Germany, and elsewhere.[6] This was linked to a discourse on the sciences that presented them as autonomous, quite separate from society and prepared to be put into practice for the common good;[7] (2) the *industrial world of enterprise*, interested in efficiency and planning, but also in integrating the sciences and the workforce into the company *as long-term partners* – a norm that was in agreement with the ideology and practice of the welfare state; (3) the *hierarchical world of tradition*, the top-down modes of conceiving authority and power. Together, these three worlds of social justice defined a dominant way of being in society, stressing respect for labor and merit, for integration and social protection, for 'progress' and the 'rational development of Nations'. Then, on the other hand, organized around quite a different set of reference points concerning justice, was the *market world*, with its appetite for freer movement of goods and funds, a preference for *laissez-faire* legislation and opportunistic action, but which still, then, accepted state regulation and contractually organized social relations.

Over the last decades, a new *cité de justice* has emerged, a new set of values for social behaviour and justice has progressively reordered most social beliefs. In this new 'common universe', *the market world* became the dominant form of regulation (I am not essentializing it, nor claiming that its formatting was not profoundly altered between 1960 and 2000 – on the contrary *the market world* was profoundly transformed during that period).[8] In the new 'common universe' that finally took shape, personal engagement, responsibility, transparency and participation became some of the values that mattered – along many others that are not central to me in this paper. Derived from that universe are the notion of *governance* – advocated by major international institutions like the World Bank and by most think tanks and European governments; the engagement of business on behalf of 'social and environmental responsibility' – so in vogue since the late 1980s; but also most programs in the underdeveloped world when carried out by NGOs funded by the major agencies of the North.

---

4 For a more systematic analysis, see my book (Pestre 2003), and (Pestre 2005). Add (Abelès 1999); (Angell, Relman 2002); (Beck 2002); (Boltanski, Chiapello 2000a,b); (Dahan, Pestre 2004). (Dasgupta 2000); (Dejours 1998); (Demeritt 2001); (Dubet 2006); (Dupuy 2005); (Ehrenberg 1998); (Falk 1999); (Hall, Soskice 2001); (Kalaora 1999); (Lessig 2004); (Milani, Arturi, Solinis 2001); (Poulin 2005); (Sennett 2006); (Veltz 2000); (Waring 1991).

5 In this section, I follow (Boltanski, Thévenot 1991) even though I have modified the name they give to the various *cités de justice* they identify. I also made their analysis a description of the specific world of the first Cold War (their sources are an argument to do so) and not an analysis of the *cités de justice* 'in general'. These points would deserve a detailed justification that I cannot offer here.

6 For the specificity of continental Europe with respect to the United States, the best book is (Hall, Soskice 2001).

7 For more on this and what is usually called the "linear model", see (Edgerton 2004).

8 Details in (Pestre 2003: chapter 3).

This new universe of justice, this new "spirit of capitalism" as Boltanski and Chiapello say, now organizes most of what constitutes today's values.[9]

To employ quite a different vocabulary, we have passed from one form of hegemony to another, from one way of regulating individuals and the social to another. Two things lie at the heart of this change. First, a new conception of the government of both people and things, a new practice of *governmentality* stressing, on the one hand, the duty of self-governance but developing, on the other, an unprecedented set of rationalizing procedures, most of the time of an economic and managerial kind, which bear on many kinds of *subjects* – consumers, patients, workers, professionals, etc. Second, a dismantling of the regulations put in place during the preceding century (like the welfare state) in favor of an extension of property rights, a fight against 'commons' – as exemplified, in science, by the privatization of a central part of knowledge through changes in patent policy.

This situation has built upon, has helped invent and has given legitimacy to new social and political demands – like human rights – even if it happened that these rights were often instituted at the cost of other rights (rights to social protection, for example) and were often granted without precisely defining for whom they were intended (which does not mean they have no social efficiency!).[10] The process itself, as with most social processes, exhibits contradictory tendencies. If I only consider the links between knowledge, participation and democracy, new forms of biopolitical control developed at the very moment when individuals received far more possibilities for free and independent action (through techno-scientific development for example, but not only) and when demands for renewed forms of democracy exploded in 'civil society'.[11] The process appears contradictory on at least one other level: because of the growing demands for (and the success of) more democratic, participatory and dialogical behaviours when assessing techno-scientific products, people in power developed strategies to cope with such attitudes. In fact, learning *to manipulate* the demand for democratic debate has now been at the heart of company strategy and politics for nearly two decades.[12]

2. Six extreme statements proposed to open discussion

I would now like to flesh out these preliminary remarks. I will start with what I consider an interesting symptom – the tendency, in part of STS, to 'withdraw' to procedural questions, questions over the *best procedures* society should adopt in order to *decide* in contexts of *uncertainty*. This question is of decisive relevance but it does not exhaust the question of how best to cope with techno-science in democratic societies, and it does not constitute the core of what we, as scientific intellectuals, should take into consideration. If I were to put that point with the most extreme severity, I would say that we should not limit ourselves to defining the best modalities of social and political engineering. Let me refine these ideas by proposing six major statements.

---

9 (Boltanski, Chiapello 2000a). Add (Banthien et al. 2003); (Kalaora 1999); (Milani, Arturi, Solinis 2001); (Moreau Defarges 2001); (Osmont 1998); (Padioleau 2001); (Roman 2000).
10 (James to be published), (Mandami 2000).
11 (Blondiaux, Sintomer 2002); (Sintomer 2001); (Callon, Lascoumes, Barthe 2001).
12 See below statement 5. Add (Jamison 1996).

Statement 1: Dialogic and participative democracy is not, and perhaps cannot become in our market-based democracies, central to the regulation of techno-science; techno-scientific knowledge and products are primarily regulated by other (competing or parallel) institutions that are outside the dialogic order: markets, administrations, courts of justice, etc.

A part of the work done in the STS field around science and the political has been oriented toward responses based on the ideal of participatory and dialogical democracy. This approach, which underlines the necessity of guaranteeing the expression of diverse points of view, is intended to facilitate the emergence of socially 'robust' solutions and of precautionary attitudes. In a kind of Habermasian move, emphasis is placed on the principles that all deliberations should be open and public, that they should be adversarial and include non-specialists, and that the decision-makers should be held publicly accountable for their choice – because then well-conceived and well-conducted, systematic, open and well-managed confrontations are likely to lead to the 'best' solutions, in technical as well as in social terms.[13] The resulting research on the most reasonable arrangements, on the ideal forms of organization for democratic exchange and expertise is extremely important. It brings out an essential path to follow – and I remain actively engaged in the struggle for its refinement.[14]

This approach is based on an idealization of social reality, however, and on simplifications of the processes by which decisions are actually made. The first widely acknowledged fact, although one that is usually quietly forgotten, is the highly uneven distribution of social and symbolic capital in society that implies a very uneven distribution of the capacity to 'participate'. The capacity to create an association or an expert group, to mobilize funds or the media, is strongly determined in social and economic terms, as well as in the relations between North and South (Lash, Szerszynski, Wynne 1996). It is often the same social groups that lose out, the same 'new groups' that emerge as the winners and the same questions that are forgotten.[15] I know that this kind of discourse sounds a little old-fashioned in our networked and post-modern societies which have moved beyond the old dichotomies of rich and poor, owners and dispossessed, or, as Boltanski and Chiapello write, that distinguish between "work and leisure, paid and unpaid, public and private, profit-sharing and volunteer work, business and ethics". Nevertheless, it might well be that not dwelling on this issue amounts either to being blind to a monumental reality, or simply acting in bad faith. These inequalities and injustices are simply too large and influential for us not to consider them with the utmost attention.[16]

The second point is that the complex game of institutional politics is greatly underestimated in these approaches. This is the case with the making and formatting of 'publics' (through management techniques or 'evidence-based medicine' today), with the relationships between political institutions and the media, with the very many modes of regulating

---

13 (Stirling to be published in ST&HV).

14 The most complete and systematic presentation of this point of view is (Callon, Lascoumes, Barthe 2001). For a more general and theoretical introduction to the question of forms of deliberation, see (Blondiaux, Sintomer 2002). See also (Hatchuel 2000); (Joly 1999); for Habermas' place (Sintomer 2001). I also intend to write a paper on the role played by the European Commission in the last ten years to reform its own 'governance'.

15 See among many (Sennett 2006); (Dupuy 2005); (Dejours 1998).

16 The quotation is from (Boltanski, Chiapello 2000b).

techno-scientific products in our 'complex societies' – a variety which cannot, and perhaps should not be reduced too quickly. These many modes range from national and international expert bodies (to daily monitor the dozens of thousands of chemical products available for example) to the elaboration of new regulations by administrations, 'agencies' (think of the FDA in the United States) and of course courts. Each mode has its own bent (it presupposes certain frames and tends to favor certain solutions) and 'deciding' which one runs the show (in each case) is a choice with often already-implied consequences.[17] Let me also mention the logic at play in the world of professional politics when confronted with dialogical democracy. Because politicians are regularly exposed to electoral sanction (perhaps not an unimportant notion) and because they are concerned with their re-election, they often demand the right to make the final decision and consider any dialogical process (always a very local and partisan set) as one way among others of grasping societal demands. More generally – since I guess we are not ready to totally abandon representative democracy, whatever its shortcomings – the question seems to be that of the articulation between *many forms of democratic activity*, from regularly organized elections to 'citizen juries' and lobbying through numerous forms of *post-hoc* controls, equilibrium of powers, administrative bodies checking how accepted regulations are applied, court, jurisprudence and legislative acts, political activisms of all kinds, from NGOs to associations and political parties – an assemblage we have perhaps not considered seriously enough.[18]

Let me consider one particular institution, the European Commission. This is a particularly interesting case since the Commission is probably the single institution in Europe that has done the most in terms of transparency, participation and good governance, that has tried to include the largest range of 'stakeholders' in its decision-making processes. One main reason for this inclusiveness might be that the Commission is an institution that drastically lacks democratic legitimacy since it never faces any direct election. It also had, in the mid-1990s when these steps were taken, an image as an excessively technocratic body that imposed abstract, general solutions that took no account of local complexities. It therefore, perhaps, had no other option that to present itself as open and attentive to the opinions arising from 'civil society' – which led it to promote a new role for itself in Europe. Like most economic and political elites, however, the Brussels-based Commission seems also to believe that 'the public' (in singular) does not really understand the true stakes involved in making technological choices. 'The public' tends to overestimate the dangers and does not understand when there is no alternative to certain developments. It is not surprising, therefore, that the commissioners, whatever their good-will and engagements in favour of participatory politics, often blaze a trail of decisions that go against the opinion of the majority (as with GMOs) or that, when matters are absolutely central, they bypass their own participatory rules. Key decisions are then made without the involvement of participatory democracy – a situation that could be easily shown, for example, through the case of open software and software patenting.[19]

---

17 A PhD thesis by Jean-Baptiste Fressoz is to come out soon on this question.

18 On the necessarily complex relationships between interest groups and politicians, see (Lochard, Simonet-Cusset 2003), notably the very pertinent remarks of Jean-Michel Belorgey in (Lochard, Simonet-Cusset 2003: 129-133). For a longer-term perspective on the emergence of deliberative processes today as well as on the forms and style of justifications for democratic decisions, see (Blondiaux, Sintomer 2002).

19 This work, which I intend to expand very soon, is based on the reading of texts found on the website of the European Commission.

Finally, and what is undoubtedly the most important element, decision-making in market democracies depends on a far greater variety of logics than *rational debate* or *the exchange of arguments* alone. The sphere of the political is not so central to decision-making and regulatory processes because a decisive role is played by economic actors who produce, innovate, buy and sell, because decisions of major importance (in social and environmental terms) are simply being made 'on markets'. Put more bluntly, democratic procedures are often only *reactive* and occupy a subaltern position in relation to the *de facto* situations created by financial and economic actors.

These actors constantly reconfigure the social by inventing and distributing techno-scientific products, by reshaping techno-industrial systems – a point that has often been discussed in technology studies since Langdon Winner's famous article of 1980: *Do artifacts have politics?* (Winner 1980) Even though 'civil society' does not lack means to fight back (I am thinking of the very dynamic 'form of life' of Linux users and their capacity to oppose the demand of Microsoft and others for the private ownership of source codes (Lessing 2004)), acting as though the main regulations were arrived at through participatory or democratic procedures is at best naive, at worst misleading.[20] These approaches forget that democratic societies rely in an essential way on not obeying any one unique set of rules, on not depending on any one single decisional system. Decisions are made according to many parallel and different logics that undoubtedly operate within the political arena, but that are also arrived at through financial and economic action.[21]

On a higher level, we can say, echoing Ulrich Beck, that the main actors of the world economy today constitute a meta-power largely disassociated both from states and from the sphere of the political. Their strength resides in their capacity to *do* things, to invest where they deem appropriate and to pull out of any country that contests their approach too vigorously. This meta-power is not in need of democratic legitimacy even if it redefines the political – as moral engagement, for example – and efficiently reconfigures the legal infrastructure via the WTO, *cabinets d'affaires* and so-called, for example, 'independent' central banks.[22] This meta-power helps to undermine the classic forms of democratic accountability, substituting such processes with infra-political practices – political correctness, for example, or superficial transparency. It has constructed a universe of regulation around itself characterized both by ever increasing litigation (it is in the courtroom that rules are often fixed and disagreements resolved), by the introduction of voluntary codes of ethics and by the exceptional role played by certain international, not citizen-controlled institutions, like the World Bank (which governs large parts of the South through an interesting new mix of words, knowledge and financial lending power) or the WTO and its way of settling disagreements.[23]

---

20 Concerning countries in the global South, see (Goldman 2001, 2005); (Woods, Narlikar 2001); (Hickey, Mohan 2003); (Onana to be published).

21 For more on this, see (Pestre 2003: chapter 3) and (Fitoussi 2002). On the differentiations to be made between decision conceived as a conversation or an organised deliberation, see (Manin 2005) and (Urfalino 2005).

22 (Beck 2002). According to Beck, this meta-power is cosmopolitan. It is also decisively based in the North, however, with support from the dominant states (G8). Its interests are only pacifist and cosmopolitan as a first approximation, since it is always ready for war to save civilization (Huntington 1997, 2004).

23 (Goldman 2005); (Ballet, de Bry 2002); (CCE International 2004); (de Ravignan 2002); (d'Iribarne 2001).

A central question thus arises – and it carries particular force in the current world where market-based regulations have attained an amplitude unknown since the last third of the nineteenth century. The question is whether the political, as the site of democratic debate, is still authorized to rule, at least to some extent, on the proliferation of techno-scientific products when most people judge their effects to be irreversible, dangerous or of no interest for the common weal. The question is whether or not all commerce is justified *a priori* when it radically affects societies – whether limitations should be imposed on the right to commerce. The tendency today – at least until the next financial crisis – is to say no, that the political cannot and even should not exercise such a power except, perhaps, in the areas that touch our moral tradition, such as cloning or pedophilia (although not the sex and prostitution industries, which are enjoying an enormous boom).[24] But doesn't this evolution represent a major regression for democracy? After all, in the industrial capitalism of the mid-20th century, the market world accepted the possibility of such limitations, as well as other forms of collective regulation and redistribution. This development might also represent a fundamental regression for the field of STS, since such questions could (and, I believe, should) constitute major areas of research.

Statement 2: Trying to develop technical democracy and hybrid fora should not undermine the right to dissent; we should also recognize that the proposals at which citizen juries arrive at may 'travel' (in social terms) as poorly as those that emanate from the world of representative politics.

What is forgotten is that a democracy is not a political regime free from conflict, that "conflict is neither an accident nor a misfortune" but "the expression of the undecidable nature of the public good".[25] Only by forgetting this could one tout the belief that, if the procedure is good and properly engineered, a more rational, optimal and legitimate consensus will emerge. Our societies are not comprised of one unique world of justice, of one unique universe of reference and the decisions made, as well as the dominant arrangements arrived at, cannot avoid also being choices in favour of some and against others. The notion of consensus – in science studies as in political sciences – might therefore be misleading, and looking systematically at the choices finally made from the point of view of those who will be most affected by them might be a helpful and heuristic tool, in terms of mental and social hygiene. The fact that collectives of people establish themselves, that they demonstrate the limits of representative democracy, or the violence done to them, is more than positive. The fact that proposals are made on the basis of different *cités de justice*, that new forms of empowerment emerge in the social body is a significant step forward – but it is equally important to be attentive to the question of who is most in need of help in order to precisely arrive at some form of empowerment.

In fact, we must recognize that decisions remain acts of authority – even when they are the outcome of the most open process. The central function of politics is to learn how to live with conflict and to turn any potential enemy into an adversary; but this does not mean that

24 On this aspect, see the strong argument developed in (Poulin 2005), who insists on the enormous explosion of the sex and prostitution industries following the liberalization and globalization process.
25 (Ricœur 1991); see also (Affichard, Foucauld 1995).

once a dispute is settled, antagonism just disappears. As Paul Ricœur has remarked, "Political discussion is without any conclusion, even though it is not without any decision" (Ricœur 1991: 166-167). Because the game of politics is a blend of collaboration and conflict, and because it does not constitute the central *locus* for resolving questions in market democracies, the right to dissent has to be defended – and one must always retain the right to contest both the decision and the *états de fait*. If we forget this right, if we behave as though everyone had the same possibility to participate, as though the procedure would allow us to arrive at the optimal solutions – or more precisely: if *we deny the right of the victim of a decision to* rationally *persist in his or her refusal*, cases of injustice will continue to arise, carrying with them the risk of populism (Mouffe 1999; Young 2001).

More importantly, since most decisions in techno-scientific matters imply fundamental choices about the society one wants (what kind of agriculture, for example?), about ethics (what kind of humanity?) and about the environment (what kind of sustainability?), their acceptability always remains problematic. This is certainly the case with politicians (notably in Europe) caught between the Scylla of the demands of their electorate and the Charybdis of what they see as necessary to insure the international competitiveness of their national economy. This situation might also hold true for participatory procedures, however wonderfully they might be conceived. Once they leave the circle in which they were originally elaborated, the resulting propositions might no longer possess any greater legitimacy, in the social body at large, than those elaborated by politicians (this resembles in some ways the argument Hobbes made against Boyle and the new natural philosophy, whose conclusions he regarded as being *a priori* unacceptable since they were decided in a closed space, a space not accessible to public scrutiny – in that case the Royal Society of London). There is no obvious reason why these propositions should not suffer a similar fate and be ignored by large sectors of the population. The convictions arrived at by these 'panels' mounted in the restrained spaces of consensus conferences or other citizens' juries perhaps do not 'travel' any better than those of the politicians.

Statement 3: Discourses of participation and good governance are to be found everywhere now; they have become central elements of a new form of governmentality combining calls for individual government of the self and extended regimes of rationalisation and of scanning of populations. These new forms of governmentality should become central objects for our investigation.

Governance has now become the standard way of conceptualizing the political. Much has already been written about what this word implies in 'ideological' terms, as well as on the chronology of its appearance. The concept was initially elaborated in business management circles at the end of the 1960s when managers were less and less capable of properly running the shop floor (at Fiat, for example, or at General Motors in the United States) (Boltanski, Chiapello 2000; Waring 1991). The concept spread in a first wave at the beginning of the 1980s following the demands for transparency made by managers of pension funds and because of the renewal of shareholders' involvement in business. For them, transparent governance of their assets was essential (Pestre 2003: chapter 3). Next, 'good governance' moved into politics via the discourses of international institutions such as the World Bank; it then started to appear in the rhetoric of NGOs working in the South to make public life more 'accountable' (Kalaora 1999). The 1990s stage of this evolution was characterised by the notion of 'global governance', in part theorized at the London School of Economics, that aimed at establishing common values for the management of a collective, and ultimately reconciled, future (Moreau Défarges 2001).

As Moreau Défarges and others have suggested, the vocabulary of governance conveys the idea that the world of politics, as it was invented and has been practiced for more than two centuries, is *de facto* obsolete. Not only because it is based on an overly conflictual understanding of the social, but also because it relies too much on the State and on the formal procedures of representative democracy. Since our societies now have global questions to solve (in environmental terms, for example), since our societies have been pacified from the outside (we have finished with the Westphalian model and no longer need to fear wars of aggression – in the North, of course) and since they have been pacified from within (we live in wealthy societies) – in sum, because there are no longer any real conflicts of interest or war and because we no longer have to worry about the redistribution of goods (expanding globalized markets are the solution) – there is no longer any important choice for society to debate and politics and the State can fade away and at last allow civil society to auto-organize freely, as it sees fit.[26]

The only remaining questions are procedural and managerial in nature – with maybe the help of some kind of international police acting against a few groups that perturb this globally happy world. In this universe of good will that has been ideologically reconciled and is without enemies (which evidently has nothing in common with the assertion of various clashes of civilizations – Huntington – or with the battle of Good against Evil), the political question of representation is no longer a live issue (Huntington 1997, 2004). Representative democracy is essentially bankrupt, and everything has become the responsibility of everybody in a world where each individual is responsible facing everyone else. The only task that remains is supervisory – or, more precisely, the invention of rules for a new game whose optimal solution is the target. In that sense, governance appears as a way of forgetting the question of legitimacy and power – and that deserves serious consideration from scholars. Having people collectively invent the most efficient solutions, favoring the multiplicity of sites where democracy could flourish are vital ideas – but not considering with the utmost care the role of electoral sanction and of the organized sphere of the political in public life is a mistake.

In short and without too much nuances, we STS practitioners should certainly return to a more informed political philosophy. Not that the question is easy, far from it. But it would be most worthwhile to consider again how we conceptualise contemporary societies and their many ways of regulating themselves. For example which functions are performed by States, agencies, NGOs and international bodies today? what are the relations between procedures, substantive questions and values? and between representative, participatory and oppositional politics? We could for example come back and discuss what was at the core of the modernist constitutions; the way more cynical traditions describe social relations; the media as forms of power (and the transformations recently introduced – think of Fox News); and reconsider in which way the metaphor of the tribunal (which we have used a lot) misled

26 Add (Falk 1999). See also (Manent 2001).

us when talking about public life.[27] If industrial and financial techno-science has become a major power today, why not, as de Vries and Halffman suggested, turn to good old Montesquieu and think of "a constitution for risk societies" to counterbalance its powers. "Legal guarantees of a [new] constitutional state" should be on our agenda, they write – "the freedom of citizens in a decent society" to be added to "the integration of interests" through various forms of dialogic activity, and to the social awareness of a very unequal distribution of powers. In that, as Halffman writes, "there is a gap on the political theory-flank of constructivist STS the size of a library", and that gap needs to be filled.[28]

We can probably go a step further and say that today 'governance'[29] constitutes the new, more elaborate and dominant form of *governmentality* (following Foucault, governmentality is considered here the art of governing, the set of administrative knowledge and technologies "by means of which we are able to constitute, define, organize, and instrumentalize the strategies that individuals, acting freely, can have with respect to one another" (Foucault 1994: 728)). These new forms of governmentality are not the exclusive property of Nation States or international organizations. Indeed, as I have already alluded to in my brief chronology of the notion of governance, this approach was initially conceived in the world of business and finance – and they are now at the core of any shop floor management. I have in mind the way in which medicine has evolved over the last decades, becoming "increasingly managed, and in details, by the state or insurance companies" at the moment patients evolved into consumers and activists. As John Pickstone says:[30] "Technical analysis is extended from diagnostic classifications to the cost-effectiveness of procedures, and the resultant power is exercised over doctors rather than patients." I also have in mind, to take a quite different case, the way migrant people are now treated by state agencies (Fassin, Memmi 2004). In Paris, they are required to be free subjects in charge of themselves, to describe to immigration officers their own itinerary and objectives – the 'charters of corporate responsibility' being a kind of identical/symmetric object. Proposed by most companies nowadays these texts, in which managers lay out their social and environmental commitments, serve multiple goals[31] – namely: To affirm the moral rectitude of the company, to present a brand image, to inspire confidence and to protect it against challenges from activist groups that might launch smear campaigns or boycotts (indeed, the timing of the announcement of such charters often supports the hypothesis of this last motivation). Another aim of such charters is, by engaging the level of voluntary commitments, to escape restrictive legislation that might be imposed from outside (Lubbers 2002). Rating agencies (comparable to the ones found in financial markets) have thus started to appear. The idea behind certain of these agencies is to assess what companies actually *do* in social and environmental terms, to compare their declared engagements with the reality of what they do – and to make these assessments public. These agencies claim that doing so will help give more weight to voluntary busi-

27 On this last aspect, see (Hatchuel 2000); (Halffman 2003).

28 (Halffman 2003: 476-477). In a more abstract but stimulating form, see also (Latour 1999).

29 There is an immense literature on governance. I refer the reader to (Osmont 1998); (Kalaora 1999); (Roman 2000); (Milani, Arturi, Solinis 2001); and, in particular (Hermet 2001) and (Moreau Defarges 2001).

30 In a short note prepared for a workshop held in Brussels on Science and the Political, 29-30 June 2006.

31 They also vary according to countries and cultures. For example (d'Iribarne 2001) and (d'Iribarne 2003) showed that the differences are quite striking between America and the more 'catholic' France.

ness commitments, that it will make companies effectively more accountable, that it will help convert these charters into *public* contracts (Pestre 2003).

Taking these engagements of 'good governance' at face value and checking up on their implementation constitutes astute politics vis-à-vis the new forms of governmentality I am trying to understand – and they should be backed. We should not forget, however, that these business engagements form part of larger strategies put in place over the course of at least the last two decades. These strategies were conceived with the help of image consultants and advertising companies. For example, E. Bruce Harrison, author of the 1993 book *Going Green*, is not alone in offering the following advice to companies. He suggests that companies be "model(s) of openness", he urges them "to submit (their) activities to public scrutiny" and "to keep up a constant dialogue with everyone interested in (their) performance" (Rowell 2002b: 34-47). He also advises them to get down to the work of managing this dialogue directly. Initiating such a dialogue allows the company to limit the number of subjects to be dealt with, to define the rhythm of meetings and, for example, *to debate endlessly without ever restricting the company's vital interests*. Regularly reviving this dialogue allows the company to choose its partners, to sponsor certain (notably environmental) activities in areas that are of marginal importance while still being essential for the company's image and at the same time to isolate those who do not wish to enter into a 'reasonable' discussion. It is also central – and this is a decisive piece of advice – to learn to speak the language of the objectors, to show the company's openness by using this very language (here, we might reflect on the generalized use of the term 'sustainable' – to which I will come back soon; during the Johannesburg summit, BMW advertised their cars in the French press as 'la voiture durable'). In the cases where the companies are not in control of the negotiation process, these same communications consultants advise companies to create their own 'interest groups' (with suitable names, such as the Global Climate Coalition set up by the oil industry), to always answer critical reports with new reports, to regularly demand the reopening of the discussion while continuing to act on other levels, in the political arena for example. And in this way, they can stall the dialogical process, permanently if necessary – which is the most efficient way to decide and go on doing business without taking any formal decision.[32]

In case you think that this all comes out of a second-rate spy thriller, let me assure you that these practices have been theorized by professional people. You can find them on the web and in libraries – and the scenario I just presented largely corresponds to the history of the debate over climate change as managed by the oil-industry over the last fifteen years. Even if these practices invalidate neither the importance of dialogical democracy nor the necessity to promote it, they remind us not to be too naïve – and to recognize the various ways our contributions and actions can be used by other people. Maybe the lesson is also that our scientific work might often have a political dimension despite itself – and despite us.

---

32 (Rowell 2002a & b). The quotations on communication advice from Harrison are taken from (Rowell 2002b).

Statement 4: New categories are at the core of our common thinking – sustainable development or precaution for example. They might cover up more than they reveal concerning what is at stake; deconstructing them in a systematic manner, and building genealogies of their deployment and acceptance, might be most helpful.

The discourse of 'good governance' is often combined with the discourses of sustainable development and of precaution, two major concepts of the last thirty years. Here again, while the intentions are noble, these notions might well cover up a major lack of analytical precision, they might be obscuring debates and using euphemisms to hide problems – they might be tools helping us to forget, to render formerly well-developed forms of knowledge invisible. Indeed, it is interesting to note that the current ecumenical discourse of sustainable development has replaced a tension that used to pit the discourse of unlimited technological development in the service of resource-intensive growth against those who argued for limited growth, for another idea of development, who placed the problem of the sustainability of unfettered growth at the heart of their analyses. Over the last two decades, this opposition seems to have vanished – in part thanks to the increasing polysemy of the expression 'sustainable development' itself. As an extreme and recent example, I cannot resist quoting the definition of the term given by the CEO of BP France, Michel de Fabiani on October 11 2001:

> "Sustainable development is first and foremost about producing more energy, more oil, more gas, and maybe more coal and nuclear fuel, as well as more renewable energy, of course. At the same time, we need to make sure that this production is not achieved at the expense of the environment."[33]

In more precise terms, the debate of the 1940s to 1960s has shifted ground and now voluntarily limited growth is less and less seen as an option among official institutions. No doubt some of these questions are raised (through the interesting thematic of industrial ecology for example (Erkman 1998), or in the alter-globalization movement) and we can hope, despite statements like Fabiani's, that things might be changing among economic and political actors; we could also hope that Dasgupta will be proven wrong when claiming that markets are not the most appropriate institutions for protecting the environment.[34] Overly positive views concerning sustainable development might however allow the fusion of fish and fowl, reassuring everyone of their own environmental and economic good sense. They allow us to lead a comfortable existence in a happy and reasonable world since *we now know* how to progress in a healthy manner, how to think about 'sustainable development'. Thanks to Hans Jonas and Ulrich Beck, we are now 'aware' of the profound uncertainties generated by technological development and are able to assign both this uncertainty and our ignorance to their appropriate places. Thus, we can see ourselves as having conceptually mastered the situation, as having finally discovered the means to reconcile what an ideology

---

33 As far as I know, one of the first articles presenting this line of thought with great insight is (Dovers, Handmer 1992). Add the exemplary book edited by (Lash, Szerszynski, Wynne 1996). The quotation from de Fabiani is taken from http://www.decroissance.org, in the section: Le bêtisier du développement durable.
34 (Dasgupta 2000) claims that environmental resources are often considered free by most techno-industrial interests and that there is little incentive to economize their use.

(evidently much too conflictual and 'political') had formerly obliged us to regard as irreconcilably opposed. All the same, despite this conceptual, ethical and post-modern leap forward, the solution may be simply a logic of unquestioned, unfettered growth without any alternative visions.[35]

Could we seriously believe that development can be sustainable without posing the question of our modes of consumption? Has the question of individual consumerism (must every car have air conditioning?) lost all sense in relation to any possible collective choice or in relation to the question of global warming? Does the market's profit-oriented technophilia naturally lead in the direction of conservation, precaution, prevention of risks and healthy development? Does the question of the redistribution of wealth have nothing to do with socially sustainable development? Should we not go back to some form of stronger regulation? While it is clear that communism is dead and gone (God rest its soul), does this mean that such questions have lost all interest for us? I fear not – even though I realize that such an outspoken comment might merit my immediate excommunication. On a more serious note, it might well be that we too regularly forget, when in the middle of our 'scientific research', that there are too many things that we do not and cannot know, and that we are busy actively producing ignorance.[36]

Statement 5: These developments should induce us to come back to two key questions: how can we talk positively about various kinds of knowledge, including science? And what are the various (positive and negative) connections we could identify between science, the political order and democracy?

The new style of governmentality that has developed around the techno-sciences over the last thirty years means a reversal of the status 'officially' accorded to knowledge. During the post-war boom and the cold war, knowledge, in the public sphere at least, was said to be free and autonomous, allowing the unfettered pursuit of the truth. That belief constituted the core of what Brian Balogh called the "pro-ministrative state", a way of managing societies, during the first cold war, by a close-knit system of professional experts and state and business leaders.[37] In the last decades, refusals and contestations of this image have become highly visible both in the social sphere – in the context of atomic energy, public health scandals, the Bhopal disaster, the affair of contaminated blood, the problems with asbestos – and in the social sciences – by feminism, science studies, cultural studies and so on. What we are seeing today, looking at how the major economic and political actors 'use' the notion of knowledge in social controversies, is a public inversion of positions: These actors have come to appropriate the ideas that all knowledge is partial and partisan, that it is normal for experts to disagree, that all the parties need to express themselves and 'negotiate' agreement about theories and facts. Indeed, they find themselves promoting this way of describing knowledge. In the spheres most exposed to polemics, notably where the interests of the world's most powerful states and companies are in play (around atomic energy, biotechnol-

---

35 See works by Andrew Jamison, for example (Jamison 1996).
36 A standard reference here is (Ravetz 1993). See also a forthcoming volume on 'agnatology' edited by Proctor & Shiebinger and (Steinsaltz, Funkenstein 1997).
37 (Balogh 1991). See also (Dahan, Pestre 2004).

ogy, agro-business and public health, for example), we see that absolutely anything and everything is said and written – under the guise of science and, apparently, in complete bad faith.

It is, for example, this capacity of the Bush administration to not just say but to affirm anything in the name of science that was the target of the petition signed by 60 Nobel Prize and National Medal of Science winners in February 2004 and by the report of the Union of Concerned Scientists in March 2004. These texts aim to expose "the suppression and distortion of research findings by high-ranking Bush administration political appointees" (concerning climate change, air quality, agricultural pollution, etc.), as well as "the undermining of the quality and integrity of the appointment process" and the "manipulation of the scientific advisory system to prevent the presentation of advice that might run counter to the administration's political agenda" (about the prevention of lead poisoning, health and dietetic questions, workplace safety and so on). No doubt we could read these texts as deliberate actions on the part of scientists to reassert their authority. While it might not be entirely wrong, I definitely believe that they *are also* symptomatic of a major social and political change that we should in no way ignore. Another example is the intervention by the Bush administration on US and UN reports on obesity – and the conclusions that should be drawn in terms of diet, food and soft drinks.[38]

The same instrumentalization has been theorized around political circles. In order to "win the Global Warming Debate", the Luntz Research Company (which advises the American Republicans on their public relations and helps them to better justify their refusal to sign the Kyoto and Montreal protocols) notes that American people believe that there is no consensus (or something near consensus) among scientists. "Should the public come to believe that the scientific issues are settled," writes the Luntz Company, "their view about global warming will change accordingly." The report draws the following conclusion: "Therefore, you need to continue to make the lack of scientific certainty a primary issue in the debate, and defer to scientists and other experts in the field." You should also emphasize, says the document, the importance of "acting only with all the facts in hand" and "making the right decision, not the quick decision". Presenting the "language that works", the memo proposes the following expressions – which were regularly used by Republican leaders: "We must not rush to judgment before all the facts are in. We need to ask more questions" or "Scientists can extrapolate all kinds of things from today's data, but that doesn't tell us anything about tomorrow's world." And rather than regulate, it is better to insist that

---

38 (Union of Concerned Scientists 2004a & b); (US House of Representatives 2003). According to articles in the *Financial Times* and the *Herald Tribune* from January 2004 (Fighting obesity (2004) and White House accused (2004)), the American Academy of Pediatrics asked for a ban on the sale of sugared soft drinks during school hours. Faced with this, the companies selling soft drinks replied by contesting the data while promising to withdraw their distributors from schools. Simultaneously, William Steiger, special assistant for international affairs at the Department of Health and Human Services of the American administration, challenged the findings of a parallel WHO report arriving at the same conclusions. He wrote the following: "The assertion that heavy marketing of energy-dense food or fast-foods outlets increases the risk of obesity is supported by almost no data." He also questioned a WHO assertion that increased fruit and vegetable consumption reduces the risk of obesity and diabetes. And he finally saw no necessity to withdraw distributors. Add the declarations of the CIO of Coca-Cola in the *Wall Street Journal Europe* (Terhune: 2004).

"what we need to do is to put American creativity and American innovation to work". That way, the memo concludes, you will have a greater chance to not get an international agreement that would threaten the economy – and the United States will be able to show that they help "other nations save their forests and build safer energy sources" (Luntz Research Company 1994).

It is important not to misunderstand my point: I am not so naïve as to doubt that all knowledge is always and already caught up in the social, that it is always 'situated' and cannot become a view from nowhere. Knowledge, and scientific knowledge in particular, is always enmeshed in unending networks of reference and I know that no such thing as 'objectivity' exists or could exist *in abstracto* – however much we might wish this were the case in today's context. On this point, I follow Jasanoff and Wynne against Collins and Evans in the debate they had in *Social Studies of Science* in 2002. Supporting the multiplication of institutions of knowledge, favoring debate, banking on encounters as learning processes – in sum, *acting on the idea that knowledge and ignorance go together and are socially distributed* – is absolutely central. It is always better to back the diversity of places of knowledge production and maximize confrontation.[39]

I also know two other things, however, two other things that are just as important. First, there is such a thing as deliberate manipulation and this practice is becoming the rule – *precisely to destroy any collective learning process.* Second, the control of legitimate knowledge is politically decisive, and those in power can easily remain satisfied with a view that 'everything is relative' *when they can alternatively produce facts on the ground.* I do not pretend to have a simple solution to offer for the dilemma that results from this situation, but the fact that it has recently become a lot more urgent means that it simply cannot be ignored. Whether we think about the evaluation of drugs, comparisons between GMOs and non-GMOs, the causes of obesity, or the health effects of bad working conditions, the present tendency is overwhelmingly in the direction of marginalizing 'objectifying' expertise, of privatizing it or suppressing it.[40]

What results is a situation where the weakest (or those who are furthest removed from the reigning orthodoxy or the mass media) are unable to make their version of knowledge heard. As surprising as it may seem, therefore, I propose that rather than exercising the ever-present possibility of re-opening the black boxes of knowledge – as we have done so efficiently in science studies and history of science when confronting scientism – we instead (or also) consider the possibility of *helping keep certain black boxes closed (at least for a while) against their being rashly and artificially reopened for instrumental purposes.*

These observations lead me to make three proposals. First, we should return to a theoretical consideration of what makes knowledge different, what made it a discourse different from, say, political discourse. The fact that Jasanoff and Wynne rather than Collins' and Evans are right in terms of principles, in the polemics I have just alluded to, does not remove the basis for Collins and Evans' remarks – for example that while science-studies scholars claim that politics and science are organically intertwined and in continuity, these scholars also seem to know how to distinguish them. To say that science and politics are organically dependant does not imply they are identical, nor that all claims have the same value

39 For more precise analyses of these points, see (Pestre 2005).
40 On the instrumentalization of results concerning GMOs, see *La Recherche*, November 2003.

– whether assessing the effects of a pesticide on health or debating, between historians, what was at stake for Pasteur at Pouilly-le-Fort. Telling what science is *not* does not mean we are able to articulate a coherent discourse on it in a positive manner. Whatever the difficulty of the task, I believe we need to put that question back at the core of our theoretical concerns. My second proposal would be that we, as STS scholars, should accept the limits of our own science, be ready to water down our certainties and, for example, introduce other 'differences', differences we did not consider in the past. The fact that we agree that there is no such thing as 'objectivity' does not mean, for example, that we cannot fight for it, fight for the promotion of attitudes that re-affirm *the duty of objectivity* – as we fought for the promotion, in expert processes, of procedures to help guarantee some form of independence. Norms differ from practices, but norms matter and have their own efficiency. At that level, the key is to imagine the means to translate a norm into institutional terms – which is less than easy. Finally, we should be as pragmatic as possible since we are faced with a real paradox and powerful actors. It is likely that we will have to work with strongly context-dependent propositions and with considerations that are theoretically unsatisfying and contradictory. But if the world we are facing is itself contradictory, perhaps we cannot escape such solutions (in fact, even STS is a human construction and has some limits!). In other worlds, I believe that it is urgent that we, as historians or practitioners of STS, accept withdrawing ourselves from the reassuring beachheads that we have constructed and return, along with everybody else, to the cold waters of the new world that is taking shape in front of (and with) us.[41]

Statement number 6: Today, we are living in a time of maximal tension between two great historical dimensions of the process of 'modernization', one linked to democracy and its extension, the other to human invention and technophilic business.

I would now like to gather some of the strands of my argument and propose a more global reading. My initial argument rests on a proposition made by Ravetz about the paradox in which we find ourselves today (Ravetz 2001).

My thesis can be summarized as follows: the large-scale social disputes in Europe around techno-scientific-industrial products, around GMOs, stem cell research or cloning, around nuclear power plants and nuclear waste, or around the environment, have to do with today's sharpening of one of the historical tensions of the 'modernization process' that has shaped our societies over the last two centuries.[42] That tension could be described (even if I know how partial this presentation is) on the one side as a process of *democratization*, a process of *dialogue* that places the political at its center, and, on the other side, as relying intrinsically on technological *progress*, human *ingenuity* and invention. On one side lie the demands (the

---

41 As far as expertise is concerned, it is now admitted that independence could function as a norm. Procedures, including sanctions, are sometimes instituted for those who do not respect the rules. I propose we try and imagine similar 'rules' for objectivity. On expertise, see (Hermitte 1998) and (Joly 1999).

42 These tensions manifest themselves when decisions have to be made about techno-scientific-industrial matters perceived as possibly affecting us in an irreversible way (us as individuals, or us as social groups and/or us as an ecosystem); and when decisions have to be made in a context in which the consequences are not known or uncertain – or in any case hotly debated (in particular among various kinds of scientists); and when stakes are considered high (as with global warming) or in which certain key values are at stake (they might be ethical, social, political or environmental).

rights) for individual and collective recognition and control over oneself and one's life; on the other lie the demands for (the rights of) science-and-technology-based activity (and notably business) to promote innovation for the good of all. On one side is 'the political' as the space in which the *polis* collectively decides its future; on the other side are the values of the free individuals, the free market and of the entrepreneurs who rely on scientific and technological innovation to propose new products. Both sides are based on a 'rational' understanding of the situation. On one side is the political rationality of *precaution*, a rationality that wants to debate the individual and collective consequences of actions before accepting the new techno-industrial-scientific order and its products; on the other side is the technological and commercial drive that up to now has always been under control and has been profitable to mankind (are you opposed to new drugs?). Note, finally, that both aspects simultaneously made our modern (and post-modern) world – *and that both dimensions might be crucial*: if societies were governed solely by market rules, if nothing else were there to counterbalance them, societies would probably disintegrate rather quickly and be exposed to constant violence. On the other hand, a world where the allocation of goods would be carried out solely by political leaders (even if democratically elected) would certainly lead to a disaster just as dramatic (Fitoussi 2002). No doubt democratic debate is integral to modernity but twentieth-century history has made us very cautious, and we now know how well-identified minorities can be systematically excluded by totally democratic procedures.

Many things have contributed to the emergence of this new tension over the last decades. One key factor I want to mention, however, is the drastic change in proprietary rights in the United States in the very early 1980s – organically connected with the possibilities that science and its associated technologies have opened up in terms of practical action: the possibility of recomposing DNA, of large computational means, etc. This redefinition of property took shape through a series of deliberate and political moves, as we know, but also through a set of largely independent moves – the reorganization of Nasdaq and changes in pension funds regulations, for example – a complex sequence of decisions I tried to document in a book I published in 2003. Of course we cannot forget the weakening of the nation state as the central locus of democratic decision, particularly in the South; of course, we should consider the growing 'technicization' of decision-making processes and the multiplication of expert advice (from economics to interdisciplinary activity around the environment), the proliferation of major techno-industrial failures (as in the case of mad cow disease) and decisive sociological changes (the expansion of groups with high-level educational capital and the growth of individualism and/or of identity politics). One element remains central, however, the possibility opened in the early 1980s to privatize what was, up to then, defined as open, public knowledge. The core of the new world we now live in is there, *around a new definition of property* – the consequences being new modes of regulations at the international level (largely through markets and the control of the 'technologicalfrontier'), massive redistribution of wealth, major threats to forms of life – and that explains why the tension exists again between democratic politics and the market's technophilia as the two major spheres of reference.[43]

---

43 Discussions in WIKO, notably with Zhiyuan Cui, helped me get to the point more sharply than in (Pestre 2003).

So, what is at stake in the debates that occupy large segments of our societies is an opposition between (at least) two different definitions of the collective good, two different emphases on what is important in terms of political regulation *today*. Of course, many preoccupations are *de facto* intertwined in such a process: the question of trust in *science*, in *expertise*, in administrative and economic *regulations*, in political and economic *elites*, in the kind of progress the *techno-industrial world* promotes when opening one Pandora's box after the other and bringing about '*faits accomplis*' through market systems – but the core is largely around property, and we should examine the way more complex notions and practices of property could emerge.[44]

3. On the importance of expanding our agenda

What conclusions can we draw from these remarks about our way of working and acting, about what we can do as intellectuals and social actors? My first suggestion is that we need to escape our excessively irenic apprehensions about the world and, for example, broaden our understanding of the growing demand for the democratization of techno-scientific-industrial choices. Many things are at stake, not only political issues but also practical and theoretical issues. My second suggestion is that we abandon the quest for responses that are universally applicable, and that we instead constantly alternate the levels of approach, from theoretical to pragmatic and back, from the order of principle to that of compromise, from the universal to the local, from the propositional to the critical, from the constructive to the deconstructionist, and *vice versa* – that we learn to be mobile in terms of our positions, presuppositions, objectives, tools, and preoccupations.

This could mean, for example, accepting at least in part the Habermassian version of the picture and helping public debate flourish where it might lead to positive reappropriations and a productive learning process – but this could mean simultaneously maintaining a strong critical stance toward moves aimed at manipulating these debates or, to take the movement against biopiracy, to be part of the *substantive debate* and come up with a more complex understanding of property rights than those that have emerged over the last 20 years.[45] This might also simultaneously mean launching a critique against both governance and liberal forms of governmentality by pointing out the policies of the World Bank or the inconsistencies of the 'social and environmental' governance of business. We need to analyze many crucial questions – the effects the tangled power structures have on agro-alimentary politics and what the *Codex Alimentarius* has become following its arbitrary transformation by the WTO, as well as the possibilities of alternative legal frameworks that would serve other ends. Similarly, it is most urgent to analyze the effects of these global power structures on

---

44 A last word before concluding. We all know that the "modernization process" could be described (and should be described) in many other ways, but my point is not the "modernization process" as such – a very loose expression anyway, an expression of minor importance to me in this context.

45 The easiest approach to this immense field is to look at the websites of Third World Network, Biowatch, Science and Development Network, the Convention on Biological Diversity of the United Nations Environment Programme, etc. See also (Hermitte, Kahn 2004).

health – and ask how we can assess the consequences of the new Private-Public Partnerships put in place in Africa over the last decade.[46]

This also implies describing how one governs today by means of words, how battles over governance have largely become battles conducted on a global scale to win over minds.[47] It also implies a reflexive look at ourselves, at the roles we play, whether consciously or not, in the globalization process. Here, I am thinking of the transformation that anthropology has undergone as a discipline since being regularly called upon to help with the introduction of good governance in the South. This situation has sparked lively debates within the anthropological community about how it has been co-opted by good governance, what the effects are for the populations that are supposed to be reordered with the help of anthropologists, and the consequences for 'anthropology' as a discipline. Obviously, a similar self-reflection could be started in the STS field.[48]

In my view, we really need to address these questions, debate them in the manner of questions to be grappled with, difficult and contradictory questions, and not restrict ourselves to issues of what is instrumentally realizable. It is quite probable that this kind of reflection will not result in any solutions that can be put into action in a non-problematic way – but this does not mean that we should not do it. Contradiction is a constituent element of the world, and not facing up to it does not make it disappear. Reflection in the social sciences requires accepting that one cannot always be in a position to offer a solution. The alternative would mean to give oneself over to the ideology of positivist science. Our duty, on the contrary, should not be limited to making operationalizable propositions; it is also to accept thinking about what is difficult to think about, to speculate on alternative futures, on radically different modes of being. Utopia remains an essential concept in the social sciences, just as it is in the political realm.

46 On these last two points, see (Romi forthcoming) and (Reich 2002), notably (Buse, Walt 2002).
47 On this decisive question of the battle of words, I suggest starting with the 1999 issue of Ethnologie française, particularly (Abelès 1999); (Kalaora 1999), and (Bellier 1999).
48 Here, I am thinking of the very interesting reflection that has been launched, through seminars, by the Association Française des Anthropologues in Paris on the practice of ethnology in the context of globalization.

Claus Zittel

The Politics of Cognition: Genesis and Development of Ludwik Fleck's 'Comparative Epistemology'*

1. The Myth of Fleck's Relativism

Many scientific theorists consider Fleck to be quite obviously one of the most important pioneers of the present-day "radically social constructivist relativistic trends" in scientific theory.[1] This is contradicted, however, by the fact that at central points throughout Fleck's entire work, an attempt at creating an 'Über-science' [*Überwissenschaft*] recurs, which can hardly be brought into harmony with relativistic arguments. Moreover, Fleck's concept of science changed over the years, which is why one can only with great difficulty ascribe to him a consistently uniform position, but must rather differentiate his thinking into different phases. This therefore casts into doubt the widely-held impression of an epistemological relativism with Fleck. It can further be feared that the interpretations which are based on such dominant preconceptions have until now exerted a too one-sided influence on the historic as well as systematic discussion of Fleck's work. Why?

Advocates of relativistic positions in scientific theory claim that the natural sciences, rather than being distinguished from the cultural sciences through a privileged access to nature, as is otherwise mostly advanced by philosophers of science,[2] are deeply embedded in variously different cultural systems of interpretation. Accordingly, they further argue, the natural scientific ideals of objectivity, exactitude and evidence are also culturally variant and reveal themselves when more closely examined to be standpoint- and context-dependent concepts with regulative function. They go on to claim that in particular Ludwik Fleck was one of the first to recognize clearly the socio-cultural dependence of science. One could refer here for example to his slogan: "In science, just as in art and in life, only that which is true to culture is true to nature" (Fleck 1979: 35; Fleck 1980: 48). What Fleck makes particularly attractive for scientific theory is that with him, the cultural milieu of science is not applied from without, but instead arises out of the natural science's perspective itself. And indeed, what is fascinating about Fleck's work today is not only the concrete demonstration of how research practice and thought style are intertwined with each other in different cultural and social contexts, but also the lucid descriptions of how scientific imagination is constituted in the face of resistance with the aid of material objects and practices according to defined social preconditions. All of this appears to be a strongly relativistic program.

---

* For their helpful discussions and comments the author wishes to thank: Moritz Epple, Falk Müller, Florian Schmaltz, Reinhard Schulz, Evelyn Fox-Keller, Johannes Fehr, Ilana Löwy, Peter Galison, Ian Hacking, John Pickstone, Cornelius Borck and Sylwia Werner.

1 Compare for example (Löwy 2000); (Cohen, Schnelle 1986: 421-444); (Bonah 2002: 187-207); (Borck 2004: 447-464); (Löwy 2004a); (Löwy 2004b).

2 Compare (critical of this) (Rorty 1993) and on the background (also with respect to Fleck) (Zittel 2002).

2. Cracks in the Standard Story

The impression, that Fleck's program is relativistic, doesn't at all conform to the way in which Fleck himself declared and theoretically based his methods and goals. For Fleck, as is well known, characterized his epistemology as "comparative epistemology" ("vergleichende Erkenntnistheorie").[3] But such a "comparative thought style research" (Fleck 1979: 27f., 40, 51, 64, 101, 108; Fleck 1980: 41,56, 70, 85, 133, 142), as postulated by Fleck, lays claim to the possibility of taking a superior standpoint, from which it would be possible to distinguish differing thought styles and to arrange them in historical order, so that they could be organized by grouping corresponding concepts from different thought styles together and framing certain thought lines by starting with particular leading metaphors (proto-ideas). Here Fleck observed "own *historical regularities* in the naturally-fated course of ideas" ("eigene *historische Gesetzmäßigkeiten* im Ablaufe der Ideenschicksale"[4]).[5] At the same time Fleck didn't shy at all from the quite frequent use of apodictic formulations in order to clarify his point of view.

Here are some initial examples; others follow below. Fleck holds

> "the postulate 'to maximize experience' as the '*supreme law*' of scientific thinking. Thus, once the possibility of such comparative epistemology arises, it becomes duty" [...] (Fleck 1979: 50; Fleck 1980: 70). "A rule of thought that allows one to make use of more details and more compulsory connections, *as the history of science teaches us*, deserves to be emphasized." (Fleck 1979: 22; Fleck 1980: 34). "The concept of syphilis *must* be investigated like any other case in the history of ideas [...] Many, very solidly established scientific facts *are undeniably linked*, in their development, to [...]" (Fleck 1979: 22; Fleck 1980: 35). "In comparative epistemology, *cognition must not* be construed as only a dual relationship between the knowing subject and object [...]".[6]

---

3 Before its publication, the title of his main manuscript "Die Entstehung und Entwicklung einer wissenschaftlichen Tatsache" was, for example, "Die Analyse einer wissenschaftlichen Tatsache. Versuch einer vergleichenden Erkenntnistheorie" ("The Analysis of a Scientific Fact. Attempt at a Comparative Epistemology" – unpublished letter of Fleck to Schlick, September 5, 1931).

4 Emphasis added (Fleck 1979: 9; Fleck 1980: 15).

5 Because of such positions, some scientific historians refer to Fleck's comparative epistemology in order to make greater structural comparisons of thought styles in the history of science. That in doing so, a context overarching meta-standpoint must also be possible is assumed to be unproblematic. Compare (Fuchs 2002). Such references to the history of science in Fleck's name require a still further precision: Fuchs compares only the thought style of two individual thinkers and isn't interested in a thought collective. But Fleck makes here quite a rigid distinction: "The Presence of thought style makes it necessary and, indeed, indispensable to contruct the concept 'thought collective'. Whoever might nevertheless prefer to eliminate the thought collective must introduce value judgements or dogmatic faith into the theory of knowledge. In so doing he would only succeed in creating a particular dogmatic type of epistemology out of the general comparative type" (Fleck 1979: 41; Fleck 1980: 57).

6 (Fleck 1979: 38; Fleck 1980: 53) (emphasis added). Cf.: "das Postulat vom Maximum der Erfahrung für das *oberste Gesetz* wissenschaftlichen Denkens. Zeigt sich nun einmal die Möglichkeit einer vergleichender Erkenntnistheorie, so wird sie zur Pflicht" (Fleck 1980: 70). "Ein Denkprinzip, das mehr Einzelheiten und mehr zwangsmäßige Koppelungen wahrzunehmen erlaubt, verdient, *wie die Geschichte der Naturwissenschaften lehrt*, den Vorzug [...]. Der Begriff der Syphilis *muss* wie ein denkhistorisches Ereignis [...] unter-

How relativistic is such a theory, that speaks in many places of 'law' and 'duty' in imperatives? Or we could ask the question the other way around: When, how and why has Fleck tried to go beyond relativism to establish a context-transcending standpoint? How can we frame Fleck's studies theoretically without suddenly getting trapped in the position of taking a meta-standpoint?

3. Further Questions on Fleck's Work

Sometimes this conflict between the different statuses of Fleck's claims and descriptions has been noted by historians of science without it being considered crucial.[7] But, this discrepancy brings up systematically and historically interesting questions that not only allow us better to follow the development of Fleck's epistemology, but also lead us into the heart of Fleck's thinking – the questions of the commensurability and the translatability of knowledge situated in different contexts. These questions have been recently discussed under the label of epistemological contextualism, which considers questions such as what the conditions of understanding are for other or archaic cultures and how we can at all formulate adequate criteria of interpretation in the face of such cultures. The crucial question of epistemological contextualism is then: Is it required to presuppose a basic universal rationality that in principle cannot be relativized and contextualized, in order to get at least a vague understanding of the thought styles of other thought collectives? One must also ask if the possibility of comparison of thought styles in Fleck's comparative epistemology implies a basic anti-relativism with respect to a universal rationality which we are tacitly to presuppose.

How does Fleck's position exclude itself from this? Fleck, in any case, requires that he adequately reconstruct according to given criteria other systems of knowledge which he presents. Even further: For Fleck's comparative investigations, as already hinted at by the above quotation, a significant *descriptive, understanding and explanatory optimism* is characteristic, which implies *astounding generalizations* with respect to the objects being investigated, as well as to the historical conclusions. This suddenly becomes clear when one, for example, more closely examines Fleck's concept of epochs (which he obviously allows the cultural history to define) or when one considers that he assumes for each epoch a type of general development logic for thought styles (Fleck 1979: 23; Fleck 1980: 35). Questions that necessarily follow are, for example: Does Fleck present *the* history of syphilis or a story to which one can imagine adding many other alternative histories? How generous, according to Fleck, are the measurements of thought styles to be conceptualized? What relation does his description of the evolution – and accumulation – of knowledge within individual thought styles (here Fleck thoroughly acknowledges advances) have to the anti-evolutionary consequences which follow from the comparison on a meta-level of many thought styles? What, for example, is with his concept of proto-ideas? He claims that they are substantiated

sucht werden" (ebd.). "Viele wissenschaftliche, bestbewährte Tatsachen verbinden sich durch *unleugbare Entwicklungszusammenhänge* mit vorwissenschaftlichen [...] Urideen" (Fleck 1980: 34f.). "Vergleichende Erkenntnistheorie *darf Erkennen nicht* als zweigliedrige Beziehung des Subjektes und des Objektes [...] betrachten" (Fleck 1980: 53).

7 Compare, for example (Stachel 2004: 351-382, 370).

differently during different epochs – that sounds like the application of a holistic-structuralistic insight (Fleck 1979: 23f.; Fleck 1980: 36). Why are they then according to Fleck still the same ideas? Why does one in the application of his method find that “in the development of ideas, primitive pre-ideas often lead continuously to modern scientific concepts” (Fleck 1979: 100; Fleck 1980: 131)? Doesn’t such a discussion of a proto-idea or a pre-idea contradict the relativity of thought styles? Doesn’t Fleck himself show quite clearly, when he writes on the same page that a thought A and its opposite B, which surface as a contradiction within a thought style, indeed don’t do this when they are advanced by advocates of different thought styles, for thereby “it will just *not* be *the same* thought” (“so ist es eben *nicht derselbe* Gedanke” Fleck 1979: 100; Fleck 1980: 131, emphasis added). According to Fleck, research on thought styles requires the application of a sociological method in epistemology. Is the sociological method, however, historicized or is it the new ‘Über-science’?

On the other hand Fleck claims, with a view toward ‘archaic’ forms of thinking: “but we have no standpoint available to us from which we could uniformly understand and investigate them. We are also unable to understand the personalities of archaic thinkers” (Cohen, Schnelle 1986: 80f.; Fleck 1983: 86). Doesn’t Fleck unexpectedly adopt such a viewpoint, in order to make this claim? And what does ‘uniformly’ mean here? Why should we at all want to understand them all *uniformly*? As will be shown, Fleck wants to perform exactly this. And moreover: The labeling of a thought style as ‘archaic’ or ‘primitive’ is of itself normative and orients itself on an advanced model of knowledge or on an art-historical concept of a circle of styles (from an archaic to the classical period, then comes the Enlightenment, the Hellenistic period or the Baroque period). We must therefore take care not to see already in Fleck a prelude to Lévi-Strauss’s cultural relativism. Some thought styles are characterized as incommensurable with one another (Cohen, Schnelle 1986: 80; Fleck 1983: 87): How are we then to evaluate Fleck’s own possibilities for understanding in his historical research? How is his own thought style to be described and is it, too, to be relativized? In short, a strictly relativistic reading of Fleck’s theory of science would in light of such passages encounter considerable difficulties.

After the phase, in which Fleck’s fascinating scientific and research analyses have been brought out of the long shadow of Thomas Kuhn and their main ideas established, successfully comes to a close – the volume of collected essays we have before us and the conference from which it sprang are also evidence therefore – it is time to examine more closely Fleck’s theoretical fundamentals.

4. The Politics of Epistemology: An Alternative History of Fleck’s Thought Path

Over the decades Fleck expanded and generalized his ‘comparative epistemology’ more and more. The genesis and development of Fleck’s contextual epistemology can be divided into five distinct phases. I would therefore like now to show that Fleck’s project began from a specifically standpoint-bound way of thinking (1). Then, he expanded his insights into a *theory* of thought-styles (2), which he in the next step elevated into a science (3). Finally, he built out this science into a general synthesis of methods (4) and in the end even into a fundamental science (5).

A Standpoint-Related Way of Thinking

Fleck begins initially with a contribution, in which he emphasizes the distinctiveness of medical thinking compared to that in the natural sciences: In *Some Specific Features of the Medical Way of Thinking* (1927), Fleck argues that medicine is a much weaker concept of science than natural science. Its exceptional position rests on the fact that medicine does not deal with average cases but rather studies deviating phenomena. One could therefore say that medicine, according to Fleck, acts principally in an anti-Aristotelian manner. Secondly, medicine looks, above all with help from statistical methods, for laws to identify phenomena that do not conform to accepted theory ("Gesetze für nicht gesetzmäßige Phänomene", (Cohen, Schnelle 1986: 39; Fleck 1983: 37)). But this would not be sufficient. Fleck's third point holds that empathy and intuition have to come into play in order to be able to perceive an incidence of illness at all. Fourthly, note must be taken of the resulting far driven process of abstraction in medicine in producing concepts of species "whose fictitiousness is considerably greater than in any other field of science". The result is a "characteristic divergence between theory and practice" (Cohen, Schnelle 1986: 42; Fleck 1983: 40). In contrast to other disciplines, medicine does not have at its disposal a basic theory, as chemistry has with the atomic theory (Cohen, Schnelle 1986: 44; Fleck 1983: 43), but rather only temporary and dynamic central ideas. Lastly, if the subject of investigation, the disease, can only be observed in its states of progress, then historicity and temporality are specifically inherent to the concept of that disease. Fleck's considerations culminate in an exceedingly broad-minded comparison: "then medical thinking differs in principle from scientific thinking in that it uses Gauss's coordinate system, while the latter uses the Cartesian system" (Cohen, Schnelle 1986: 45; Fleck 1983: 45).

Obviously at this time Fleck himself still had a great admiration of scientific logic, which he later derides. He identified scientific thinking not with Bacon and the empirical disciplines, but rather conceived of a caricature of Cartesian mathematical fundamentalism. Here he reserved epistemological relativism and the social causality of science exclusively for medicine. Later he dared to expand these insights gleaned from medicine to the other natural sciences, whereby a careful restraint in regard to the choice of examples cannot be misinterpreted, since he sticks closely to the medical evidence. Examples from physics, as well as from the arts, are comparatively rare. It remains that we believe that he conceived his basic scientific-theoretical ideas within the framework of a counterformulation to an – in his eyes, objective – scientific thought style. To doubt would be the further question of whether a Fleck critic could play Fleck's earlier analyses off against his later ironical ones, in that he or she contextualized his theory of thought styles and simply declared it to be a medically-specific special case.

The Theory of Thought Styles

The transition from an epistemology arising from a medically-specific perspective to a general theory of cognition was achieved by Fleck in his article *On the Crisis of Reality* (1929). This text is probably the most relativistic and most radically constructivist contribution to the philosophy of science that Fleck ever wrote. In it is contained, for example, the oft-quoted statements from Fleck:

> "Thus every thinking individual, being a member of some society, has his own reality in which and according to which he lives. Everybody has even many, sometimes contradictory, realities: the reality of everyday life, a professional, a religious, a political, and a small scientific reality [...] Each cognition, each system of cognitions, each entry into the social realm has its own corresponding reality" (Cohen, Schnelle 1986: 49; Fleck 1983: 48).

But then follows the adjunct: "This is the only correct standpoint" ("das ist der einzig gerechte Standpunkt", Cohen, Schnelle 1986: 49; Fleck 1983: 48). This formulation appears remarkable to me. Fleck doesn't write, this is the only 'true' position. The usual English translation here, leads to confusion, since the German word 'gerecht', which translates to 'correct', has in contrast to the English word 'correct' absolutely no epistemological connotation, only normative. Fleck switches admittedly to a meta-level, but he doesn't justify this switch in this place epistemologically, rather he characterizes it as normatively or politically motivated – it's a political decision!

Furthermore, the scientific way of thinking is now correspondingly described in political terms and is generally qualified as democratic (Cohen, Schnelle 1986: 50; Fleck 1983: 49), the sharp opposition between theory and practice is extended to all science as characteristic (Cohen, Schnelle 1986: 51; Fleck 1983: 50) and – now pluralistically – the "dependence on purpose" of all "scientific truths" is observed (Cohen, Schnelle 1986: 52; Fleck 1983: 51).

The crucial impulse for this broadening obviously came less from Kurt Riezler's article (*Krise der Wirklichkeit*, 1927),[8] to which Fleck's text is presumably the answer, but rather from a contribution in the previous issue of the same journal, *Die Naturwissenschaften*, penned by Niels Bohr, *Quantum Theory and Newer Development in Atomics* (*Das Quantenpostulat und die neuere Entwicklung in der Atomistik*[9]). Fleck refers here and repeatedly later to this article by Bohr,[10] whose presentation of quantum mechanics obviously opened for him the possibility of broadening his medicine-related point of view in the direction of other sciences. The process had begun.

The Third Step: The Invention of a New Science

Either in 1932 or through the Polish translation of 1933, Fleck was familiar with a lecture by Erwin Schrödinger titled "Ist die Naturwissenschaft milieubedingt?"[11] ("Is Science a Fa-

---

8 (Riezler 1927: 705-712). For the neo-Kantian Riezler and the in this time ever present discourse on 'crises' see: (Borck 2004); for further cultural background see: (Harrington 1996).

9 (Bohr 1928: 245-257). Ilana Löwy believes though, that Weigl and the Polish school of medicine influenced Fleck to generalize, to science as a whole, his view that diseases are socially constructed. But she admits that it is difficult to say just how familiar Fleck had been with the work of this school. I am equally skeptical, in particular because Weigl's concepts remain within medical thought. Furthermore, it is sufficient to consider the sources whom Fleck himself quotes, in order to be able to make plausible the opening of Fleck to a general, critical view of science in a collective framework. Compare (Löwy 1990: 217).

10 Compare (Cohen, Schnelle 1986: 66; Fleck 1983: 67).

11 (Schrödinger 1932); translated into Polish (Poznanski 1933). Also printed in (Mayenn 1994: 295-333). Here also in the publisher's preface, there are hints to the divided reception of the Schrödinger lecture. Remarkable in particular is that the title chosen for a criticism was "Der Stil der Physik" ("The Style of Physics" ibid.: 11, Fn 15).

shion of the Times?"). This lecture offered Fleck a further, important confirmation of his thoughts from the perspective of physics and, years later, during his controversies with Bilikiewicz[12] (1939), Fleck referred to this text. Schrödinger, also, contests that the sciences study special objects reserved only for them and may therefore claim an exclusive access to nature. He asserts rather the thesis that modern physics shares characteristic features with modern culture. The changes in modern physics are therefore determined by cultural factors. In particular, Schrödinger recognized relationships between the artistic environment of the Weimar Republic and the development of modern physics. In his opinion, quantum mechanics, with its casting in doubt the causality dogma of classical physics, shared the then current artistic need for upheaval and disorder, and, in the artistic movement of the new unemotionality, he recognized parallels to the Puritanism in physics which strives to eliminate superfluous concepts. Furthermore, in art as well as in physics there is a trend toward statistics and mastery of masses. All in all, Schrödinger offers a constellation of concepts which we also find with Fleck, whereby however with Schrödinger the social analysis of the cultural commonalities observed by him is missing.

Quite probably encouraged by this, Fleck's "comparative epistemology" (Cohen, Schnelle 1986: 67; Fleck 1983: 69) appears for the first time in his 1935 article, *Scientific Observation and Perception in General*, under the definition of a *science*:

> "a theory of knowledge based on the sociology of thinking and on the sociological history of the development of science, [which is the] science which compares different thought styles and investigates the circulation of thoughts within various thought-collectives" (Cohen, Schnelle 1986: 72; Fleck 1983: 75).

The discrepancy between relativism of thought styles and the general claim of the sociologically ordered meta-method clearly comes forward now. But when Fleck states, as he did in his contemporaneously published main work (Fleck 1979), that it would be about time now "to assume a less egocentric, more general point of view and to speak of comparative epistemology" (Fleck 1979: 22; Fleck 1980: 34), he seems to be unclear about what he means by that. It is not surprising then, that his statements just a few years later provoked philosophical contention. The Polish philosopher Izydora Dąmbska, a student of Twardowski and who, as Fleck, came from Lwów, brought the problem of Fleck's basic assumptions precisely to the point in her critique of Fleck:

If science is "ex definitione an intersubjective entity [...], accessible to the control of many ordinary people, speaking a common language, [then] out of the negation of the possibility of intersubjective cognition, the denial of the possibility of science"[13] therefore follows.

It would be neither correct nor appropriate, as recently happened, simply to ridicule Dąmbska's objections and to sacrifice Fleck's counterpolemic.[14] For what Dąmbska makes

---

12 (Fleck 2007: 264-288). Compare (Symotiuk 1983: 3-4, 569-582) and (Löwy 1990).

13 (Dąmbska 1937: 288-294) translation by C. Zittel.

14 I consider in particular Birgit Griesecke's characterization of the Fleck-Dąmbska controversy to be unfair. As we have seen, Fleck's argumentation of exactly the point with which Dąmbska takes issue is not very stringent. Furthermore it should be considered that also Dąmbska represents in pure culture a quite definite thought style, namely that of the Lwów-Warsaw school. In this way, the thought styles of Fleck

clear is that one either has to give up sharp relativism as a consequence of Fleck's concept of thought styles or has to dissolve the traditional concept of science, including its *claim of objectivity*, and conceptualize science as art. Fleck, however, in contrast to his relativistically inclined successors or predecessors (such as Nietzsche), never was willing to accept this consequence or to abandon his own scientific ideal, and it remains to be asked, for what reasons Fleck so dearly clung to this ideal. For only apparently did he make concessions in this question: Only at the very end of his response to Dąmbska and, markedly, within a philosophical debate, does he feel obliged to minimize the status of his epistemology:

> "The theory of thought styles liberates us from many outdated prejudices and uncovers huge fields worthy of investigation. In this sense, that is, for its liberating and heuristic role, I think, it is true."[15]

But in the earlier sections he takes a different tone. Again his ethical and political motivation clearly comes forth, for Fleck expects from his concept of thought styles that it leads to tolerance.16 Furthermore it is exclusively *his* concept which opens up the possibility of common understanding among humanity (!) and gives hope to the overcoming of totalitarian claims of power:

> "One cannot examine any beliefs independent of the entire culture of a given society in a certain epoch. The point of departure must be a community of people living together, and such a method leads above all to a sociology of thinking which, because it can be a comparative science, encompasses the most general beliefs. In the epoch which we approach, the epoch of synthesis and of the disappearance of particularisms, it will be unavoidable. The specialization and the differentiation within society is growing and will grow further. One cannot speak of an ongoing (lasting) effect of the brutal attempts to control the minds of men. Agreement is only

match those of Dąmbska in their incommensurability, and that would be completely in the sense of Fleck. Compare (Griesecke 2005: 21-28).

15 (Fleck 1938: 192-195). A german version, translated by Sylwia Werner, is published in (Choluj 2007: 439-473).

16 Also the last of Fleck's essays, "Crisis in Science", culminated in his swearing to the normative power of his epistemology: "Comparative studies on styles will make the students more tolerante towards strange styles, and prepare them for co-existence" (Cohen, Schnelle 1986: 157; Fleck 1983: 180). Compare in particular also Fleck's statements in his discussion with Narcyz Łubnicki and Tadeusz Tomaszewski (1948): Ludwik Fleck: Streszczenie referatu i dyskusji: Problem obserwacji naukowej, in: (Fleck 2007: 295-297): "Die Denkkollektive besitzen einen besonderen Denkstil, eine besondere soziologische Struktur und eine besondere historische Entwicklung. Nur durch die vergleichende Soziologie des Denkens kann man eine rationale Wissenschaft über das Erkennen schaffen. Indem wir einen vergleichenden Denkstil schaffen, befreien wir uns aus der Bindung der Denkkollektive und der Epoche, ähnlich wie uns das wissenschaftliche Wissen aus dem Chaos der Einmaligkeit der individuellen Erlebnisse befreit [...] Die vergleichende Wissenschaft über die Denkstile und Denkkollektive wird in der gleichen Bedeutung objektiver als die einseitige Theorie des Erkennens, die nur auf einem Stil gestützt ist; wie auch das Erkennen objektiver ist, das auf den Aussagen mehrerer als eines einzelnen Menschen beruht. Die Wissenschaftler können sich nicht vom Denkstil ihres Kollektivs einzig innerhalb der Probleme ihrer Wissenschaft befreien. Außerdem kann ein Wissenschaftler z.B. religiös sein, d.h. in manchen Bereichen nach einem mittelalterlichen Stil denken. Einen vergleichenden Denkstil zu schaffen, bedeutet zugleich, ein neues Gebiet des Denkens zu schaffen, in dem die Mitglieder verschiedener Stile in einem Bereich gleichzeitig denken können." (Translation by Sylwia Werner).

possible based upon the method of comparison: Only so does one create a common thought collective, free through criticism and common through tolerance. Therein lies the content and worth of the principle of thought styles, and from this standpoint one must, as I believe, investigate them. These are neither sophisms nor relativized truths" (Fleck 1938: 192-195).

In contrast to the more reticent plea for an heuristic theory of thought styles that breaks open the traditional ways of seeing, Fleck went in 1935 further into the opposite direction, and thereby further strengthened exactly the generalizing thrust of his epistemology, so that with it he additionally claimed to be able to give a *unified* explanation for all styles of thought.

The Attempt at Synthesis: The Theory of Thought Styles as a Universally-Unified Science

In *The Problem of Epistemology* (1936), Fleck says that "up to now" we have lacked a point of view that has allowed us globally to comprehend and investigate the "intellectual personality of ancient thinkers" or "archaic forms of thinking" (Cohen, Schnelle 1986: 80f.; Fleck 1983: 86). But such a global method from a superior viewpoint would be established by the new "comparative science of thinking" (Cohen, Schnelle 1986: 79; Fleck 1983: 85) or "science of cognition" (Cohen, Schnelle 1986: 81, 85; Fleck 1983: 87, 92).

For Fleck, therefore, the point is in no way *to recognize one's own limits of understanding and to relativize one's own standpoint in the face of extrinsic patterns of thinking*. For Fleck, the problem of incommensurability arises only on the horizontal level, not for the olympic position of his theory. For *solely* on the basis of his comparison of thought styles is, as quoted above, general commensurability possible, between differing thought styles, however, not at all or barely. On this horizontal level, that is, from the inner perspective of individual thought styles, he distinguishes therefore between groups within which communication is possible and groups in which "people [...] are not in the least able to agree with or understand each other, as if they belonged to different thought-groups (thought-collectives) and communities" (Cohen, Schnelle 1986: 81; Fleck 1983: 87). Three major thought styles, each with their own evolution, are differentiated, and this, by the way, entirely not with local but rather completely generous parameters: the philosophical, the scientific, and the mystical. Only within a collective is communication possible.

There seems to be no problem for Fleck that his own claim of understanding already transgresses the boundaries of communication which he qualified as absolute. Consequently, he would thus have to define and relativize his own sociological perspective (that was Nietzsche's consequence) and/or admit that a common rational basis exists for understanding different thought styles. However, by taking such a position, his perspective could not turn out as a global synthesis but as a pluralistic manner of explanation. According to this perspective, one could hypothetically try to alternate between different perspectives without synthesizing them. This could be a back door to the dilemma of justification. Fleck however, propagates the possibility of a synthesis:

> "Should the science of thought styles be a science capable of development, useful and rich in a substantial way, it ought to broaden the range of its interests. It must not be limited to the study of the domains and stages of science which are officially rec-

> ognized at the given moment, but, taking into account the variety of thought-styles and the multiplicity of thought-collectives, it must become a comparative science. [...] Consequently it ought to include psychological, sociological and historical methods. Its subject will be the whole of cognitive life, its organization, fluctuations over time and developmental peculiarities, local features, properties of its various forms [...]. Epistemology thus understood is a science of thought styles. When comparing thought styles, studying their historical genesis, their development, the social forces which produce and maintain them, the methods of introducing them into the thought-collective, epistemology [it] will earn a positive opinion from contemporary 'official' scientific knowledge [...]" (Cohen, Schnelle 1986: 97f.).17

Fleck specifies on several occasions *expressis verbis* that his position should not be understood as relativism or skepticism: "This view outlined here should not be construed as skepticism. We are certainly capable of knowing a great deal" (Fleck 1979: 51; Fleck 1980: 70). In the controversies with Bilikiewicz,[18] he makes clear that "no cognitive relativism results from the theory of thought styles. 'The truth' as a current state of transformation of thought style is always only one: it is completely determined within one thought style" (Löwy 1990: 268).[19] Fleck's epistemology is therefore not relativistic, but rather relationistic.

---

17 "Soll die Theorie des Erkennens eine entwicklungsfähige Wissenschaft sein, nützlich und reich an konkretem Inhalt, muß sie den Bereich ihrer Interessen erweitern. Sie kann sich nicht auf die Untersuchung der im gegebenen Moment offiziell anerkannten Wissensbereiche und -etappen beschränken, sondern muß, indem sie die Mannigfaltigkeit der Denkstile und Vielheit der Denkgemeinschaften in Betracht zieht, vergleichende Wissenschaft werden. [...] Sie muß also *psychologische, soziologische* und *historische* Methoden umfassen. Ihr Gegenstand wird die Gesamtheit des Erkenntnislebens, dessen Organisation, zeitlichen Fluktuationen und Entwicklungseigenheiten, lokalen Merkmale, die Eigenheiten seiner mannigfachen Formen [...]. Eine so gefaßte Theorie des Erkennens (Ökonomie, Technik, Kunst, Politik, Mythologie und Psychiatrie werden einbezogen) ist die Wissenschaft über die Denkstile" mit ihr gelangt man "zu einer geeigneten Anschauung des heutigen offiziellen Wissens" (Fleck 1983: 107f.).

18 Bilikiewicz, to Fleck's surprise, reads Fleck's concept as transcendental-philosophical. Fleck rejects this decidedly, however he allows his comparative epistemology to be counted among the neo-Kantian attempts at advancing a perspective of perspectivism, such as with Hans Vaihinger (*Philosophie des Als Ob,* 1911) and then later with Friedrich Kaulbach (*Wissenschaft und soziale Umwelt,* 1939). Fleck refers to Bilikiewicz's *Die Embryologie im Zeitalter des Barock und Rokoko*, an otherwise unremarkable investigation in which, in a very general way and with reference to Wölfflin's and Joel's style concept, a relationship is constructed between the collapse of absolutism and new embryological insights of the Animaculists about the independent individual life of spermatozoa. Fleck demands an investigation which more specifically captures the collective character of the scientific process of discovery. Bilikiewicz reacts with annoyance, and accuses Fleck of a transcendental idealism in the neo-Kantian manner, in the epistemological view of which the cleft between natural science and cultural science disappears. The second objection is that the medical basis is too small for epistemological generalizations. Thirdly, Bilikiewicz is convinced that historical facts are independent and do not underlie any cultural shift.

19 Because Ilana Löwy gives often quite free translations of the Bilikiewicz-Fleck-controversy, I quote always also the polish source and the unpublished german translation of Sylwia Werner: "Z teorii stylów myślowych nie wynika też żaden relatywizm poznawczy. 'Prawda' jako aktualny etap przemian stylu myślowego jest zawsze tylko jedna: jest ona *stylowo* bez reszty zdeterminowana." (Fleck 1939b: 169, Fleck 2007: 282) "Aus der Theorie der Denkstile resultiert auch kein kognitiver Relativismus. Die 'Wahrheit' als eine aktuelle Etappe der Denkstilumwandlung ist immer nur die eine: Sie ist *stilbedingt* vollständig determiniert. "

One can speculate over the reasons for this reobjectivization of his theory of knowledge. Perhaps Fleck was still so obligated to the scientific ideal of objectivity of the scientific community of the Viennese Circle, which he otherwise criticized, that he was afraid of the reproach of being unscientific and relativistic.

But more convincing seems to me another conjecture, which rests on what has already been noted by Wittich and Stock (in objection to Fleck's concept), and recently (in defense of Fleck) by Cornelius Borck.[20] They point out that some of the first critical reviews of *Genesis and Development of a Scientific Fact* came from fascists (Haeberlin, Kroh, Fischer and Peterson),[21] who tried in the name of Fleck to add "race" as a further category, in order to distinguish a German thought style in the sciences from a Jewish one. The emergence of such a kind of fascistic relativism in Germany, with the purpose of discriminating against Jewish scientists, must have shocked Fleck and caused him to sharpen his constraints. In light of the new, unhealthy alliance of positivism of power, antisemitism and relativism in Germany, it seems therefore more than likely that Fleck now even more had to establish a normative or at least more general standpoint, in order to preserve a remnant of rationality at all.

Four years after the publication of his main work, and shortly after the appearance of the above-mentioned criticisms, Fleck began the work of refining these constraints. In his article *Science and Social Context* (1939), Fleck now describes the new relativistic approach to science as double-edged: we would be, he wrote, at a "marvelous turning point on the path of expansion of science [...], from which unfolds vistas of an entirely inconceivable and exotic landscape" (Löwy 1990: 250).[22] But this viewpoint holds dangerous perspectives as well: Since science depends on its social environment, it can be, from both left and right, politically abused. Fleck's scientific optimism, which resulted from the feeling of freedom from the pressures of old theories, is broken and he desperately swears the explanatory power of the effect of his own theory. Fleck is taken mostly as not quite a political thinker, and indeed the following passage shines another light on him. I quote therefore in detail:

> "Whereas 'well-educated', that is conservative scientists view with some anxiousness the emerging new aspects, clever politicians process as rapidly as possible the newly-acquired information into demagogic slogans. The sociological, collective nature of knowledge was first turned into a political slogan involving the social and class conditioning of science, and then the competing political trends created the spirit of nation and race to provide a mythical world view propagated through the ages [...] we will fabricate evidence to obtain politically necessary and predicted results. We will introduce planned management of thought, free creation will be replaced by

---

20 Compare (Stock 1980); (Wittich 1983). A rich source for the understanding of the societal role of the thinking of (Kuhn 1978: 105-113); (Wittich 1981); (Borck 2004: 8, 12).

21 Petersen wrote, for example: "In eigentümlicher und von dieser Seite unerwarteter Weise schließt sich Fleck damit unserem neuen deutschen Denkstil an, der eine voraussetzungslose, absolute Wissenschaft verneint" (Petersen 1936: 239). Otto Kroh wishes that "die rassische Bestimmtheit des Denkstils miteinbezogen wird" and Fischer finds, the "Volksgemeinschaft wird vermißt" (Kroh1936); (Fischer 1936).

22 "Jesteśmy naprawdę na cudwnym zakręcie drogi rozwojowej nauki, z którego otwiera się widok na zupełnie niepojęty egzotyczny krajobraz" (Fleck: 1939a: 151, Fleck 2007: 265). "In der Tat befinden wir uns an einer wunderbaren Kurve auf dem Entwicklungsweg der Wissenschaft, von der aus sich ein Ausblick auf eine ganz und gar unbegreifliche exotische Landschaft öffnet", transl. by Sylwia Werner.

bureaucratic centers, and propaganda will replace the autonomous penetration of society. [...] But, first of all, the following danger is imminent: a new generation is growing up of future scientific workers who believe that truth in the good, old professional sense of this word, does not exist anymore. Some became fanatics because they have been deprived of confidence in reason: the others became cynical because they see that there is no absurdity big enough not to be believed, if only it is promoted by shrewd and intensive propaganda. That is why the study of the dependence of science on environment and epoch is particularly important today" (Löwy 1990: 250f.).[23]

Fleck wanted his theory to be understood as a kind of ideology-critique. He warned against confusing his science of thought styles with the aesthetic approach of the new generation of scientists *who had already begun to think in relative terms as well*:

---

23 "Podczas kiedy "dobrze wychowani", więc konserwatywni uczeni z pewnym lękiem zamykają oczy na nowość powstających aspektów, sprytni politycy co prędzej przerabiają wyłapane wiadomości na hasła demagogiczne. Z faktu socjologicznej, zespołowej natury poznania zrobiono naprzód polityczne hasło socjalnego, klasowego uwarunkowania wiedzy, a potem konkurencyjny kierunek polityczny stworzył ducha narodu i rasy, by peprzez epoki snuł mit światopoglądowy. [Ponieważ każda wiedza jest zależna od środowiska, więc trzeba proces odwrócić: do zmienionych sztucznie środowisk dorobić odpowiednią wiedzę. Przecież i tak nie ma obiektywnej nauki! Więc co prędzej trzeba "zrobić" lewicową lub prawicową, proletariacką lub narodową fizykę, chemię itd.] Do politycznie potrzebnych, z góry przyjętych wyników dorobimy dowody. Zaprowadzimy planową gospodarkę myślową, biurokratyczne ośrodki zastąpią wolną twórczość, propaganda zastąpi autonomiczne przenikanie społeczeństwa.
Zabiegi te byłyby śmieszne, gdyby nie były poniekąd groźne. Nieuk, czy raczej pół-nieuk, który co słyszał o wynikach sztucznej hodowli ras końskich, próbuje wyhodować konia ze skrzydłami...] Lecz jedno przede wszystkim niebezpieczeństwo grozi: rośnie pokolenie przyszłych pracowników naukowych w tej wierze, że nie ma już prawdy w dobrym dawnym sensie fachowym. Jedni stają się fanatykami, bo odebrano im zaufanie do rozsądku, inni stają się cynikami, bo widzą, że nie ma tak wielkiego głupstwa, które nie uzyskałoby uznania dzięki sprytnej intensywnej propagandzie. Oto dlaczego zagadnienie zależności wiedzy od środowiska i epoki jest dzisiaj szczególnie aktualne" (Fleck 1939a: 151, Fleck 2007: 266). "Während 'die gut erzogenen', also konservativen Gelehrten, mit einiger Furcht ihre Augen vor dem Neuen der gerade entstehenden Aspekte verschließen, verarbeiten schlaue Politiker schnellstens die erhaschten Informationen zu demagogischen Parolen. Aus der Tatsache der soziologischen, gemeinschaftlichen Natur des Erkennens wurde zuerst eine politische Parole eines sozialen, klassenbedingten Wissens gemacht und dann schuf die gegnerische politische Richtung den National- und Rassengeist, um durch die Epochen einen weltanschaulichen Mythos weiterzuspinnen. Da jedes Wissen von der Umwelt abhängig ist, muß also der Prozeß umgedreht werden: zu der künstlich veränderten Umwelt muß ein entsprechendes Wissen beschafft werden. Denn es gibt sowieso keine objektive Wissenschaft! Umso schneller muß eine linke oder rechte, proletarische oder nationale Physik, Chemie usw. 'gemacht' werden.] Zu den politisch nötigen, vorher festgelegten Ergebnissen werden wir die Beweise beschaffen. Wir werden eine Planwirtschaft für das Denken einführen, freies Schaffen wird durch bürokratische Zentren ersetzt, Propaganda wird an die Stelle des autonomen Durchdringens der Gesellschaft treten [...] Diese Maßnahmen wären lächerlich, wären sie nicht auch gefährlich. Ein ungebildeter oder eher ein halb-gebildeter Mensch, der etwas von den Ergebnissen der künstlichen Zucht von Pferderassen hörte, versucht ein Pferd mit Flügeln zu züchten [...] Doch vor allem droht die folgende Gefahr: Die Generation der zukünftig wissenschaftlich Arbeitenden wächst im Glauben heran, daß es die Wahrheit im guten alten Sinne des Fachs nicht mehr gibt. Einige werden zu Fanatikern, denn man nahm ihnen das Vertrauen in die Vernunft, – die anderen werden zynisch, denn sie erkennen, daß es keine noch so große Dummheit gibt, die nicht dank einer schlauen intensiven Propaganda Anerkennung fände. Das ist der Grund, weshalb das Problem der Abhängigkeit des Wissens von der Umwelt und der Epoche heute besonders aktuell ist" (Translation by Sylwia Werner).

> "However, the approach employed by the majority of authors is artistic and literary rather than scientific; it consists in intuitive feeling of similarities. [...] Such intuitive approaches are not yet sufficiently elaborated to be studied. The approach is still too literary and indeterminate: the statements, when extracted from rhetoric formulations [...] do not convince anyone. [...] [but] an adequate starting point for a positive study of the effect of the epoch on science can be found *only in the general sociology of thought*" (Löwy 1990: 251).[24]

Within the context of this argument, Fleck was able to explain more precisely the problem of understanding other cultures. In his view, it is namely more an act of empathy than a task of translation. It is not so important how one *translates* old concepts into modern terms, but how to get a feeling for the coloring of styles and the foreign concepts in their various contexts.

> "The style aura of concepts undergoes change and is followed by changes in the concepts themselves. So, first one should examine the atmosphere of concepts and their style-coloring, which is reflected in the linguistic use of certain words, and one should pay particular attention to their metaphorical use. Only this opens the way to the study of the thought style of a given epoch" (Löwy 1990: 253).[25]

At first glance it seems that this combination of empathy, thick description and linguistic approach could avoid the epistemological consequences that otherwise necessarily derive from the idea of translatability. But Fleck mixed epistemological with political-aesthetic argumentation in the attempt to connect his descriptions to an exceptional superior science which would justify them from a higher level:

> "We learn the style-dependent aura of concepts: archaic, incomprehensible sentences acquire a sense, and we find how the present meaning has been derived from an old one. In this way, artistic impressions, intuitive conjectures, and subjective feelings can be changed into relations *derived from independent laws of the sociology* of

24 "Lecz takie ujęcie, jakie przeważnie stosują autorowie, ujęcie raczej artystyczne, literackie niż naukowe, polegające na intuicyjnym wyczuwaniu podobieństw [...] i związków nie nadaje się jeszcze do badania. Za dużo w nim literackiego, dowolnego: wyrwane z pięknego tekstu, oglądane na zimno, zdania te nie przekonywują. [...] Dlatego zdaje mi się, że punktem wyjścia dla pozytywnych badań nad wpływem epoki na naukę może być tylko ogólna socjologia myślenia" (Fleck 1939a:151-152, Fleck 2007: 266). "Doch ein solcher – eher künstlerischer und literarischer als wissenschaftlicher Ansatz, der von den Autoren hauptsätzlich verfolgt wird und der auf einem intuitiven Erfühlen von Gemeinsamkeiten [...] und Zusammenhängen beruht, ist für die Forschung noch ungeignet. Zu viel Literarisches und Beliebiges ist darin: Aus einem schönen Text herausgerissene Sätze überzeugen, kalt betrachtet, niemanden [...] Deswegen scheint mir, daß der Ausgangspunkt für ein positives Erforschen des Einflußes einer Epoche auf die Wissenschaft nur die allgemeine Soziologie des Denkens sein kann" (Translation by Sylwia Werner).

25 "Stylowa aura pojęć zmienia się, za nią zmieniają się poglądy. Otóż trzeba przede wszystkim badać tę aurę pojęć, ich stylowe zabarwienie, odbijające się w zwyczaju językowym odnośnie używania pewnych słów, zwłaszcza metaforycznego ich używania. To dopiero otwiera drogę do badania stylu myślowego epoki" (Fleck 1939a: 154, Fleck 2007: 268). "Die Stil-bedingte Aura von Begriffen unterliegt dem Wandel, ihr folgen die Veränderungen der Anschauungen. Daher muß man zuerst die Aura der Begriffe untersuchen, ihre Stilfärbung, die sich im sprachlichen Gebrauch beim Verwenden bestimmter Worte, insbesondere ihrer metaphorischen Verwendung, widerspiegelt. Dies erst eröffnet den Weg zum Erforschen eines Denkstils in einer gegebenen Epoche" (Translation by Sylwia Werner).

thought and theory of the development of thought. We [...] attain a science of cognition rich in details and capable of cultural growth" (Löwy 1990: 252f.).[26]

With the intention of defending the scientific status of his insights, Fleck switched to a meta-level and in the process combined epistemological and normative justifications. One could admittedly object that possibly every epistemological justification is inextricably coupled with normative interests. Nevertheless it would remain necessary to distinguish between different ways of confusion. One could therefore, following Nietzsche's example, conclude a subjective or individualistic 'moral' of cognition from the theory of thought styles, whose 'objective' status is regulated by power struggles.[27] But, specific for Fleck is

---

26 "Poznajemy stylową aurę pojęć, archaiczne zdania niezroumiałe nabierają sensu, stwierdzamy, jak dziesiejsze znaczenie powstało z prymitywnego. Artystyczne impresje, intuicyjne domysły, subiektywne wyczuwanie daje się tą drogą zamienić na związki wyprowadzające się z odrębnych praw socjologii myślenia i rozwoju myślenia. [...] Otrzymujemy zdolną do rozwoju i bogatą w szczegóły naukę o poznawaniu" (Fleck 1939a: 153, Fleck 2007: 268). "Wir lernen die Stil-bedingte Aura von Begriffen kennen, archaische, unverständliche Sätze erhalten einen Sinn, wir stellen fest, wie die heutige Bedeutung aus einer primitiven entstand. Artistische Eindrücke, intuitive Vermutungen und das subjektive Erfühlen lassen sich auf diesem Wege in Zusammenhänge umwandeln, die sich aus den unabhängigen Gesetzen der Soziologie des Denkens und der Entwicklung des Denkens ableiten. Wir [...] erhalten eine Wissenschaft über das Erkennen, die ausbaufähig und reich an Details ist" (Translation by Sylwia Werner).

27 The greatest proximity to Nietzsche is admittedly found in Fleck's earlier article "On the Crisis in Science" ("Zur Krise der Wirklichkeit"). Here Fleck relativizes the concept of truth that: "there exists no fundamental difference between 'appearance' and 'truth', the difference is one of development" ("Es gibt keinen grundsätzlichen Unterschied zwischen 'Schein und Wahrheit', sondern nur einen Entwicklungsunterschied" (Cohen, Schnelle 1986: 56; Fleck 1983: 56), thereby moving closer to Nietzsche's criticism of knowledge. Fleck's optimistic prediction of a new science is unmistakably colored by a Nietzschean tone: "At the present time we are so fortunate as to witness the spectacle of the birth, the creation, of a new style of thinking. Let us give free rain to the creators, the experts! [...] Education for life will be different; life and art will be given a different form. A new, up to date reality will be created. What is the use of akward metaphysics, if tomorrow's physics will transcend all phantasy? Let us leave free hand to the experts and reserve room in our thinking for the future" (Cohen, Schnelle 1986: 57); "Wir haben gegenwärtig das Glück, das Schauspiel der Geburt, der Erschaffung eines neuen Gedankenstils zu erleben. Man lasse dem Schaffenden, den Fachleuten, freie Bahn! [...] Man wird anders zum Leben erziehen, Leben und Kunst anders gestalten. Man wird eine neue, zeitgemäße Wirklichkeit schaffen. Wozu plumpe Metaphysik, wenn die Physik von morgen jede Phantasie überflügeln wird?" (Fleck 1983: 57).

"Truth is not 'relative' and certainly not 'subjective' in the popular sense of the word. It is always, or almost always, completely determined within a thought style. One can never say that the same thought is true for A and false for B. If A and B belong to the same thought collective, the thought will be either true or false or both. But if they belong to different thought collectives, it will just *not* be *the same* thought! It must either be unclear to, or be understood differently by, one of them. Truth is not a convention, *but rather (1) in historical perspective, an event in the history of thought, (2) in its contemporary context, stylized thought constraint.*" ("im historischen Längsschnitt denkgeschichtliches Ereignis, im momentanen Zusammenhange: stilgemäßer Denkzwang" (Fleck 1979: 100; Fleck 1980: 131). It is now admittedly not at all possible to see why this basic structural insight of Fleck shouldn't also hold for his proto-ideas, whose timelessness, along with the lines of thinking constructed with their help, would also be contextually dissolved.

For the distinction between science and art is also with Fleck only a quantitative one: "The artist translates his experience into certain conventional material by certain methods. His individual freedom is, in fact, limited; by exceeding these limits, the work of art becomes non-existent. The scientist also translates his experience, but his methods and materials are closer to specific (scientific) tradition. The signs (i.e., concepts, words, sentences) and the ways in which he uses these signs, are more strictly defined and are more subject to the influence of the collective; they are of more social and traditional character than those used by

the attempt to establish a non-subjective science. In this light, perhaps it doesn't appear entirely so odd that Bilikiewicz attacked Fleck, much to Fleck's surprise, for his "transcendental idealism".[28]

The Final Step: The Ideal of a Mathematics-like Fundamental Science

The last stage of elaboration of Fleck's epistemology appears in the late essay *Crisis in Science* (1960). It was already being said that Fleck, in his earlier work, had classified science as democratic. Amazingly in any case, Fleck saw the democratic state of contemporary science in that the criterion of truth lies in generality, in great quantity, from which a general provability of scientific research proceeds (Cf.: 105; Fleck 1983: 117). This characteristic was in this point a result of Fleck's (purposefully) optimistic meta-science, because these questions of power disappeared to a great extent and were reduced to simple questions of understanding between various thought collectives. The problem with this determination is that it basically conflicts with his theory of thought styles, for the more plural the thought styles are conceived, the more difficult the understanding becomes and the more difficult the control of the ever more esoterically and incomprehensibly appearing science. One should perhaps also evaluate this lack of diligence on Fleck's part against the backdrop of the rising totalitarianism of the time.

In his later essay *Crisis in Science*, Fleck rectifies his earlier insight that natural sciences are democratic, by stating in the introduction that the influence of politics and financiers

---

the artist [...] The collective of science is much more dense than the collective of art" (Löwy 1990: 271f.). Why then does Fleck still adamantly defend the boundary between science and art, between scientific and nonscientific statements?

Also Nietzsche had already spoken "Against positivism, which halts at phenomena – 'There are only facts'" – and countered: "I would say: No, facts are precisely what there are not, only interpretations" ("nein, gerade Thatsachen giebt es nicht, nur Interpretationen" 7 [60], Nietzsche 1988: 12, 315). Science therefore, just as for example art, has initially a global *meaning* (compare *Jenseits von Gut und Böse* No. 14 and No. 22, Nietzsche 1988: 28, 37) and has as such its right. The scientific interpretation of the world tends however to misjudge its status of interpretation and to declare itself absolute, holds Nietzsche further. On the contrary, according to Nietzsche, "Above all, one should not wish to divest existence of its *rich ambiguity*" ("Man soll es [das Dasein] vor Allem nicht seines *vieldeutigen* Charakters entkleiden wollen" (*Die Fröhliche Wissenschaft* No. 373, Nietzsche 1988: 3, 625), therefore – and now follows the place where one still hears Fleck's echo – we must "become *physicists* in order to be able to be *creators* in this sense [...] Therefore: long live physics!" ("Wir müssen *Physiker* sein, um in jenem Sinne *Schöpfer* sein zu können [...] Und darum: Hoch die Physik!" (Nietzsche 1988: 3, 563f.). The comparison with Nietzsche is clarifying, insofar as he can demonstrate which theoretical escape routes were also at that time historically possible. Fleck, with his concept of a creationist science, approached Nietzsche's philosophy without, however, sharing Nietzsche's epistemologically skeptical premises and relativistic conclusions. The differences are in the end too great, that one could under the omen of philosophizing with possibility – designs place Fleck in a line from Nietzsche to Musil. Nietzsche brings the different, although completely aesthetically designed systems of knowledge into antagonistic opposition and views their relations to one another as relations of power. And this, before a background of a theory of culture which states to what extent the general erosion of values in times of decadence leads also in science to a loss of traditional certainty, and how self-reflection, perspectivization and relativization of the beliefs in the objectivity of sciences are undermined by this itself. From this, Fleck stands quite apart.

28 Compare footnote 18.

leads to opportunism in research. His diagnosis of causes as well as his therapeutic advice are disarmingly simple: The gap, which he posited earlier, between holy ideals of objectivity and truth on the one hand and the concrete practice of research on the other, is identified as a cause of misery. Fleck hoped however that the demystification of those ideals as simply temporary guidelines would restore freedom of research, and that this would simply be an effect of his theory. He therefore claimed that the "sociology of thinking should be developed as a fundamental science equal in its value to mathematics" (Fleck 1979: 156; Fleck 1980: 179). This later article, though, lacks much of what he successfully and impressively had managed in his earlier writings. His considerations here remain general, and simply list previous insights. One cannot simply conclude that the refusal of many professional journals to print the article came from a general resistance to the thinking of Fleck in the 1960s.

Fleck's understanding of culture and society remains rather vague and general, and concrete political constellations, ideologies, and personal non-scientific interests are mostly absent in his analyses. Differences between thought styles are stated but struggles between them are analyzed without recourse to material and societal interests and questions of power.

His dream of a "democratic" science is thus quite different from a conception of science, such as that advanced by Michel Foucault, who analyzes discourses of thought collectives always as dispositives of power.

5. A Look Forward

Can one free Fleck from the relativistic-objective trap that he set for himself? Or must we take leave of Fleck completely, as *spiritus rector* of colorfully relativistic postmodern conceptions of science, and leave him to the jaws of orthodox scientific theory? Not at all. I have already mentioned that one can sometimes understand Fleck's change to a meta-level as a normative decision, and what appears to be normatively offered need not be an epistemologically inevitable consequence. Furthermore, one could also emphasize that with Fleck, the incommensurability of thought styles isn't based on different ways of *thinking*, but on different thought *styles*. One could attempt a riveting description of other thought styles without having to translate them. This aspect could be played off against modern contextualistic scientific theories.

Epistemic contextualists like Michael Williams, Wilfried Sellars, and Bob Brandom[29] deal with the establishment of a *space of giving and taking of reasons*, that is, forming the contextual framework for claims of knowledge. This is where the references to basic conditions of rationality of understanding are obvious, for argumentative methods of proof remain decisive. Also with Fleck's successors, who speak of *styles of reasoning* (Hacking 1992), the orientation on rational procedures is unmistakable. Fleck's case is different since, according to his epistemological premises, he needn't involve himself in this game of proof, and, as we have seen, his transition to a meta-level isn't epistemologically compelled, but on the contrary conflicts with central assumptions of his scientific theory. His concept of thought style, which admittedly suggests that primarily thinking would be addressed, should

29 For example: (Sellars 1997); (Brandom 1994); (DeRose 1992); (Williams 1991).

rather be accentuated on the second part – the style – since this is by no means a superfluous addition, but rather a factor of its formation. One cannot overlook that Fleck's concept of style is neither simply a logic category nor primarily a social one, as Porus or Harwood would emphasize,[30] but one of aesthetics, which owes thanks to the carefully transfer of art-historical style categories (such as those developed by Wölfflin or Riegl) into a history of science.[31] Non-propositional components such as feeling, mood and perception of "Gestalt" have, with Fleck, a much greater share in it. But is necessary to make here a further distinction: Bilikiewicz as well refers even explicitly to Wölfflin's concept of style. But in contrast to Bilikiewicz, who uses the concept of style for constructing simultaneous existing general analogies between science and culture, Fleck investigates the distinct historical development of thought styles. 'Intuition' by Fleck doesn't sustain as by Bilikiewicz the seeing of analgies but rather the distinct "Einfühlung" into strange thought styles. But through this strong share the respective thought styles actually become incommensurable and the possibilities of understanding are either highly limited or entirely dispensed with. It seems to be difficult for some Fleck scholars to see that point.

It is worth noting in this context for instance Jonathan Harwood's critique of Fleck's concept of constraint of style. For him, this can only be understood in the way that a social collective, with the aid of a dictated style, executes constraint which style itself could not enforce (Harwood 1986: 183). Fleck has a different opinion. He states, "an artistic painting also exhibits its own constraining style [...] every product of intellectual creations has relations 'which cannot exist in any other way.' They correspond to the compulsory, passive links in scientific principles. These relations can be, as it were, objectified and regarded as expressions of 'beauty' and 'truth'" (Fleck 1979: 101; Fleck 1980: 132).

Such non-rational factors and aesthetic elements in the formation of knowledge can be brought to light by Fleck's conception, whereas most of the current contextual theories of knowledge lack this ability. In light of Fleck's insights, the narrowness of these contextual theories of knowledge becomes transparent.

If we, however, think with Fleck, against Fleck, and beyond Fleck, we attain a differently geared contextual epistemology: Then, Fleck's historical-sociological analyses act against the intentions of their creator, as an antidote to widespread universalism, anti-historicism, and rational reductionism in the theory of science.

30 (Porus 1983: 3-4, 545-560). (Harwood 1986: 16, 173-187).
31 Compare (Wessely 1991).

Mary Jo Nye

Science and Politics in the Philosophy of Science: Popper, Kuhn, and Polanyi

The names of Karl Popper (1902-1994), Thomas Kuhn (1922-1996), and Michael Polanyi (1891-1976) are well-known among scientists and historians and philosophers of science. During the 1960s, in particular, their work became the focus of considerable discussion because of their independent attacks on empiricism, inductivism, and positivism. They each rejected the philosophical tradition that uses simple empiricism to differentiate the sciences from the arts, from religion and metaphysics, and from ideology. They each published a *magnum opus* around 1960: Polanyi's *Personal Knowledge* appeared in 1958; the English translation of Popper's *Logic of Scientific Discovery* came in 1959; and Kuhn's *Structure of Scientific Revolutions* debuted in 1962.

Kuhn's *Structure of Scientific Revolutions* has sold over a million copies in two dozen languages (Heilbron 1998: 505). This is a resounding success for a scholarly book with footnotes, and it marks the appeal of Kuhn's interpretation of the nature of science for a very broad public audience. Like Kuhn's audience, Popper's was a large one, and many scientists regard Popper as a giant among philosophers. Peter Godfrey-Smith recounts the instance of attending a lecture by a Nobel prize winner in medicine and hearing mostly about Popper, rather than about virology (Godfrey-Smith 2003: 57). The virologist might have been Sir Peter Medawar, who often is quoted for saying: "I think Popper is incomparably the greatest philosopher of science that has ever been."[1]

In comparison to Kuhn and Popper, Michael Polanyi likely is the least well-known in the general public. However, Polanyi's name has iconic status among some philosophers and sociologists of science, who cite Polanyi's *Personal Knowledge* for it anti-reductionism, its formulation of the idea of tacit knowledge, and its originality in recognizing and describing science as social practice, rather than as formal method.[2] Polanyi is honored by scientists, too, because he was a scientist with a long career of distinguished scientific contributions at the forefront of physical chemistry. The Polanyi Medal is awarded every two years by the Royal Society of Chemistry in the field of Gas Kinetics. Michael Polanyi's son John Polanyi was Nobel Prize laureate in Chemistry in 1986.

I have written elsewhere and in some detail on Polanyi's scientific work and career in Budapest, Berlin, and Manchester from the 1910s to the late1930s (Nye 2000; Nye 2001; Nye 2002b; Nye 2007). Here I examine how politics shaped the agenda of Polanyi's writings about science, as it did the work of Popper and Kuhn. Popper, Polanyi, and Kuhn all argued or accepted the belief that science can flourish only in Western-style democracies, or

1 E.g.: http://www.takingchildrenseriously.com/the_education_of_karl_popper (downloaded 21-XII-06).
2 (Golinski 1998: 17): "Kuhn adopted the phrase 'tacit knowledge' from Michael Polanyi to characterize a large part of what the scientist learns." For references to Polanyi in the literature of science studies, see (Thorpe 2001: 19-20). An example is (Collins 1974: 165-186).

what Polanyi called the republic of science. This proposition has come under increasing criticism among historians of science in the last decade, as exemplified, for example, in Loren Graham's book *What Have We Learned about Science and Technology from the Russian Experience?* and in the volume *Politics and Science in Wartime: Comparative International Perspectives on the Kaiser Wilhelm Institute,* edited by Carola Sachse and Mark Walker (Graham 1998; Sachse, Walker 2005: 17-19).

In order to analyse the political agendas of their philosophical writings, I sketch the family backgrounds, formal education, and early intellectual and political concerns of Popper and Kuhn, turning, then to Polanyi, and concluding with some final comparisons of the political dimension in the philosophical work of these three twentieth-century figures.[3]

Karl Popper vs. Positivism and Historical Determinism

Karl Popper, Thomas Kuhn, and Michael Polanyi were men of different generations. Polanyi was the oldest of the three, born in 1891 in Budapest. Popper was born in Vienna in 1902. Kuhn, twenty years younger than Popper, spent his early years in New York City and Croton-on-Hudson before going to private schools in Pennsylvania and Connecticut (Andresen 1999: 43-67). All three were from originally Jewish families of some financial means and intellectual ambitions. None of the families were religiously observant. Popper's parents converted to Lutheranism in 1900 (Brudny 2002: 97).

When Michael Polanyi decided to leave Hungary for Germany in 1919, he received a baptismal certificate from the Roman Catholic Church, but he did not participate afterwards in Catholic sacraments.[4] Both Popper and Polanyi married Catholic women, Anna Josefine Henniger and Magda Kemeny (Brudny 2002: 62-63; Scott, Moleski 2005: 74).

Karl Popper was the youngest of three children and the only boy. His father headed a law firm located adjacent to the large family apartment which contained a library of some 10,000 books. Simon Popper was a scholar, social reformer, Freemason, and radical liberal. Friends of the family included Sigmund Freud's sister Rosa Graf, Joseph Breuer, and, among Karl's friends, Konrad Lorenz and Rudolf Serkin.[5] Young Karl left home at sixteen, toward the end of the Great War, and tried out some classes at the University of Vienna (Popper 1976: 19).

---

3 All three have been the subject recently of long biographical studies, including (Hacohen 2000) and (Brudny 2002); on Popper (Fuller 2000) and on Popper and Kuhn (Fuller 2004); (Scott, Moleski 2005) on Polanyi. Also on Popper (Magee 1973); (Popper 1976) and (Watkins 1997). On Kuhn (Andersen 2001). On Polanyi (Wigner, Hodgkin 1977).

Kuhn has been the least fortunate in biographers, given the viciousness of Fuller's attack against his biographical subject in (Fuller 2000), a diatribe continued in (Fuller 2004). Fuller's wrath largely has to do with what he sees as incorrect politics in Kuhn's philosophy of science, in comparison to Popper and Polanyi, while also attributing to Polanyi (in a description of Polanyi and his milieu that often is erroneous) the origins of many of Kuhn's insights; e.g., see (Fuller 2000: xiv-xvi, 5-7, 139-140), and (Fuller 2004: 121). On Kuhn's philosophy of science (Hoyningen-Huene 1993); (Bird 2001); (Nickles 2002); and (Marcum 2005).

4 (Scott, Moleski 2005: 55). Polanyi received a formal certificate of baptism from Vienna, where the baptism was registered. Regarding the certificate, letter to Michael Polanyi (Misi) from Károly Pollacsek (Uncle Karl), Vienna, 7 January 1920, MPP: Box 1, Folder 6.

5 (Popper 1976: 9); (Brudny 2002: 52); (Watkins 1997: section on Vienna).

The end of the First World War brought events that appear pivotal for Popper's mature orientation. Sympathetic with the goals of socialism, Popper briefly turned to the Communists in 1919 until he witnessed the police shooting of unarmed young Socialists during a demonstration orchestrated by the Communists. Popper later said that he was horrified by Marxist demands that this kind of conflict be intensified in order to bring about the historical destiny of the rule of the proletariat, and he abandoned Marxism and the Communists for the Social Democrats (Popper 1976: 33-35).

In 1919 he was in the audience at a lecture by Albert Einstein. Popper later claimed that Einstein's lecture demonstrated to him, on the spot, the secret of the scientific attitude. Einstein, he recalled, recognized that experiments were capable of showing that the relativity theory was untenable, but they could not conclusively establish it as true, and this was the real scientific attitude (Popper 1976: 38). Popper took it upon himself to learn about standards of truth through studying mathematics and physics, pursuing university lectures during the 1920s (Brudny 2002: 59). On the practical side of things, Popper apprenticed himself to a master cabinet maker from 1922 to 1924, then took up social work for a year in Alfred Adler's child guidance clinics, and passed qualifying exams for teaching mathematics, physics, and chemistry in secondary school (Popper 1976: 7-8; 39-41).

After Popper began taking classes in 1925 at Vienna's new Pedagogic Institute, his interests shifted from the psychology of discovery to the logic of discovery.[6] He began reading works by Ludwig Wittgenstein, Rudolf Carnap, Hans Reichenbach, and other contemporary philosophers, some of whom he met at Moritz Schlick's Thursday evening private seminar which became known as the Vienna Circle.[7] Popper rejected the positivistic thrust of their discussions and the demarcation of science from metaphysics on the basis of verifiability and induction. In contrast, in the tradition of David Hume, Popper regarded objections to inductivism as insuperable. No matter how much evidence might be marshaled to link cause and effect, the statement of a law, rather than of a coincidence, is a leap of faith (Popper 2002: 6, 7, 10, 13). By spring of 1932, Popper settled on falsifiability as a criterion for a true scientific law (Hacohen 2000: 198).

Unable to find a publisher for his long manuscript *Die beiden Grundprobleme der Erkenntnistheorie (The Two Fundamental Problems of a Theory of Knowledge),* he wrote a shorter version called *Logik der Forschung (The Logic of Investigation*), which Schlick and Philipp Frank published in 1935 in their new positivist book series. Members of the Vienna Circle called Popper their "official opposition," but ironically the *Logik* became identified with logical positivism until its re-publication in English translation in 1959.[8]

About the time *Logik* first appeared in German, Popper completed the main outline of another book, *The Poverty of Historicism.*[9] He was hoping for a job in England, but it was Canterbury University College at Christchurch in New Zealand that offered him a refuge in spring 1936, a year before Germany's occupation of Austria. He published the *Poverty of Historicism*, which appeared in three parts in the London School of Economics journal *Economica* during 1944-1945. *The Open Society and Its Enemies*, written in New Zealand, was

6 (Watkins 1997: section on Vienna); (Popper 1976: 72-78).
7 See the manifesto in *Verein Ernst Mach in Wien* 1929.
8 (Watkins 1997: section on Vienna); and (Popper 1976: 87).
9 (Popper 2004: xi); and (Watkins 1997: section on Vienna).

the other arm of what Popper called his "war effort."[10] It appeared in 1945, with detailed critiques of Plato, Hegel, and Marx that were adumbrated in the earlier *Poverty* manuscript. It was these political books that made Popper's critique of scientific positivism and inductivism first widely-known.

Popper's main target in these political writings was the doctrine of historical determinism, which he characterized as sheer superstition. While he thought it was legitimate to interpret history in metaphors or themes of the struggle of classes, races, or ideas, he rejected the notion that there can be theories of historical evolution based in observations of historical facts and the discovery of rhythms or patterns or laws (Popper 2004: 140, 143). In order to argue that history is not a science, Popper had to describe what science is and what scientists do.[11]

Science is not a method of beginning with observations, wrote Popper, and historical laws cannot be based in induction of so-called patterns from facts. Descartes was partly right in describing the deductive and systematic nature of scientific knowledge, but Descartes was wrong in taking its foundations to be secure and self-evident initial premises. Rather, the starting point of scientific thinking is conjecture or hypothesis that must be refutable in principle (Popper 2004: 90, 120; n. 3, 120-121; n. 45, 83). Falsifiability is both a criterion of scientific method and a criterion of demarcation from pseudo-science.

> "We make progress if, and only if, we are prepared to *learn from our mistakes* [his italics]: to recognize our errors and to utilize them critically instead of persevering in them dogmatically. Though this analysis may sound trivial, it describes, I believe, the method of all empirical sciences" (Popper 2004: 80).

Progress in science depends on free competition of thought and thus on political freedom.

While Popper directed the *Poverty of Historicism* against totalitarianism and the claims of historical destiny by fascists, Nazis, and communists, his critique of Soviet communism was considerably more muted than was his critique of Plato's epistemology and politics. Popper withheld judgment on Soviet Russia, as did many liberals and socialists, and he praised Marxist intellectuals for wanting to solve urgent social problems.[12] Popper wrote in his autobiography that he demurred from publishing anything anti-Marxist until 1935 because anti-Marxism was so closely associated in the 1920s and 1930s with fascism (Popper 1976: 36).

In 1946 Popper came to England, just after the *Open Society* was published, to teach at the London School of Economics, where he became Professor of Logic and Scientific Method in 1949. By 1951 his LSE seminar was well-known for its court of admirers, but also for Popper's authoritarian and egocentric demeanor, which many people found insufferable. Michael Polanyi presented a paper in Popper's seminar in March of 1952 on "The Stability of Beliefs," in which Polanyi compared belief-systems of the African Azande people to modern belief systems, including Marxism and modern science, and he emphasized the difficulty of breaking out of any belief system in which one is educated. According to John

---

10 (Popper 2004: ix); (Popper 1976: 115) and (Hacohen 2000: 383).

11 On the question of historical laws, see (Agassi 1963); (Donagan 1964); (Dray 1964); (Danto 1965); (White 1973); (Dray 1973); (Gardner 1974).

12 (Hacohen 2000: 396-397); (Brudny 2002: 129-130).

Watkins, Polanyi emerged from the seminar gravely offended at the way Popper had chaired the session and had treated him (Watkins 1977: section on Early Days at LSE).

In 1950 Popper gave the William James Lectures at Harvard University, where he met George Sarton, Bernard Cohen, James Bryant Conant, and the young Thomas Kuhn (Popper 1976: 128). Kuhn and Popper met again at Kuhn's home in Berkeley in 1962 and at a 1965 conference at Bedford College in London (Heilbron 1998: 510). Popper did not much care for Kuhn's views, saying, in particular, of Kuhn's notion of normal science that it was a recent phenomenon and "a danger to science" (Hacohen 2000: 532). Let me turn now to Kuhn and to some political foundations of Kuhn's philosophy of science.

Kuhn and the Goals of General Education in a Democracy

Thomas Kuhn's father, a consulting engineer, had studied at Harvard University, and his son entered Harvard in the fall of 1940 with the aim of majoring in physics. He graduated in three years and immediately went to work for the war effort in radar, joining John Van Vleck in the spring of 1943 at the Radio Research Laboratory on the Harvard campus. During 1944 and 1945 Kuhn was assigned to Great Malvern in Worcester and then to the 9th Air Force in Rheims. He was in Paris on 25 August 1944, the Day of Liberation.[13]

In September 1945 Kuhn entered the physics graduate program at Harvard, working with Van Vleck in solid-state theoretical physics, completing his thesis in 1949 on electronic binding energies of metallic solids (Kuhn 1949). In the meantime, he took philosophy courses. At a cocktail party, a young woman asked him what he did, and, as his later student John Heilbron relates, the party chanced to fall silent so that everyone heard Kuhn's answer. "I just want to know what truth is," was his reply (Heilbron 1998: 506). Kuhn was not sure, however, that physics was his route to the truth. He was finding group theory and solid-state physics dull in comparison with the work of contemporaries who had been at Los Alamos. By his own account, he was not convinced that he could shine in his current field (Baltas et al. 2000: 272-273).

As an undergraduate writer and editor for the *Harvard Crimson*, Kuhn had voiced strong support in the student newspaper for the policies of Harvard's President, the organic chemist and wartime science defense administrator James Bryant Conant. In particular, Kuhn applauded Conant's calls for using American higher education as a tool in total war, and he found appealing Conant's argument for the importance in a democracy of public understanding of science.[14] When Conant invited Kuhn to assist him in developing a new general education course on science using historical examples, Kuhn saw this as an opportunity to do something interesting and important while finishing his physics dissertation (Hufbauer 1998: 9). In the summer of 1947, as Kuhn studied texts for a unit on the history of mechanics from Aristotle to Galileo, he experienced the joy of discovery. "To my complete surprise, that exposure to out-of-date scientific theory and practice radically undermined some of my basic conceptions about the nature of science and the reasons for its special success" (Kuhn 1962: V.). For Kuhn, who also was reading psychology, it was a Gestalt switch in his

13 (Heilbron 1998: 506); and (Hufbauer 1998: 7).
14 (Hufbauer 1998: 6; also 10-11); (Harvard University 1945); (Conant 1947); on Conant (Hershberg 1993).

conceptual framework (Baltas et al. 2000: 293). Kuhn decided to follow the history of science as a profession.

In laying the groundwork for history of science in the new General Education curriculum at Harvard, Conant was using some of Michael Polanyi's work. Kuhn later recalled,

> "I liked it quite a lot – I don't remember just what it was, except that I kept feeling terrible at those points where [Polanyi] [...] spoke as though extrasensory perception was the source of what the scientists did" (Baltas et al. 2000: 296).

Polanyi's little book *Science, Faith, and Society* had been published in 1946, and it is not surprising that Conant would have noticed it, since Polanyi was a chemist, like Conant. Polanyi also was writing essays in the 1940s for the newly organized Society for Freedom in Science, and his pamphlets on the "Rights and Duties of Science," "The Planning of Science," and "The Foundations of Academic Freedom" can be found in Harvard's Widener Library. Kuhn may have read the pamphlets; but in any case, he recalls in a letter to William Poteat having read the 1946 book *Science, Faith, and Society* and Polanyi's 1951 collection of essays *The Logic of Liberty*.[15]

In the spring of 1951, Kuhn gave a series of eight lectures at Boston's Lowell Institute. In his opening lecture, he announced that his lectures would show that the empiricist view of the "ideal scientist [as one who] abandons all prejudice [...] is altogether wrong." As outlined in an essay by Karl Hufbauer, Kuhn argued that firm preconceptions are woven into patterns of scientific research, and that this is a good thing. He denied the possibility of dispassionate neutrality in the working scientist, but not the objective validity of scientific results. He used the words "anomaly" and "paradigm" in these lectures, over a decade before publishing the *Structure* book.[16]

While on a five-year faculty appointment at Harvard teaching history of science, Kuhn applied for a Guggenheim Fellowship with a proposal for two books, one on "The Copernican Revolution" and another on "The Structure of Scientific Revolutions." The Copernican Revolution proposal grew out of a segment in the course he taught with the chemist Leonard Nash and from Kuhn's preoccupation with the problem of scientific revolutions as shifts in intellectual systems, particularly as described by Alexandre Koyré. Kuhn intended to argue in the *Structure* book that scientific theories serve the function of ideologies that both direct future work and constrain the creative imagination.[17]

By the time the Copernican Revolution book appeared in 1957, Kuhn was teaching at Berkeley (Kuhn 1957). He spent the next academic year at the Center for the History of the Behavioral Sciences in nearby Palo Alto, on recommendation from Robert Merton and Edward Shils.[18] In November 1958 Polanyi visited Palo Alto, and Kuhn heard him give a talk on tacit knowledge. Of that talk, Kuhn later reminisced "I liked the lecture all right, and it's possible that it helped me to get to the idea of paradigm, although I'm not sure" (Baltas et al.

---

15 (Polanyi 1975); (Polanyi 1964); (Polanyi 1951); (Polanyi 1945); (Polanyi 1946); (Polanyi 1947). See Letter from Thomas S. Kuhn to William H. Poteat, 28 February 1967, cited in (Scott, Moleski 2005: 246).
16 (Hufbauer 1998: 14-16). Also see (Andresen 1999: 62-63).
17 (Hufbauer 1998: 20-21, n. 86 on 64). Hufbauer includes the text of Kuhn's Guggenheim Fellowship application on 42-51.
18 (Merton 1977: 95-100, 122), cited in (Hufbauer 1998: n. 88 on 64).

2000: 296). Polanyi's new book *Personal Knowledge* had just appeared. Kuhn looked at it, he later said, but decided at the time that "I *must* [his italics] not read this book now. I would have to go back to first principles and start over again, and I wasn't going to do that (Baltas et al. 2000: 296). Later, when Kuhn did try to read Polanyi's *Personal Knowledge*, he did not like it, although he cited the book and Polanyi's notion of tacit knowledge in a footnote in *The Structure of Scientific Revolutions*.[19] Kuhn's *Structure* book appeared in a series edited by former members of the Vienna Circle Rudolf Carnap and Philipp Frank.

The notion of scientific "dogma" was the theme of a paper that Kuhn presented at a symposium on "Scientific Change" at the University of Oxford in July of 1961, shortly before the *Structure* book was to appear. This symposium was organized as a major event by the Division of the History of Science in the International Union for the History and Philosophy of Science, and Kuhn and Polanyi met once again on this occasion. Polanyi, who had been a Senior Research Fellow at Merton College in Oxford since 1959, had been invited to comment on Kuhn's paper, along with A. Rupert Hall of the University of Cambridge. Kuhn summarized his ideas on paradigms, normal science, anomalies, crisis, and scientific revolutions. As a commentator, Polanyi followed up by saying that "The paper by Mr. Thomas Kuhn may arouse opposition from various quarters, but not from me." Polanyi had been trying for years to call attention to scientists' deep commitment to established beliefs.[20] Let us turn now to Polanyi.

Michael Polanyi and the Problem of Soviet Science

Michael Polanyi was born in Budapest, the youngest of five children and the son of a civil engineer turned railroad entrepreneur who died in 1905. His widowed mother Cecile was at the center of a group of intelligentsia that included poets, painters, and scholars, most of them sympathetic with bolshevism and socialism (Scott 1983: 11; Szapor 1997: 43-54). His elder brother Karl would acquire an international reputation in the history of economics by publishing *The Great Transformation* in 1944, in which he argued for the inevitably socially divisive nature of the capitalist market economy (Wigner, Hodgkin 1976: 413).

Polanyi completed a medical degree in 1913 and a Ph.D. in physical chemistry in 1917 at the University of Budapest, having studied physical chemistry in Karlsruhe with Georg Bredig and Kasimir Fajans. After working briefly with Georg de Hevesy in Budapest, Polanyi returned to Karlsruhe in 1919, and in 1920 he moved to Berlin to a position at the Kaiser Wilhelm Institute for Fiber Chemistry. In 1923 he became director of the chemical-kinetics research group in Fritz Haber's Institute for Physical Chemistry and Electrochemistry, where the Fiber Institute was housed.

From the 1910s through the 1930s Polanyi pursued several physical chemistry research programs, often simultaneously. One focus was the adsorption of gases on solid surfaces, which was the subject of his Ph.D. thesis within the classical framework of thermodynamics. In 1930 he and Fritz London offered a new and generalized explanation for adsorption

19 (Baltas et al. 2000: 297); and (Kuhn 1962: n. 1 on 44), referring to chapters 5 and 6, in particular, in Polanyi's *Personal Knowledge*.
20 (Polanyi 1963: 375). Also (Kuhn 1963: 347-369); (Hall 1963: 370-375); and discussion in (Crombie 1963: 380-395).

forces using quantum mechanics.[21] Another focus of Polanyi's work was x-ray diffraction and solid-state studies of natural fibers and metals (Nye 2000; Nye 2001). His best-known scientific work is in the field of chemical kinetics and dynamics, and he developed, with the American physical chemist Henry Eyring, the theoretical approach that came to be called the "semi-empirical method" for predicting activation energies for chemical reactions. Their derivation of the first potential energy surface for chemical reactions is considered one of the foundations of the modern field of chemical dynamics.[22]

Following Adolf Hitler's appointment as German Chancellor, the civil service laws of April 1933 forced Polanyi and most of his fellow Jewish scientists to leave Germany. Polanyi took up a position as Professor and head of the Laboratory of Physical Chemistry at the University of Manchester, where he gradually spent less and less time in scientific research. In 1948 he exchanged his chemistry professorship for a chair of social studies created explicitly for him. In 1959 he retired and moved from Manchester to Oxford, having published five books since 1940 that were economic, political, and philosophical in nature: *The Contempt of Freedom* (1940), *Science, Faith, and Society* (1946), *Full Employment and Free Trade* (1948), *The Logic of Liberty* (1951), and *Personal Knowledge* (1958).[23]

Polanyi's very first political publication appeared during the First World War in 1917 as a Hungarian pamphlet addressed "To the Peacemakers." It was written after Polanyi had been retired from active duty as a medical officer in the Austro-Hungarian Army due to ill health. The essay appeared in the progressive journal *Huszakik Század (Twentieth Century),* edited by his friend Oscar Jászi, who had been head of the progressive and westernizing Radical Party in Hungary since 1914 (Scott, Moleski 2005: 46-47). In a call for peace, Polanyi warned against the "materialist prejudice" that convinces people that war is in their interest, and he criticized the "tacit presuppositions" and nationalisms that subvert the goal of a peaceful Europe united by internal cooperation and by a system of law and order, including the establishment of a joint army (Polanyi 1997c: 16, 22, 27).

When the Hungarian Republic was formed in November 1918, Michael's brother Karl was general secretary of Jászi's Radical Party, and Michael became a secretary in the Ministry of Health (Scott, Moleski 2005: 48-49). In early 1919, before a Communist coup took place in March, Michael Polanyi wrote another article on political issues, this time in the journal *Szabadgondolat (Free Thought),* edited by Karl and published by the progressive Galileo Circle. Michael and Karl both were active in the Galileo Circle, a Hungarian progressive group advocating science as a tool for social reform. They also participated in meetings of a group called the Sunday Circle, which brought them together with Karl Mannheim, George Lukács, and other intellectuals in the home of poet Béla Bálazs.[24] In his 1919 article "New Scepticism", Polanyi warned scientists and artists not to fall subserviently before the demands of politicians, but to fulfill their mission of exploring "truth" and to analyze political notions such as "class interests" and "ideologies" as problems, not as solutions, for the current political troubles (Polanyi 1997a: 53-56).

21 For details (Wigner, Hodgkin 1976: 421-448); (Scott, Moleski 2000); Nye in (Hook 2002); and Nye in (Nye 2002a).

22 See the symposium "Fifty Years of Chemical Dynamics" 1981.

23 (Polanyi 1940); (Polanyi 1946); (Polanyi 1984); (Polanyi 1951) and (Polanyi 1958).

24 (Scott, Moleski 2005: 22-23, 41). See (Cash 1977: 2, 8). Also (Congdon 1991); and (Palló 1998).

After Polanyi left Hungary for Germany in late 1919, Polanyi devoted himself fairly singlemindedly during the early 1920s to his scientific work and career in Berlin. In 1928, however, with Leó Szilárd, and John von Neumann, Polanyi began following a seminar in economics led by Jacob Marschak.[25] Marschak had emigrated in 1919 from Russia to Berlin, where he became a leading socialist Marxist theorist and writer on economic policy.

In 1928 the state of affairs in Russia was of more than marginal interest to Polanyi, whose mother was originally Lithuanian. There was much talk of new scientific and technical jobs that were opening up in the Soviet Union, and his niece Eva and her husband Alex Weissberg would move to Kharkov in 1931, where Weissberg would work at the Physical Technical Institute.[26] Polanyi made his first visit to the Soviet Union in 1928 at the invitation of the physical chemist Abram Joffé.[27] While there, Polanyi collected information about research facilities and about the general situation of prices, wages, and living conditions, writing his sister Laura ("Mausi") of his pessimistic impressions of the economic system and commenting in a notebook on the "brutal and stupid fanaticism" of the Soviet state.[28]

Upon his return to Berlin, with his second child (John) due in January, Polanyi ruminated on his own successes and failures in the Berlin scientific world during the past eight years. He was busy and prolific in 1928, publishing eighteen singly or co-authored chemistry papers in that year alone. He had many worries, however, among them the financial support for his laboratory and its workers. He worried, too, about his own prospects for truly making a mark in science, ambitiously thinking of the examples of his elder Berlin colleagues Albert Einstein and Max Planck. Could he yet make a great discovery or inaugurate a new theory? As he confided to his diary, could he even keep up mathematically and conceptually in the new quantum mechanics with younger colleagues such as Eugene Wigner (1929 Diary, 9 January (MPP: 44: 4))?

In December, Polanyi participated in the ceremonies for the 60th birthday of Fritz Haber. Throughout the history of science, Polanyi told his audience, there have been innovators and traditionalists, revolutionaries and conservatives. Einstein, Planck, and Rutherford are among the revolutionaries and the great heretics of science. Polanyi praised Haber as one of the upholders of scientific tradition whose science is not just one of knowledge and enlightenment, but of decisive action. As a great scientific leader, Haber knows that science must control its own life as a family of scholars, knowing how to *rule itself* [my italics] in relation to the state and the economy (Polany 1928: 1316-1317). Note Polanyi's themes here: the tension between tradition and change, between dogma and heresy; the scientific life as a life not only of intellect but also of conduct; and the constitution of science as a family, which

25 Interview of Michael Polanyi with Thomas S. Kuhn, 15 February 1962: 10-11. SHQP.

26 See Livia Polányi, "Cornucopias of History: A Memoir of Science and the Politics of Private Lives" (reprint, courtesy of Pnina Abir-Am, no page numbers), whose grandfather was Adolf Polanyi. She writes that she is one of fewer than ten living individuals born with the name Polanyi and that none live in Hungary.

27 Letter from Fritz Haber to Generalverwaltung, KWG, 21 November 1928. MPG: Abt. I Rep. 000 1A #1165. Also see letter from Haber to Stefan Bogdandy, 16 November 1928. MPG: Abt. I. Rep. 000 1A #1165. Also (Scott, Moleski 2005: 108-109); (Palló 1996: 39-43).

28 Quoted in (Scott, Moleski 2005: 109), from Letter from Michael Polanyi to Laura (Mausi) Polanyi, 29 April 1928; and Tatra Notebook August 1928 (MPP 44:3).

Polanyi later would term a "republic," characterized by self-rule within the larger political economy.

Also in December of 1928 Polanyi read the French intellectual Julien Benda's new book *Le Trahison des Clercs (The Treason of the Intellectuals)* (Scott, Moleski 2005: 109). What he found in Benda were some of the same ideas expressed in Polanyi's own youthful political essays of 1917 and 1919. Benda, like Polanyi an assimilated Jew, condemned the racial, class, and national passions sweeping Europe, including Zionism (Benda 1969: 3-4, 11-12). Similarly, too, to Polanyi, Benda lamented the preeminence of what he called the "religion of the material" and of political movements that claimed to be carrying out the spirit of history (Benda 1969: 120-121, 27-28). Benda wrote of the passing of a

> "[...] class of men whom I shall designate 'the clerks,' [*clercs*, in the medieval sense of the word] by which term I mean all those whose activity essentially is *not* [his italics] the pursuit of practical aims, all those who seek their joy in the practice of an art or a science or metaphysical speculation [...] and hence in a certain manner say: 'My kingdom is not of this world'" (Benda 1969: 43).

These sentiments were hardly new and bring to mind the Munich lecture of 1918 on "Science as a Vocation" by Max Weber, in which Weber decried the waning of asceticism and of devotion in German scientific life, challenging scholars and teachers to rededicate themselves to their calling (Weber 1948: 34, 138-139).

Early in 1930, Polanyi organized a group in Berlin, which met for dining and discussion at the Kaiser Wilhelm Institute's new Harnack House. The group's members included London, Szilárd, Von Neumann, Eugene Wigner, and the economists Gustav and Toni Stolper. They debated whether socialism was better for science and technology than capitalism, the merits of the Soviet Five-Year Plan, and the relationship between pure and applied science.[29] Polanyi wrote an article for the May issue of *Der deutsche Volkswirt (The German Economist)*, edited by Gustav Stolper. In this short essay, Polanyi argued for government support of science even when practical benefits might not be immediately obvious (Polanyi 1930: 1149-1151). A visit to Leningrad in 1932 further cemented Polanyi's misgivings about the Soviet Union. The next return to the Soviet Union in 1935, after Polanyi had moved to Manchester, was worse. Polanyi saw his niece Eva and her husband, both of whom were arrested shortly afterwards on charges of plotting against Stalin and the state.[30] On that visit, too, Polanyi talked with Nikolai Bukharin, who was supporting central planning of science under Five-Year Plans to Polanyi's dismay (Polanyi 1966: 3-4). Bukharin was executed in 1938.

Polanyi's non-scientific publications stepped up considerably. In 1937 the American chemist Melvin Calvin, who was visiting in the Manchester Laboratory, found it hard to interest Polanyi in chemical conversations any more (Calvin 1991-1992: 40-42). T. R. Hog-

---

29 Polanyi discussed the meetings in his interview with Thomas S. Kuhn, 15 February 1962: 10-11 (SHQP). Also see letter from Polanyi to Frau Dr. Toni Stolpher, 25 January 1930 (MPP: 2:6) and handwritten note, undated (MPP: 43:2).

30 On this (Szapor 1997: 47-49) and note 14 on 49. Arthur Koestler, who had first met Eva Striker in the Budapest kindergarten run by Polanyi's eldest sister, later said that it was Eva's story that was the inspiration for *Darkness at Noon*. See (Caesarani 1998: 13, 78, 83) and (Weissberg 1952: viii-xii, 13).

ness, visiting Manchester on behalf of the Rockefeller Foundation, noted the change in the man whom Hogness called "the ace of European physical chemists."[31] Polanyi wrote a critical essay attacking Sidney and Beatrice Webb's sympathetic view that the Soviets were undertaking a great social engineering work of uncertain success (Polanyi 1997b: 59). He engaged in debates with his brother Karl from time to time, since, Karl, too, had emigrated to England and was lecturing for the Worker's Educational Association in London and Oxford, before embarking for North America in 1940.[32] Polanyi more and more worried about the possible application to British science of ideas about central planning. He immensely disliked all the writings on science by British Marxists, especially the biologist Lancelot Hogben's book *Science for the Citizen* published in 1938 and the physicist J. D. Bernal's book *The Social Function of Science* published in 1939 (Polanyi 1975: 1-26).

Polanyi's book of essays *Contempt of Freedom* appeared in 1940, earning wider attention for his political and economic views.[33] In August 1941 Polanyi gave his typist Olive Davies a typescript of twenty-five pages titled "Beginning of a Book on the Scientific Life." He rarely could be found now in the laboratory, itself sparsely populated because of the war (Scott, Moleski 2005: 186, 187). He wrote an essay called "The City of Science," a harbinger of the later 1962 article in *Minerva* on "The Republic of Science" (Scott, Moleski 2005: 188-190). Another crisis point came in 1943, when Polanyi read that the internationally respected Russian geneticist Nikolai Vavilov had used, as his last defense in his battle for Mendelian experimental genetics, its practice by Western scientists. The opposition headed by Trofim Lysenko rejected Vavilov's argument, insisting instead on the necessity for biological science to offer immediate practical results in agriculture and claiming the insufficiency of scientific evidence for the existence of the gene. For Polanyi, the trial of Vavilov was a challenge to the legitimacy of the entire Western tradition of science. On what grounds, he asked, do we base our confidence in our western scientific knowledge? How can we defend it (Polanyi 1964: 9)?

Drawing upon his notes and essays, Polanyi pulled together a systematic answer to these questions in his Riddell Memorial Lectures at the University of Durham in March 1946. Oxford University Press published the lectures under the title *Science, Faith and Society*, the volume read by Thomas Kuhn in the late 1940s. Its major ideas reappear in *The Logic of Liberty* and *Personal Knowledge.* Thus, Polanyi's big book in the philosophy of science, like Popper's, appeared after his political writings had already introduced his main themes in philosophy of science to a broad readership.

Science, Faith, and Society begins with the question of "what is the nature of science" and moves to the question of legitimate grounds for demarcating magical from naturalistic interpretations of events. Polanyi found observation too weak a reed upon which to base the

---

31 Fellowship Report of T. R. Hogness (University of Chicago), 16 August 1937. RFA: Folder 110, Box 13, Series 717D, Record Group 1.1/401D.

32 After 1947, Karl Polanyi held a professorship at Columbia University, but he lived in Canada because his wife could not obtain a U.S. visa due to the McCarran Act. See (Bohannan, Dalton 1965: 1508-1511). Karl and Michael had both personal and intellectual differences that caused some tension between them in the 1930s. See (Scott, Moleski 2005: 153-154).

33 (Scott, Moleski 2005: 183). He sent the book to Blackett in October of 1941, who found in the work a hostile attitude not toward the Soviet Union alone, but, as Blackett put it, "to all that is generally called progressive politics – 'progressive obsessions,' in your words." See (Nye 2004: 38-39).

demarcation. The method of scepticism, too, was inadequate as a demarcator. Scientists are not prone to doubt, nor do they abandon working hypotheses in the face of conflicting evidence, he said, citing his own professional experience that:

> "In my laboratory I find the laws of nature formally contradicted at every hour, but I explain this away by the assumption of experimental error. I know that this may cause me one day to explain away a fundamentally new phenomenon and to miss a great discovery [...] yet [...] if every anomaly observed in my laboratory were taken at its face value, research would instantly degenerate into a wild-goose chase after imaginary fundamental novelties" (Polanyi 1964: 31).

Science is characterized, Polanyi continued, by "schools of research" in each of which there is a master-pupil relationship within a wider institutional and material domain consisting of periodicals and textbooks, research grants and salaries, and facilities for teaching and research. The system is coherent and self-governing, with no central authority. The pupil learns from the master in the way that a child learns a language or a student learns to play the piano. The master chooses the problems, selects a technique, reacts to clues and difficulties, and keeps speculating all the time (Polanyi 1964: 43, 47-48, 49-51, 45, 43, 44). It is a system of apprenticeship rooted in a common tradition, and no revolutionary scientist can succeed without appealing to this tradition and authority. Claims for discovery and innovation are based on commonly-held scientific values of validity, profundity, and intrinsic interest. The process of choosing one system over another is a process of "conversion" influenced by intuition and "finally conscience." The outcome is one that achieves the just sanction of the scientific community (Polanyi 1964: 51-52, 49, 67, 63).

Thus science has a social nature, as well as well-established normative values and rules. But, what, Polanyi asks, is its claim on truth? The scientist's search for the solution of problems, he writes in 1946, is analogous to the "prayerful search for God [...] the urge to make contact with a reality, which is felt to be there already [...] waiting to be apprehended" (Polanyi 1964: 34-35). The "novice" scientist gives himself to the service of an ideal. To accord validity to science is to express a "faith" which is upheld within the community. Here are echoes of Weber and Benda, but also of the Berlin scientists whom Polanyi most admired. Max Planck had spoken publicly in 1937 of

> "[...] the steady, ongoing, never-slackening fight against skepticism and dogmatism, against unbelief and superstition, which religion and science wage together. The directing watchword in this struggle runs from the remotest past to the distance future: 'On to God!'"[34]

Over the course of the next few years, Polanyi worked on his *magnum opus* which became the book *Personal Knowledge,* published in 1958. Negotiating a middle way between objectivity and subjectivity in characterizing scientific epistemology, Polanyi settled upon what he called personal knowledge by way of emphasizing the personal commitment, passion, and faith in the experience of the scientific calling. He spoke of the "fiduciary" nature of the scientist's quest for truth. In Michael Polanyi's view, science could not be done without personal faith in a transcendent truth (Polanyi 1958: 268).

---

34 (Polanyi 1964: 64, 73, 82), (Planck 1944: 306), quoted in (Heilbron 1986: 185).

The Political Dimensions of Post-Empiricist History and Philosophy of Science

In a widely-discussed lecture of 1994 on "Science as a Weapon in Kulturkämpfe in the United States during and after World War II," David Hollinger rooted a shift from talk about scientific method to talk about scientific communities in the anti-communist and anti-centralization concerns of scientists and other intellectuals during the 1940s and early Cold War period. He especially notes the roles of James Bryant Conant, Vannevar Bush, Warren Weaver, Edward Shils, and Robert Merton for whom the scientific community became a model of organization for a free and democratic society, just as the self-governing autonomy of the scientific community was crucial for the advancement of science itself.[35]

From Hollinger's perspective, Conant's introduction of the history of science into the Harvard General Education curriculum, Popper's popular writings on the open society and science, and Polanyi's essays in support of freedom in science and the pursuit of pure science all used the history of science as a history of liberal humanism. Both science and government operated best through freedom of action and discussion in which decisions are made by a community of interacting peers. This was a political agenda in the context of the Cold War, formulated within the broad scope of defending democratic states against advances of authoritarian states and international communism. This was an agenda, too, to establish the principle of scientists' autonomous control over their research independently from the bureaucratic control of agencies that funded them.

Anna-K. Mayer notes the similar postwar triumph in England of liberal humanists in the history and philosophy of science profession. Before the war, she notes, Joseph Needham and Walter Pagel had founded the Cambridge History of Science Committee. They both were scientists. While doing research in embryology, Needham wrote a history of embryology, and while working as a pathologist, Pagel completed a study of religious motivations in seventeenth-century medical biology. Karl Mannheim and J. D. Bernal were among the guests they invited to speak in their lecture series.[36] However, by the time Needham returned to England after the war, historian Herbert Butterfield, theologian Charles Raven, English professor Basil Willey, and members of the Society for Freedom in Science, including the physicist William Dampier, the paleobotanist Hugh Hamshaw Thomas, and the historian Michael Postan had become leaders of the History of Science Committee. Butterfield and his circle rejected materialist and economic interpretations of history and emphasized liberalism as the embodiment of science and vice-versa. They taught the history of science as the history of ideas, placing emphasis on civic notions of truth and virtue (Mayer 2000: 676, 679-683).

I have examined above the strong and quite specific political roots of the philosophical writings about science of Popper, Polanyi, and Kuhn, and we have seen that Popper and Polanyi first introduced the philosophical themes for which they later became known in their political writings. Popper was a social democrat at heart, appalled in his youth by the ex-

---

35 (Hollinger 1996: 155-174); also see 80-96 for (Merton 1973: 267-268). More broadly (Jones 1988). More tendentiously (Saunders 1999). The argument for the autonomy of the scientific community within a system of governmental support is Vannevar Bush's classic text in United States Office of Scientific Research and Development 1945. On the scientists' political actions in the U.S. (Smith 1965).

36 (Mayer 2000: 666-671, 671-672, 674-675). See (Needham, Pagel 1938).

cesses and self-confidence of European communists and fascists in Europe who arrogantly and wrongly, in Popper's view, claimed the mandate of historical destiny. He settled on the principle of doubt, not certainty, as the fundamental characteristic of modern science. The scientific process is one of conjecture and refutation, of trial and error, just as should be true in the political realm of the open society. For Popper, social engineering and scientific planning in the state can never result in perfection or utopia, any more than scientific investigation and explanation can result in a theory true for all time.[37]

Educated like Popper in Central Europe, Polanyi was a professional scientist with a long scientific career and considerable interest in political and economic affairs from his youth. The year 1928 marked an intellectual turning point for him, reinforced by his later visits to the Soviet Union and by the professional and personal disruption caused by his forced immigration in 1933 from Germany to England.[38] How could his science be differentiated from the lies and falsehoods of Nazi and Soviet science? How could it be protected in the world of the future? One answer lay in resistance to central planning. Another lay in challenging demands from a central government for science to serve the needs of the people and the state. Science, Polanyi argued, is not a recipe for new inventions and applications. It is not a set of simple rules for logical reasoning and empirical methodology, but a longstanding tradition and a set of consensual dogmas and tacit practices that are the common currency of a community of peers, who educate their successors through a system of apprenticeship in an autonomous community adhering to the values of its scientist members. The British theoretical physicist John Ziman has perspicaciously analyzed the nature of science and the scientific community. Of Polanyi, he writes "The keyword is 'community.' For Polanyi, science is more than the defining activity of a group of individuals; it is the product of their coordinated activities [...]. This is a simple, yet devastating insight. [...]. It is devastating, for it puts out of business most efforts to cerebrate [sic] the production of scientific knowledge as if it were a mechanical process driven by logical clockwork" (Ziman 2000: 21).

The political roots of Thomas Kuhn's philosophy of science are considerably less personal and intense than Popper and Polanyi's. However, as David Hollinger suggests, the *Structure of Scientific Revolutions* is a book fully resonant with the aims of Conant's general education program and with the liberal humanist politics of American intellectuals in the postwar period.[39] As we have seen, Kuhn drew specifically upon the early writings of Polanyi in working out his own reflections on the nature of science. Kuhn's philosophy of science provides an implicit argument against central planning since Kuhn characterizes science, as does Polanyi, as a closed, or at least normally closed, community of self-governing peers. Kuhn uses an array of historical cases that largely support the argument that scientific ideas are important independently of their practical benefits to society. Like Polanyi, Kuhn recognized the inconsistency of Popper's emphasis on skepticism with the actual behavior of scientists, and Kuhn knew that scientists do not abandon immediately their theories when faced with falsifying evidence.

Kuhn's view on the truth of scientific theories lay neither with Popper nor with Polanyi. For Kuhn, the unit of scientific achievement is the solved problem rather than a theory or

---

37 On these points, see (Godfrey-Smith 2003: 59, 68).

38 From a lecture given in 1944 on BBC and reprinted in (Polanyi 1964).

39 (Fuller 2000: chapter 5); (Fuller 2004: 121).

paradigm that is closer to the truth (Kuhn 1962: 169-170). Science is progressive because new paradigms increase the numbers of problems solved. In contrast, for Popper, confidence in truth of a theory can be increased, but never wholly achieved. For Polanyi, no scientist would continue working without faith that there is a reality and truth to be known, and the scientist's calling, in the meaning of Weber and Benda and of Planck and Einstein, is the search for that reality.[40] In their views of scientific truth, Polanyi and Popper were more old-fashioned than Kuhn, demonstrating their deep roots in Central European culture, in contrast to Kuhn's roots in American pragmatism. Still, it should be noted, in contrast to Kuhn, that the writings of the great American pragmatist Charles Sanders Peirce came to be permeated with theism in his later years and with the conviction that scientific method demanded elements of faith. These views were shared by William James and other members of the Harvard group in the 1860s and 1870s known as the Metaphysical Club.[41]

Both Popper and Polanyi conceived of what they called three worlds of knowledge, or *Wissenschaft*, which bear striking similarities to the philosophical views espoused in the 1930s by many European scientists, including Max Planck, that hesitant father of the quantum revolution and director of Polanyi's Kaiser Wilhelm Gesellschaft. Planck's speeches and essays sometimes distinguished what he called realms of the senses, mathematical structures, and the physicist's world picture. Similarly, Popper differentiated what he called the three worlds of physical processes, psychological states, and the abstract products of the intellect. Polanyi divided knowledge into three kinds: objective, subjective, and personal knowledge, in which personal knowledge requires a commitment of belief – the very leap of faith that Hume had recognized as the common sense solution to the problem of inductivism. As noted, Planck, like Polanyi, spoke of the scientist's faith that wages war again skepticism and superstition: "for the religious man, God stands at the beginning, and for the scientist, at the end, of all thinking."[42]

Karl Popper was knighted in 1965 for his defense of freedom and rationalism. Michael Polanyi received an honorary degree at Princeton University in 1946 with a citation that described him as: "A physical chemist who has devised new tools to determine how fast atoms react; a veteran campaigner against those who would take from science the freedom she requires for the pursuit of truth."[43] Kuhn's philosophy of science received no official honors tied to political statements, although his philosophy proved consistent with liberal humanist political arguments that the scientific community flourishes best by ruling its own affairs, even in times of intellectual anomaly and crisis. The real radicalism of Kuhn's work lay in his abandonment of the kinds of truth claims made by traditional science and scientists. Like Popper and Polanyi, however, Kuhn had no sympathy with the anti-intellectual and anti-scientific uses to which all three philosophers' anti-positivist critiques have been put in more recent political times. In their personal and political commitments to scientific values, Popper, Kuhn, and Polanyi were post-positivist, but not postmodern.

---

40 (Nickles 1998: 51-85); also (Ruse 1999: 13-36).

41 (Fichman 2002: 77-79) and (Menand 2001).

42 (Hacohen 2000: 21-22); (Heilbron 1986: 138); and quotation from (Planck 1944: 304-305), in (Heilbron 1986: 185).

43 Quoted in (Scott, Moleski 2005: 202). Among those also receiving honorary degrees on this occasion were Niels Bohr, Linus Pauling, and Reinhold Niebuhr.

Notes

I am grateful for permission to have consulted the papers of Michael Polanyi (abbreviated MPP below) at the Special Collections of the Regenstein Library at the University of Chicago, as well as papers at the Archives of the Max-Planck-Gesellschaft in Berlin-Dahlem (MPG), the Rockefeller Foundation Archives (RFA), and the Sources for the History of Quantum Physics (SHQP) at the American Institute of Physics. Some research for this project was supported by National Science Foundation grant no. SBR-9321305. I am grateful to Martin X. Moleski, S.J. for making available to me before its publication the biographical manuscript that William Scott was unable to complete before his death in 1999, revised as William T. Scott and Martin X. Moleski, S. J., *Michael Polanyi: Scientist and Philosopher* (Oxford: Oxford University Press, 2005). I thank Karl Hufbauer for permission to cite his unpublished paper "Kuhn's Discovery of History" (1998). Thanks, too, to Keay Davidson, who is completing a book on Thomas Kuhn, for helpful comments.

Unpublished Sources

MMP: Michael Polanyi Papers, Special Collections, Regenstein Library, University of Chicago, Chicago, Illinois

MPG: Archives of the Max-Planck-Gesellschaft, Berlin-Dahlem

RFA: Rockefeller Archive Center, Sleepy Hollow, New York

SHQP: Sources for the History of Quantum Physics at the American Institute of Physics, College Park, Maryland

Moritz Epple

Links and Their Traces: Cultural Strategies, Resources, and Conjunctures of Experimental and Mathematical Practices

Introduction

The title of this collection poses the question whether science can or should be conceived as a form of cultural practice. Depending on the meanings of the terms involved, this question has various ambiguities and implications. To begin with, one can easily trivialize the issue by giving narrow meanings to 'science' and 'cultural practice'. If science is strictly understood to be a corpus of transpersonal scientific knowledge, either as printed in journal articles, monographs and textbooks, or conceived philosophically as an 'established' set of propositions, then of course science is no cultural practice, in fact it is no practice at all. However, the activities involved in establishing scientific knowledge – from the first vague and perhaps unintended steps of guessing through hard work in laboratories or theoretical argument, to the communication, criticism and defense of knowledge claims in layered communities and using various types of media – are of course practices, or rather a whole set of related but different practices. Considering these and further related practices, and using a broad notion of culture such as Clifford Geertz's idea that a culture consists in the web of meanings a group or society gives to the actions of its members and to the world, the answer to the question whether science is a cultural practice is again trivial, indeed, a resounding 'yes'. Under these assumptions, *any* meaningful social practice, and hence scientific practice, is *a fortiori* a cultural practice.

In order to reach more interesting insights, then, we must necessarily complicate the matter. Based on a roughly Geertzian understanding of cultural practices (i.e., as sustained threads or fabrics of activities connected by the meanings attributed to them by their actors and/or the society within which these activities take place) the question is approached, here, from the following perspective: What can historians of science say about the relationships between *those* practices in which scientific knowledge comes into being and *other* social or cultural practices, i.e. those not commonly regarded as scientific? Or, to put it metaphorically, in what ways are practices of scientific knowledge production woven into the fabric of cultural practices at large? Indeed, I want to take a further step in framing the issue to be discussed here. It has probably always been agreed that certain practices essential for the undertaking of the sciences are, at the same time, practices of cultural importance beyond science. Practices of communication across the boundaries of expert groups (*exoteric* practices, to use the term coined for the purpose by Ludwik Fleck[1]), practices of securing patronage from institutions of power to scientists (cf., e.g., Mario Biagioli's well-known studies

1 See (Fleck 1980: ch. 4, sections 3, 4).

of Galilei's career[2]), competition for social resources and political influence in modern society (one might refer here, for instance, to the studies of scientists' roles in the French revolution and its aftermath[3]), efforts to incorporate parts of science into educational systems etc. – all of these are obviously activities which integrate the subculture(s) of scientific inquiry into the cultural fabric at large.

But what does all this tell us about what might be called the epistemic core of scientific activity, about the *esoteric* practices of generating scientific knowledge in episodes of research? In what ways and in what senses are the practices of knowledge generation themselves linked with other social and/or cultural practices? Where does an intellectual advance in a scientific practice, a "knowledge move" as we might say,[4] rely on or shape *another* cultural practice in a specific way, and in a way that can be documented on a micro-historical scale? The question to be discussed in the following is thus one concerning the various kinds of *links* between different cultural practices, one of which will always be a practice of scientific knowledge generation. In some cases, the existence of such a link is clear from certain material in the historical archive,[5] documents or other items that are, at the same time, traces of both a scientific and at least one further cultural activity. I will call such items *link traces*. In other cases, the link emerges only from an historical analysis of a whole set of historical traces, e.g. if a certain norm or value is shared between a scientific and another cultural activity. The question thus becomes, what kinds of links between scientific knowledge generation and other cultural practices do we know of, or can we expect to find through suitably-framed historical analysis.

Of course, this question is not a new one; indeed, it has been touched upon since the turn of the history of science to the analysis of scientific practices. However, one can both discuss the cultural practices of science without worrying much about the core practices of scientific knowledge generation (and thus cultivate, as it were, a happy externalism of a more recent kind), or one may discuss these core practices (in laboratories, at writing desks, ...) without bothering much about 'strange' excursions into politics, religion, literature, etc. The distribution of these historiographic leanings is certainly quite uneven with respect to different areas of science and different periods.

During the last decades, a substantial amount of historical insight of this kind has been presented for *experimental* practices in various fields of science.[6] Experiment has always been regarded as one of the basic pillars of knowledge generation in modern physical science, hence the history of experiment is definitely a crucial domain for our discussion. I will therefore begin with a review of some well-known studies on the history of experimentation in order to point out certain types of links that have been found to exist between experimental knowledge generation, on the one hand, and a variety of cultural and social practices, on the other. At least since the rise of Newtonian science, the other pillar on which scien-

---

2 (Biagioli 1993); (Biagioli 2006).

3 An entry point to the relevant literature is (Dhombres, Dhombres 1989).

4 The term is taken from (Schaanning 1997).

5 Here I understand the archive to be the collection of all (material) traces of past activities. I thus deviate somewhat from the notion developed in (Foucault 1969). It would be an interesting exercise to discuss the relation between Foucault's *enoncé* and what may be called a trace of a past activity.

6 Some of this work is present in this volume, see e.g. the contributions by Rheinberger, Schaffer, Steinle, and Wilson.

tific knowledge generation rests was thought to be mathematical argument, and, in broader terms, mathematisation. Detailed historical analyses of the practices of mathematisation and mathematical argument in science as *cultural* practices are still rare compared to the recent history of experiment.[7] Thus there is good reason to continue the discussion into this domain. This will be done in the second, main part of the paper. Along both lines, and this will lead to my concluding remarks, historical studies of scientific practice have taught us to appreciate the fluid and many-faceted character of knowledge in the making; a fluidity that prohibits the strict separation of science (as a specialized professional enterprise) and culture at large.

1. Cultural and social links in experimentation

Strategies and normative patterns: An experimental way of life?

Let me begin with some remarks on Steven Shapin's and Simon Schaffer's ground-breaking study *Leviathan and the Air-Pump* of 1985. As is well known, the subject of this study was Robert Boyle's experiments with vacuum pumps in the late 1650's and 1660's, Christian Huygens's attempts to replicate these experiments in Den Haag and the reactions of various philosophically-oriented critics, among whom Thomas Hobbes stands out. Shapin and Schaffer set out to show that Boyle and his followers successfully promoted not just a few scientific 'results' about a new phenomenon called 'the vacuum' but, in fact, an entire experimental "form of life", as the authors write in allusion to the late Wittgenstein.[8] To establish that – for instance – sound did not travel through evacuated space within the main container of Boyle's pump, that magnetism, in contrast, did, that Torricelli's void (i.e., the void in a Torricellian apparatus, now placed in the container of Boyle's air-pump) could be increased or diminished by controlling the amount of air remaining in the container, all this required Boyle – the authors argue – to operate on at least three levels which connected the experimental events with further practices both closely related to and far removed from the experimental stage: on the level of *material practices*, where instrument makers and their skills and tricks enter the experimental way of life; on the level of *literary techniques*, which were used to enable what Shapin and Schaffer termed the 'virtual witnessing' of the experiments; and on the level of *social practices* that served to create both a local community and a wide-ranging network of experts who shared a suitable pattern of norms and values, attaching value to 'experimenting' and distancing themselves from other varieties of natural philosophy or scholarship.

7 One of the few areas where cultural issues have been addressed in the history of mathematics is statistics and probability theory, see note 31 below. Several of Donald MacKenzie's studies, since (MacKenzie 1981), touch on the social and cultural role of certain (mostly applied) areas of mathematics. The same is true for Catherine Goldstein's work, see e.g. (Goldstein 2001); (Goldstein 2003); (Goldstein 2005). See also the contribution by Norton M. Wise in this volume and some of my own earlier work.
8 See (Shapin, Schaffer 1985: 15).

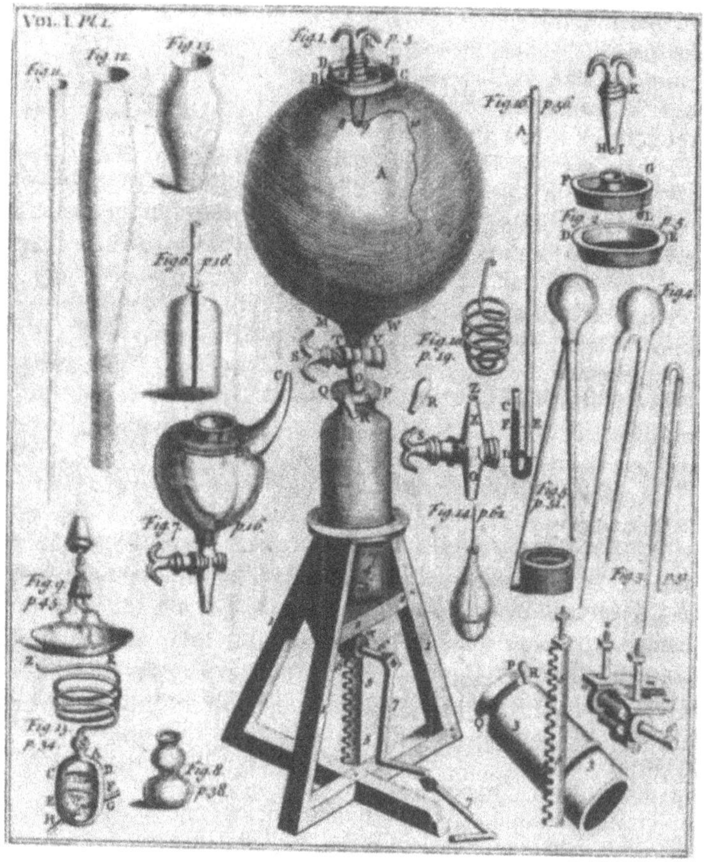

Figure 1: Robert Boyle's first air-pump, engraving of 1660 (Shapin, Schaffer 1985: 27).

In this tangle of events related to the air-pump experiments, several things turned out to be fairly fragile and unstable – and these are the points of Shapin and Schaffer's analysis that I want to recall here in particular. On the first level, the highly complex air pump(s) used in this experimental practice where fragile in a very material sense: The containers were easily destroyed, got cracks that had to be sealed, other parts of the instrument needed special care in order for the pump to work properly, etc.[9] The gentleman experimenter Boyle was not in a position to handle all this fragility alone. He depended on the intervention of instrument makers and of experienced operators of such complex apparatus. In this way, the skills and (partially tacit, personal) knowledge of these participants entered into the production of the effects described by Boyle in his written accounts. At such points, therefore, we have a link of the scientific practice of experimentation with the practices of other crafts even when no written trace is preserved. On the other hand, the remnants of the instruments themselves – appropriately interpreted – turn out to be link traces of their own kind.

9 See for instance (Shapin, Schaffer 1985: 29-30, 231-235).

The fragility of scientific practice on the second level – i.e. that of writing and communicating about the experiments – becomes clear when comparing the written accounts of Boyle's experiments with the difficulties of other researchers trying to replicate the experiments. Seemingly detailed narrative descriptions of both apparatus and experimental procedure suggested to readers that they were what Shapin and Schaffer called *virtual witnesses* of Boyle's experimentation. However, as for instance Huygens's difficulties in replicating certain crucial experiments demonstrate, the transparency and detail of the texts was deceptive.[10] Crucial details were missing (such as the kinds of sealing substances that were used at joints and moving parts of the pump). Rather than being 'proofs' of well-defined experimental phenomena, the texts gave suggestive, but more or less vague descriptions of events in a complicated experimental practice. For critics such as Hobbes it was therefore easy to point out weak points both in the apparatus and in the experimental procedures. Even the basic knowledge claims of the experiments, i.e., that the pump produced an evacuated space within the main container, remained contested. Hobbes was not completely wrong when he criticised that the operations of the pump could not be properly understood "unless the nature of the air is known first" (Shapin, Schaffer 1985: 115). The link made on this level is one between laboratory practice and practices of writing which extended far beyond the laboratory and the narrow scientific circles and crafts tied to it.

Shapin and Schaffer argue, thirdly, that Boyle's success was much less due to straightforward experimentation (as if such ever existed) or stringent argumentation than to the effective social tactics of winning followers for his experimental style. The active members of the Royal Society of London who were admitted to Boyle's experimental performances had themselves subscribed to a specific experimentalist ideal of scientific practice, to a normative pattern of scientific rationality telling them what and how to see and to do. A philosopher such as Hobbes whose pattern of rationality was different never acquired the status of an actual experimental witness, and foreigners such as Huygens were only allowed to see the experiments if they had shown themselves to be capable experimenters in accordance with the standards of the Royal Society at the time.

Whatever criticisms one may have of Shapin and Schaffer's account of Boyle's work, the overall analysis certainly points out important dimensions of science-culture links that will be encountered in other cases as well. As is well known, *Leviathan and the Air-Pump* has been discussed and criticised along quite different lines. The book has been challenged, in particular, regarding the authors' tendency to imply that the acceptance of Boyle's experimental facts was essentially the result of a successful overpowering of critics by a pressure group of experimentalists. Whatever one may think of this issue, however, it can hardly be denied that the crucial phenomenon, the 'vacuum' and its several properties, only gradually took shape in Boyle's practice, in similar contemporary practices, and in the discussions that ensued. What exactly was found in Boyle's container after pumping out a quantity of air? Rarefied air? To what degree? Another material substance, some subtle part of ordinary air? Some sort of ether? Actual, empty space? No one could tell with certainty. The basic phenomenon and the concepts used to describe it remained fluid, partially vague, subject to debate. This fluidity of the experimental phenomenon, one might say, left room for different

---

10 The same insight could be drawn from modern attempts at a replication of the experiments, see e.g. (Heering, Rieß, Sichau 2000).

epistemic perspectives, perspectives which, in turn, were determined by different cultural orientations. The glassblower, the gentleman friend of nature, the mechanist philosopher could all draw their own conclusions about what the apparatus showed during its operation. An interesting twist in the debate has been made by Peter Dear,[11] who points out that the culture of experimentation, so carefully described by Shapin and Schaffer, was in itself a rather transient historical phenomenon even within the circles of the Royal Society of London. When Isaac Newton began his career, he brought a different way of experimenting, or more accurately, a different use of experimentation, to the Society. Rather than exploring a wide range of individual experimental events, he sought to use specifically designed experiments to support particular theoretical statements, such as the idea that the sun's light was a mixture of differently coloured and variously refrangible rays rather than a simple entity. Newton's early experiments with prisms were refined until they could serve as proofs in a theoretical context, a function which was again clearly demonstrated in Newton's written and published account. His *Opticks* of 1704 was a book about an experimental (and speculative) science, but it was written in a largely Euclidean style. It was organized by series of definitions, axioms, and "proofs by experiment", rather than around experimental narratives as with Boyle.

Dear points out that this method of incorporating empirical knowledge into theoretical argumentation dates back to an older tradition running through the transformations in 16th and 17th century science; he terms it "the mathematical way in the scientific revolution". If he is right, this shows that not only particular experimental phenomena, but experimentation itself was a fluid form of scientific practice, an issue of controversy (acted out within the community of experimenters e.g. between Newton and Hooke), and subject to historical change. Moreover, the specific shift in question touches on the practices of mathematical argumentation; mathematics served, here, as a model for empirical science. From a Boylean point of view, this might be seen as an unexpected cultural role of mathematics – to which I will return in the second part of this paper.

Or, instead of transience, one might emphasise the plurality of styles of experimentation and insist that there may have been not one, but several 'experimental ways of life' during the 17th century.

Practical Resources

Let me now briefly turn to another well-known study of 19th century experimentation which in a striking way demonstrated how an experimental practice may draw on elements of social practices and cultural developments. In a paper published in 1995, Otto Sibum discusses James Prescott Joule's measurements of the mechanical equivalent of heat during the 1840's (Sibum 1995). At stake was again a controversial idea, that of the full convertibility of different natural forces (or forms of energy, as William Thomson would later call them) into one another, and a controversial practice around a fragile and sensitive apparatus. In Joule's experiments, a paddle wheel mechanism driven by falling weights stirred the water in a container. Joule claimed to have measured the precise amount of heat which was transferred to

11 See (Dear 1995b), esp. chapter 8.

the water when a given amount of mechanical work was used to stir the water. This relation between mechanical work and heat, so Joule believed, was universal: Under any circumstances whatsoever, the same amounts of heat and mechanical work would be equivalent. A few years later (first in Hermann v. Helmholtz's "Erhaltung der Kraft" of 1847), this quantitative relation would form a cornerstone in the belief that the total amount of energy is conserved in all natural processes.

Following a method very successfully developed in a research group at Oldenburg University around Falk Riess,[12] Sibum and Peter Heering replicated Joule's experiment on the basis of extant information. They could show that, in this case too, resources of knowledge and skill absent from the written scientific accounts entered Joule's experimental practice.[13] In their own practice of replication, Sibum and Heering realized that they had to set up and to manipulate the experiment in quite delicate ways if they wanted to get close to Joule's results. Many problems, in particular, resulted from the necessity to take precise temperature measurements. In the attempt to understand the historical context in which the corresponding practical knowledge and skills could have been acquired, Sibum found that Manchester's contemporary brewing tradition had been the crucial resource – as a brewer's son, Joule not only belonged to the tradition, but witnessed the radical changes that this industry experienced in the very years of his research. The increased production and consumption of beer in industrialised society required a new generation of brewers who aimed at quantitatively controlled, standardised brewing processes, at so-called "scientific brewing", as the proponents dubbed it. One particular skill both necessary and strongly advocated in these changes was the ability to measure accurately and reliably various temperatures (mashing heats) during the process of brewing. According to Sibum, it was this skill, and the general interest in quantitative relations dominating the material transformation processes in industrial production, that enabled Joule both to frame his basic idea and to achieve his experimental results.

Once again, these results and ideas were much less stable than today's textbook accounts would suggest. Even though Joule specified his measured data with surprising precision (in fact, they were quite close to modern data), and even though Sibum and Heering tried to adjust their own practice to the complex requirements of the experiment, they found it difficult, if not impossible, to get similarly precise data with replicas of the experiment. To mention but one crucial point: The experimental apparatus was so sensitive to temperature changes in the room in which it was set up that the heat of one person entering the room could disturb the measurements to such a degree that they lost their significance. This meant that Joule (and the operator he obviously had) were forced to experiment in a private space, indeed in the cool cellar of his father's brewery. In a culture of semi-public experimentation this created a problem of reception: Actual witnessing was even more problematic than in Boyle's earlier experiments. Rather than a clear-cut, straightforward experimental fact, we thus find a fragile practice balancing the strong belief of the experimenter in the existence of a mechanical equivalent of heat, on the one hand, and the exigencies of his experimental setup on the other.

---

12 A survey of the work on the replication of experiments done by the Oldenburg group up to 2000 can be found in (Heering, Rieß, Sichau 2000).

13 An interesting summary and interpretation of Sibum's and Heering's replication practice may also be found in (Pickering 1995: 104-109).

As Andrew Pickering's interpretation of this episode makes clear, its epistemic core, so to speak, consisted in an activity happening *in time*, in a complex *performance* (Pickering 1995: 107-108). It was this performance (including all material items involved in it) which allowed and actually required a link between the cognitive practice of Joule acting as a 'scientist' and the practices of his brewing environment upon which he was able to draw. Such links are particularly difficult to reveal historiographically – indeed, what are their traces in our present? Interestingly enough in the above episode, the precise values of Joule's measurement serve as a kind of link trace. In the practice of replication they were revealed to be the results of a complex activity in time that could only be understood (and, to a certain degree, re-enacted) after its extra-scientific practical links had been explored.[14]

Conjunctures

The above example calls to our attention the way research activities unfold in time on a small time scale, as well as the specific practices involved in research. This temporality has been made the corner-stone of Hans-Jörg Rheinberger's thorough analysis of experimental research in molecular biology. I would like to briefly turn to his work, in order to point out a further pattern of linking.

Drawing on a highly sophisticated case study on the *in vitro* synthesis of proteins, Rheinberger proposed to study what he calls – taking up the language of scientists – the *experimental systems* within which research is pursued (Rheinberger 1997). These systems are conceived as dynamic configurations, consisting of the material setups of experimentation, of methods, skills and techniques applied by the researchers, of intellectual components entering the research problems and the interpretation of what is happening in the experiments, etc. – in short, of everything functionally involved in "the smallest integral working units of research" (Rheinberger 1997: 28). As research is going on, this configuration undergoes a permanent change. Certain aspects of the research problems are resolved, while others move into the foreground, the material setup undergoes modifications, methods and techniques are varied, members of the research team change their views, etc. As long as an experimental system is producing new knowledge, it is thus characterised by a process of "differential reproduction", as Rheinberger points out. In this process, two major components can be discerned whose functions are complementary to each other. One component consists of the "technological objects" and material procedures which are used in experimental research as ready-made, at least temporally reliable tools for finding answers to particular research problems. (An example might be the thermometers that Joule used in his setup, when and insofar as they functioned reliably during a given time span.) The other component is that which generates, rather than answers, problems of research. Rheinberger calls this second component the "epistemic things" of research: the intellectual constructs about which scientists do not already know everything, about which they want to know more, such as 'the vacuum', the 'mechanical equivalent of heat', the 'structure of the DNA', or what your favourite example might be.

---

14 By no means do I want to claim that all precision measurements are link traces. Whether or not this happens to be the case must be established for each measurement in one way or the other, historical replication (where it is suitable at all) being merely one of them.

Seen from the perspective of knowledge generation, and this is one of Rheinberger's main points, *both* components are essential. Without the partial obscurity of the subjects of research, without their intrinsic vagueness, there would be no open problems, and consequently no future for research. Without temporarily stable procedures for generating answers, on the other hand, there would be no stable knowledge as one of the outcomes of research. As research progresses, the two components interact. The epistemic things of one research episode may become well-understood tools of further research episodes, and technical procedures which worked well in one situation may turn out to produce unexpected problems in another situation and thus themselves become epistemic things.

Now it might appear (and it has been critically remarked) that such an analysis of experimental systems is just another – albeit more sophisticated – way of telling a story about the internal dynamics of research practices, and thus it might not seem particulary helfpul in the present context. However, once a micro-historical analysis of the differential reproduction of experimental systems is undertaken seriously, it quickly becomes clear that some of their elements circulate between *different* systems, that the dynamics of one experimental system may link with that of another, that two such systems may unexpectedly correlate – Rheinberger terms this a *conjuncture* of experimental systems – they may *hybridize*, i.e. merge together, or one system may *bifurcate* into two. In this way, arrays of interrelated experimental systems come into focus; they form what Rheinberger terms an *experimental culture.*[15] Nothing prevents us, and indeed the analysis of many research episodes invites us, to carry this analysis just one step further. The circulation of elements of experimental systems such as apparatus, techniques, or specific problems is usually not limited to an exchange between one experimental system and another. It can also involve an exchange with a practice *outside* the domain of experimental science properly speaking. One example is – in Rheinberger's story – the use of radioactive isotopes for tracing the activity of proteins in cancer research just after World War II. Not only was the availability of the carbon isotope ^{14}C for medical research dependent on the existence of suitable apparatus and technologies within in the scientific and military complex of nuclear research that had developed during the war, but the easy accessibility of funding from the US Atomic Energy Commission for medical research using radioactivity contributed to the establishment of "a structural coupling" between "an experimental system for analyzing malignant growth" of cells and "the global political situation". Stretching Rheinberger's terminology somewhat we may call such a coupling a conjuncture as well – not between experimental systems, but between an episode of scientific research and an episode of a different social (and in this case, political and technological) practice. Indeed, this coupling also depended on events and historical developments that were unexpected both for the researchers and the non-scientific actors involved: neither could the medical researchers foresee before the end of the war that the Atomic Energy Commission would be willing to finance certain kinds of their research, nor could those involved in nuclear technology anticipate that radioactive tracing would develop into a core technique of biological research. In this particular case the historical traces of this conjunctural link are manifold and of quite different kinds. They range from archival materials documenting the applications for funding to the various forms in which the technique of using radioactive tracers in molecular biology are still in use.

15 See (Rheinberger 1997: ch. 9), and Rheinberger's contribution to this volume.

2. Cultural and social links in mathematical practices

Let us now turn to the second group of practices of knowledge production mentioned in the introduction, practices of mathematization and mathematical practices in general. If the recent historiography of experimentation has not only shown fascinating examples of links between scientific and extra-scientific practices but also provided interesting proposals for describing and analysing such links – what can be said about the abstract objects, the proving procedures and the research practices of mathematics, be it in the context of natural (or even social) sciences or in mathematics as a distinct discipline? Does it make sense to speak of mathematical ways of life in a similar sense as Shapin and Schaffer spoke of an experimental way of life, characterised by a distinct set of strategies and normative patterns, some of which relate activities of mathematical research with other social and cultural practices? Do we find resources beyond mathematical practice similar to those brewing practices which became a resource for Joule's work? And are there conjunctures or hybridisations between episodes of mathematical research and other episodes of cultural and social activity?

The following remarks are intended to support the idea that this is indeed the case, and not just superficially so but at the very core of crucial episodes of the production of new mathematical knowledge. As in the first part of this paper I will draw chiefly on recent contributions to the history of this particular science to illustrate the point.

An important preliminary step for such an analysis is to overcome a certain variety of anachronism that – despite all care given to the descriptions of particular mathematical achievements – still lingers in many accounts of the history of mathematics (be it accounts by historians of mathematics or accounts by other historians). I am referring to the implicit understanding that the very idea of what this science is has been more or less stable over time and in fact close to what we today conceive of as 'mathematics', a science of precise definition and rigorous proof, organising an array of fields of mathematical knowledge ranging from the venerable fields of arithmetics and geometry to the more recent branches of higher mathematics. Even a modestly sophisticated look at the historical usages of the term 'mathematics' and its counterparts in other languages, however, reveals a variety of differing meanings, both extensionally and intensionally, that should warn us against taking the modern understanding of this discipline for granted in earlier periods.[16] As a consequence, the same variation has to be allowed for the notion of 'mathematical practices', which has to be detached from modern expectations as well.

To illustrate the dangers of mistaking 'mathematics' to be a timeless notion, let me remind you of some traditional accounts of the beginnings of European science, as for instance in Dijksterhuis' classic *The mechanization of the world picture.*[17] There we read: "The true opposition between classical [i.e., early modern] and medieval science will only become completely clear if one adds to the description of mechanics as a science of motion the characteristic of a mathematical treatment."[18] Galileo's famous dictum that the book of

16 A brief summary of these variations has been sketched in (Epple 2008a).
17 The Dutch original appeared in 1950 under the title *De mechanisering van het wereldbeeld.* I am working from the German edition (Dijksterhuis 1956); an English translation was first published in 1961.
18 My translation from (Dijksterhuis 1956: V, § 7).

nature is written in mathematical symbols, the heavily mathematical character of Newtonian science, the striking rise both of mathematical calculus and of mathematical physics since the late 17th century have all been taken to support a straightforward story of the growth of physical theory, drawing, of course, on empirical data but driven by rigorous, and increasingly mathematical, reasoning.

Much of the new history of experiment has been written precisely in opposition to this simple story (which, of course, had its own twists and oddities, e.g.: when exactly was the idea of a mathematical theory of movement first conceived? By Galilei and his immediate precursors, by the Merton College philosophers, by Nicole Oresme and other medieval writers since the 14th century, or perhaps even by Aristotle?). Looking at scientific practice, so the argument was, meant going beyond mathematical theories, meant looking at the actual doings of those men and women who established modern science, meant looking at the instruments and experimental demonstrations, the practices of writing and persuasion, etc. However, by doing so the necessity of applying a similar perspective to practices of mathematisation and of mathematical reasoning themselves was often overlooked. Indeed, using and conceiving a mathematical notion or theories is itself a complex activity and not simply a matter of following a set of once-and-for-all fixed rules of definition and proof. This was not appreciated by several of the older historians of the rise of European science (including Dijksterhuis). Precisely when it came to the mathematical ingredients of modern science, these historians (knowingly) misrepresented mathematical symbolism, offered modern "reconstructions" of mathematical arguments rather than following their sources also in this regard. To apply today's mathematical language and conception of mathematics, to Kepler's, Galileo's or Newton's was thought to be a lesser vice and not actually dangerous to the historical account.[19]

As soon as historians refuse such an approach to mathematical argument, they find themselves facing the same kind of historiographical issues as in the serious history of experiment. How was mathematical language framed and employed in a particular situation, by whom, and with what aims? What exactly was meant, in a particular scientific episode, by a particular mathematical notion? What were the practical implications of using a particular symbolical technique rather than another? What constituted an acceptable definition or a convincing proof at a given moment in time or in a particular community? How were the epistemic configurations of mathematical research, its objects, language and problem solving techniques, reproduced differentially in actual research? How did 'mathematicians' (another term with a variable meaning) go about practicing their skills? And how were mathematical arguments defended in communicative settings? Only slowly are historians of science exploring the possibilities of raising such questions and thus of deconstructing the myth of a potentially timeless canon of mathematical objects and procedures.[20] Those who

---

19 In many ways one could say that the mistake was similar in kind to supposing that a detailed understanding of experimentation was inessential to a historical understanding of the new European science (a locus classicus is Koyré's introduction to his *Etudes galiléennes* of 1939). Mathematical argument – just as experimentation – was a complex tool in advancing the new science(s); a careful analysis of the forging of this tool and its historical uses is thus not far historiographically from a closer analysis of experimentation.

20 A crucial prerequisite for proceeding in this direction is to refuse immediate 'translations' of earlier mathematical language into that of our present. While it is often said that such a transposition is necessary in order to make earlier mathematics understandable, essential aspects are made invisible when earlier terms,

do have again begun to discover surprising links with cultural and social practices at the core of processes of mathematisation and mathematical knowledge generation.

Strategies and Normative Patterns: A mathematical way of life?

A few references must suffice here to make plausible my argument that it makes sense to speak of and to study 'mathematical ways of life' in the 17th century, just as it makes sense to speak of experimental ways of life. A possible starting point for doing so might again be Dear's study mentioned in the first part of this essay, or indeed much of the older (as well as more recent) literature on the new European science,[21] read in a particular way. Not only did certain strands of Jesuit learned activities emphasise a 'mathematical way' in the rise of European science, but as a matter of fact the 16th and 17th centuries witnessed the growing importance of a bewildering variety of actors referred to as 'mathematicians'. This term – used in different contexts – could mean anything from a university teacher to a clergyman or a physician, such as Girolamo Cardano, to an advocate, like Pierre de Fermat, or a merchant, engineer and writer, such as Simon Stevin. What was common to all these people leading very different, and at times very multifaceted, lives? They all engaged in a certain literary or learned activity, and one specific occupation in this context was reading and writing about mathematical problems and methods. In virtually all cases one may consider, however, this was not their only activity. Therefore, the *biographies* of early modern mathematicians themselves involve links of all sorts between the specific activity of reading and writing about mathematical problems, and other endeavours.

To engage in reading and writing was, moreover, not only a literary activity but also a social practice. Many of the mathematicians of this period became members of the correspondence networks of the self-proclaimed 'republic of letters', more than a few engaged in the efforts of publishing and printing mathematical texts, and some did so even for political authorities of different kinds. Among other things, this required them to balance the existing social order(s) with a new network of social relations emerging in the republic of letters – not always an easy task.[22]

The point to be made here is that in contrast to modern expectations about scientific disciplines, the common denominator of the varied ways of being a mathematician in the 16th and 17th century probably was much less an established core of mathematical ideas, problems and methods than a *habitus* of engaging with such matters for quite different purposes (be they religious, philosophical, technical, administrative, or possibly even 'amateurish').

---

notations and expressions are replaced by modern ones. To give just one example: It is impossible to 'translate' the word 'continuous' of the period before, say, 1850 into any of the available modern mathematical notions of continuity – for the simple reason that in contrast to *all* modern notions, the earlier uses were not part of a very precise and coherent conceptual setting. In a certain sense, therefore, it is misleading to call the replacement of earlier mathematical language by the modern one a 'translation' – much more is at stake. A more illuminating approach to these issues has been proposed in (Goldstein 1995), where the notion of 'readings' is used systematically for describing the relation between mathematical texts of a given time and later accounts, or uses, of such.

21 An entry point to the older literature is (Cohen 1994); an interesting collection of more recent studies of 'mathematical Europe' is (Goldstein, Gray, Ritter 1996).

22 See (Daston 1991) and, for an analysis of the delicate social balances involved (Goldstein 2005).

The conflicts into which such a mathematical way of life could run have been analysed in detail for the celebrated case of Galileo. In his own way, and without being as powerful as an inventor of new mathematical methods as some of his contemporaries, Galileo was clearly propagating in several of his battles rather the status and the competence of mathematics and mathematicians (understood in the broad sense of this time) than a particular system of the world.[23] That this was appreciated already by contemporaries and those engaged in similar efforts to advocate the *habitus* of mathematicians in the 18th century is documented in many early accounts of Galileo's career, not least in Etienne de Montucla's monumental *Histoire des mathématiques* written more than a century after Galileo's death. While Galileo's writings and condemnation were explicitly presented in this history as the "signal of a war about the motion of the earth that was ignited in the philosophical world,"[24] and while the narrative was composed as an account of a long-winded fight between Copernicans and Anti-Copernicans, the details of the account continually revolved around the use of mathematical argument or about empirical findings that were a 'demonstration' or 'proof' of this or that, or they commented the superiority of 'philosophy' over biblical writings, theology, and 'the authorities'. In this context 'philosophy' – a term still predominating the use of 'science' – indicated another general link between mathematical practices and the culture of the new science as well as the republic of letters in general.

At the time when Montucla's *Histoire* was first published, the varieties of a 'mathematical way of life' were the subject of open tension in the French world of *ésprit*. While many academics – most notably Bernard de Fontenelle in his *Préface sur l'Utilité des Mathématiques et de la Physique* of 1729[25] and Jean d'Alembert in the *Discours préliminaire* of the *Encyclopédie* of 1751 – were praising the striking successes of the mathematical sciences, other prominent *philosophes* expressed rather critical views of mathematics and of mathematicians in equally prominent places.[26] Some of these criticisms were quite well-founded. When Georges Buffon, in the *Discours premier* which opened the sequence of volumes of his *Histoire naturelle*, expressed strong doubts that the highly abstract knowledge produced by mathematicians would ever be of real use for humankind, and that even the attempt to mathematically understand machines still failed in most cases, he certainly had a point.[27] In 1753, d'Alembert's colleague Diderot went a step further when he wrote in his *Pensées sur l'interpretation de la nature* that mathematics was a largely valueless game of abstraction and a matter of mere convention, and that all mathematical exactness did not help in dealing with matters of sensual experience.[28]

---

23 A summary and evaluation of the relevant literature can be found in (Remmert 1998). The above remark goes well together with accounts of Galileo's career that focus on his professional status, such as (Biagioli 1993).

24 The quotation is taken from the second edition of vol. 2 of Montucla's *Histoire des mathématiques* (Montucla 1799: 295). In this edition, the account of the Galileo affair is significantly extended in comparison with the original edition of 1758.

25 See (Fontenelle 1728). On Fontenelle, see also (Gaukroger 2008).

26 To chart the debates on the value of mathematics in full detail would be a challenging but rewarding task.

27 See (Buffon 1749: 53-62); for a discussion of the troubles in hydraulics see (Epple 2008b).

28 See (Diderot 1875: Section II), on the debates between Diderot and d'Alembert see (Hankins 1970) and (Le Ru 1999).

This debate among the enlightened *philosophes* and naturalists about the reach and value of mathematics is interesting in the present context because it represents another array of links between mathematical practices and a whole spectrum of cultural roles for them. Indeed, some of the texts written for and published in this debate provide striking traces of such intellectual links, including quite surprising and strong ones when seen against a more modern background of disciplinary separation and autonomy. When d'Alembert published his reaction to the critique of mathematics expressed by Buffon, Diderot and others in the 7th volume of the *Encyclopédie* (for the last of which he officially served as a co-editor), he chose to do so in an article that discussed the persona of the *Géomètre*, i.e. that mathematician (in the wide sense of the word) who is concerned with the most esteemed branches of higher mathematics. The article is, in a striking way, a trace of an intellectual link seen by d'Alembert in the general cultural setting of the enlightenment. The study of geometry, the text states,

> "may be the only means to shatter, little by little, certain regions of Europe, the yoke of oppression and of deep ignorance under which they suffer. [...] Let there be born, if it is possible, geometers among these peoples, this is a semen which will produce philosophers over time, and almost without being noticed. [...] Soon the study of geometry will lead to that of mechanics, and this in turn will lead just by itself and without hindrance to the study of sound physics, and finally sound physics to true philosophy, which, by the general and prompt light it spreads, will soon be more powerful than all efforts of superstition. Because these efforts, however large they may be, will become useless once the nation is enlightened" (Alembert 1757: 628-629).

Once again, 'philosophy' indicates the bridge between the way of thinking endorsed by the geometers, on the one hand, and general enlightenment culture on the other. In this case the latter included an explicitly political attitude as well, as further related writings of d'Alembert clearly show.[29]

At this point, a doubt must be addressed: Are texts such as the one just quoted really traces of links between actual mathematical practices and practices of politics or of moral philosophy? Or must they instead be understood as far-fetched utopian claims that did not correspond to any social reality? The matter is too complex to be fully resolved here; suffice it to add two remarks. One the one hand, d'Alembert would not have been able to use this argument against the critics of mathematics if he could not count on the sympathy of at least some of his readers – and hence there was at least a discursive reality (or a literary strategy) in which the claimed link served a purpose. On the other hand, d'Alembert was convinced of what he took to be an empirical fact, namely that there was, in Europe, a correlation between social conditions favouring political freedom and those favouring scientific, and in particular, mathematical activity (Alembert 1757: 628).

On a more technical level, d'Alembert's optimistic views on the cultural and political force of the "esprit géométrique, c'est - à - dire l'esprit de méthode et de justesse" (l.c.) was complemented by a rather widespread belief in the power of mathematisation which can

---

29 See in particular d'Alembert's *Essai sur les éléments de philosophie* of 1759 (Alembert 1767b), covering both the moral and the mathematical sciences, and his *Essai sur la société des gens de lettres et des grands*, first published in 1753 (Alembert 1767a).

only be called utopian. This ranged from core areas of the mathematical sciences, such as emerging hydrodynamics, to seemingly remote fields like forestry. In some cases (as in hydrodynamics), 18th century mathematicians wrote down complicated differential equations which cannot be fully solved even with today's most powerful mathematical tools and computers; in others (as in certain areas of medicine) mathematical notions were 'applied' without any idea what this application might actually aid in practice.[30]

A particular field of 18th century mathematics that has been studied extensively and for which the interweavings with enlightenment culture and society have been investigated in detail is the *analyse des hazards*, the calculus of probability.[31] As this historical research clearly shows, any historiography that detaches the introduction of the basic notions of mathematical probability and the forging of the methods of probabilistic analysis from the (actual or imagined) economic, juridical, administrative, epistemological, recreational, and even religious or theological practices for which they were intended would severely limit the possibilities of understanding this crucial intellectual development in mathematics. Also in this area, the utopian component of 18th century efforts at mathematisation cannot be overlooked, and Lorraine Daston certainly has a point when she claims that these utopias form part of the "meaning of enlightenment" (Daston 2007).

Cultural Resources: Epistemic, Material

It must be admitted that some of the previous remarks are stretching the idea of an intellectual link between mathematics and other cultural practices beyond the meaning it was given at the beginning of this paper. While in the formation of probability, *specific* links between a cognitive development and social practices are obvious in many relevant texts (e.g. when Jakob Bernoulli attempts to frame a mathematical version of the legal notion of "moral certainty" in his *Ars conjectandi*[32]), an embedding of mathematical activities into the biography of an individual might not be, in many cases, a *specific* interrelation between an episode of knowledge production and an episode of some other social and/or cultural practice. The diversity of biographies of 'mathematicians' in the 16th and 17th century is thus a trace of a more general relation between the culture of mathematics and society in these centuries. However, the existence of this more general fabric of mathematical activities in the biographies of early modern scholars, merchants, engineers, amateurs, etc., may as well prepare us for seeking, and finding, traces of more specific links in the sense considered here.

Let me give one 17th century example in a little more detail. Arguably the most important innovation in mathematics in this period was the advent of analytical geometry in the first half of the century. This new set of mathematical tools, resulting from the merging of traditional geometry with new symbolical algebra, became the basis both of later mathematical analysis and – in modified form – of most uses of mathematics in the sciences for some

---

30 On hydrodynamics, see (Darrigol 2005: chapter 1), on the tensions with practice, see (Eckert 2002) and (Epple 2008b), on forestry, see (Lowood 1990), and on the 'Medecins algébristes' cf. Jean d'Alembert's critical remark in the *Discours préliminaire* of the *Encyclopédie* (Alembert 1751: vii).

31 See in particular (Krüger et al. 1987); (Daston 1988); (Brian 1994); (Hacking 1975); (Hacking 1990); among others.

32 See (Bernoulli 1713: Part IV).

time to come. In an important study, summarizing years of research, Henk J. M. Bos has presented an analysis of the very meaning of 'mathematics' and 'geometry' in Descartes' geometrical work (Bos 2001). His result can be said to prove once more the historical variability of these meanings – indeed the title Bos gave to his monograph is *Redefining Geometrical Exactness*, implying that Descartes worked out and followed his own conception of geometrical exactness, different from that or those of earlier traditions. Let me briefly sketch Bos's reconstruction of Descartes's position.

The basic objects of 17th century geometrical analysis were curves described in a plane.[33] But how could exact knowledge about such curves be reached? An indication that Descartes (just like the geometers of antiquity) found a serious problem here is his claim that it was just not possible to reach exact knowledge about all curves that had been and could be imagined. For Descartes, there were some plane curves about which exact knowledge could not be obtained. An example was the logarithmic spiral, or certain curves generated by complex mechanical motions. But which curves *were* exactly knowable and could thus form the basis of a rigorous science of geometry (these were denoted "geometrical curves" by Descartes)?

In answering this crucial question, Descartes first rejected what he thought to be a misunderstanding: The distinction between "geometrical" and "ungeometrical" curves could not be based on the fact that the latter were defined by mechanisms – in fact, he argued, even the simplest and most geometrical curves, the straight line and the circle, can be drawn on paper only by means of mechanical instruments: ruler and compass (Descartes 1637: 315). Rather, to be geometrical was a quality pertaining to certain mechanisms as opposed to others, or to some imagined constructions as opposed to others. Geometrical curves, for Descartes, were those that were generated by certain combinations of simple motions which could be "exactly measured", combinations of motions which could be described by means of certain simple, imagined, supposedly exact mechanisms.[34] The corresponding pages of Descartes' *La Géométrie* discussed several examples of such imagined mechanisms (see Fig. 2).

At the very core of one of the most consequential innovations of early modern mathematics, we thus find not a *mathematical* concept but a cultural idea, one, in fact, that guided most of Descartes' natural philosophy: the idea of mechanism.[35] The idea of *geometrical* exactness was reduced to an idea of *mechanical* exactness. This, in turn, relied on an idea of the simplicity of certain motions and of the ability to distinguish between different types of

---

33 'Curves' are to be taken here in a wide sense, including traditional plane geometric figures (composed of lines and points). Note that these objects are neither 'functions' – a notion that emerged only later – or equations (used in the 17th century to describe curves and to tackle problems related to them); to make such substitutions is one of the most common anachronisms encountered in this area.

34 See esp. the discussion on the first pages of Book II of *La Géométrie* (Descartes 1637: 315-324). Before making the famous claim that geometrical curves – "that is, those which admit of precise and exact measurement" – may be described by a (polynomial) equation which relates their points to those of a straight line (Descartes 1637: 319), he implicitly introduces the notion of geometrical curves as those arising from two motions which have a "rapport qu'on puisse mésurer exactement" (Descartes 1637: 317). See the detailed reconstruction of Descartes' arguments in (Bos 2001: chapter 24).

35 In the general thrust of founding geometry on mechanics, Newton's views on geometry are quite similar; the details, however, are different. See the discussion in (Guicciardini 2009: chapter 13, 14).

combinations of motion. One should note that neither the simple motions nor the types of combination of which Descartes speaks were defined by mathematical means. Rather, it was the other way round: These ideas defined what the very essence of geometrical exactness, and hence, what *geometry* was.[36] And while they were ideas, they were at the same time linked with *images* and a wider array of imaginings and practices shaping the 17th century understanding of mechanics.[37]

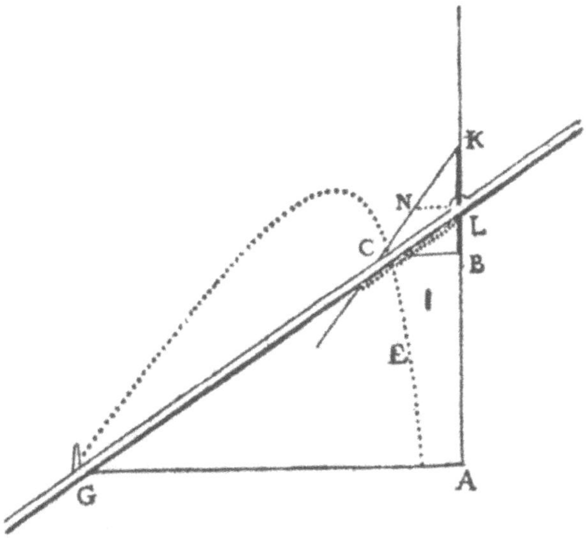

Figure 2: Instrument for drawing a "geometrical" curve (Descartes 1637: 320).

Consequently, the idea and images of mechanical exactness must be considered a cultural resource in Descartes' mathematical practice. Considering the later development of mathematics – the formation of a modern discipline around and after 1800, and the restructurings of the profession in the early 20th century – have such cultural resources lost their importance for the production of mathematical knowledge? In the next section I will present an episode of research showing that even if there should be a trend toward a stronger intellectual autonomy of at least what came to be called 'pure' mathematics, examples of very close links with other cultural activities can be found. Before doing so, however, let me point out

36 More often than not, historians of Descartes' mathematics have taken a *consequence* of this conception of exactness as its *definition*: The claim, repeatedly made by Descartes and essential to his overall conception of geometry, that geometrical curves were precisely those curves that could be described by algebraic equations. As Bos shows, this claim is 'proved' in a complex argument that relies on certain ideas about coordinated motion and construction procedures of curves. Consequently, the argument is based on a *mathematical* formulation of certain *non-mathematical* ideas (in the sense that they cannot be fully captured in formal definitions), and hence is akin to claims such as Church's thesis in the foundations of modern mathematics.
37 An insightful account of the role of imagery and images in Descartes' writings and in his conception of mechanism has recently been given by (Zittel 2009).

that even in modern, 'pure' mathematics, another type of practical resource must be considered, just as in experimental science, namely material ones.

While a full analysis of the role of material practices, of counting, calculating, writing or drawing on various supports, is still a desideratum for the historiography of mathematics, it is quite clear that the advent of electronic computers in the 20th century has brought major changes in the modes of production of mathematical knowledge. In many ways, also the history of these changes still needs to be written, for the many fields of applied mathematics as well as for pure mathematics.[38] Already know we can be certain that these changes have touched essential aspects of the self-understanding of mathematics.

One of the domains where this can be seen is the debate about the value of computer-aided proofs in mathematics – what is at stake here is the very notion of proof. The first celebrated example of such a "proof" was Kenneth Appel's and Wolfgang Haken's work on the so-called four-colour conjecture in 1976. While a sophisticated analysis of the intricacies of this problem and the setting up of a proving strategy was the result of a more traditional kind of mathematical work – the actual "proof" relied on hundreds of hours of programming, of tedious entry by hand of data into powerful computers, and of computing time, and it made the machine check a vast range of combinatorial possibilities that were impossible to survey and to deal with consecutively by a human mind – only this mental, manual and machine work taken together established the conjecture as a 'theorem'. Or did it? As Donald MacKenzie has described in an illuminating study (MacKenzie 1999), there emerged no consensus among mathematicians, computer scientists, and philosophers about the meaning of Appel and Haken's proof. A significant unease remained about whether the mathematical community should accept this kind of evidence as mathematical proof in the proper sense. This debate could not and cannot be resolved on purely mathematical grounds – or rather, what counts as 'purely mathematical' is again found to be dependent on a social (here: technological) practice. We thus encounter a similar phenomenon as with Descartes' redefinition of geometrical exactness. For the four colour theorem, another proof algorithm was eventually proposed, significantly simpler than Appel and Haken's argument, but still computer-based and uncheckable by humans.[39] Most mathematicians tend to view the issue as settled by this second proof. Several other examples of computer-aided proofs are known today. It seems that the mathematical community has more or less accustomed itself to the idea that in certain mathematical arguments, humans may (or: must) rely on automated procedures, on machines. Wherever the generation of mathematical knowledge today involves such procedures, it is inextricably linked with the technological practices and cultural images of computing.

38 There is, for instance, no major account of the advent of computational fluid dynamics (CFD), one of the major scientific developments in the 20th century. For the role of computation in pure mathematics, see e.g. (Corry 2008).

39 This new algorithm was provided in 1996 by Robinson, Sanders, Seymour, and Robin, see (Robinson et al. 1997). Meanwhile a formal verification of crucial computing code of the proof has been added, see (Gonthier 2008).

The cultural variability of mathematical objects

We have just discussed the variability of normative patterns of mathematical argument such as exactness and proof and their dependence on other cultural practices. But what about mathematical notions themselves? Is a key concept of a mathematical theory such as 'mathematical probability', 'function' or 'triangle' similarly unstable and fluid in mathematical knowledge generation as, say, the 'vacuum' in experimental science? In the actual working with such concepts, can we trace similar links beyond mathematics as in experimental science? In the final section of my argument, I would like to illustrate that this is indeed the case, using an example from my own previous work on the history of mathematical knots.[40]

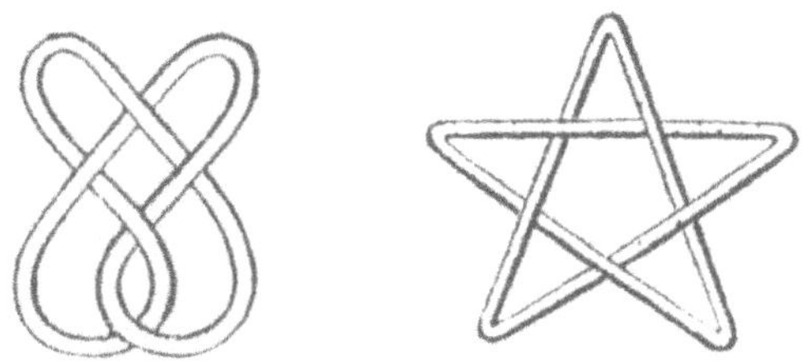

Figure 3: Knots as candidates for atoms?
Drawings in Peter Guthrie Tait's characteristic style (Tait 1877: 292).

To outsiders, this example may sound abstruse. Today's mathematicians, however, know that mathematical knots played at least two important roles in 20th century mathematics: Their investigation contributed to bringing about a major subfield of modern mathematics – topology – and the discovery of unexpected new results about knots by Vaughan F. Jones was awarded a Fields medal in 1990. Historians of physics will add that knots stood at the core of the first reductive theory of the microstructure of material atoms, William Thomson's theory of vortex atoms, which flourished in the two decades between 1867 and the mid 1880s and which motivated the Edinburgh physicist and mathematician William Guthrie Tait to work on a mathematical classification of knots for several years.[41]

But what is a mathematical knot, or rather, what has it been in history? The point is: It depends. In the practice of research, there was no fixed, ideal class of mathematical entities called "knots". To Tait, Thomson and the natural philosophers around him, a knot was either a physical object – something "constructed in cord" or a knotted vortex in ether – or an imagination of such, represented by means of diagrams on paper such as the ones illustrated in Fig. 3. In a certain practical sense these imaginations and diagrams *were* Tait's mathematical knots. All of his work was based on this conception, and indeed the mathematical arguments he developed shared its intuitiveness and vagueness. This was the perspective of a phycisist, interested both in down-to-earth experimenting (with cords or smoke rings) and in metaphysical speculation about the constitution of matter (which, Tait hoped, would turn out to be nothing but an immaterial configuration of motion in ether).

---

40 See in particular (Epple 1999a) and (Epple 2004).
41 See (Epple 1998); (Epple 1999a: ch. 4-5) and (Kragh 2002).

To early 20th century mathematicians, this outlook seemed dated and Tait's conception of knots was no longer acceptable. The knot theorists of the 1920's (such as Kurt Reidemeister in Vienna and James W. Alexander at Princeton) were trained in local environments which held mathematics to be a purely formal science, whose proofs were only dependent on logic, or, more radically, on the combinatorics of symbol systems. Accordingly, these mathematicians formed a new notion of knots. To them, a knot was just another specific entity of symbolic combinatorics, *neither* a physical object *nor* an imagination. To some extent, this was reflected even in their graphical representations of knots (which they found practically necessary as well): Rather than smooth curves, they used closed self-intersecting polygons with a finite number of sides. To their surprise, and only after hard work, they found that most results of Tait's "unrigorous" mathematics could be translated into this new framework – twenty years later, this framework would again change significantly.[42]

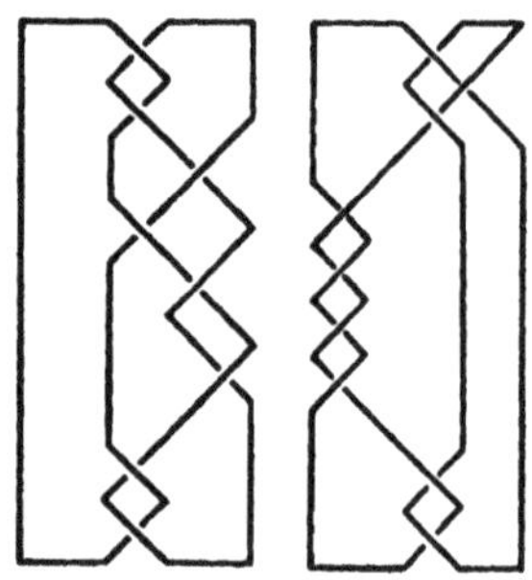

Figure 4: 'Modern' drawings of knots (Reidemeister 1932: 13).

The point of this little "biography of a mathematical object"[43] is not only that here certain objects of mathematical research turned out to be fluid, that they were redefined or refined. Rather, the point is that, as in histories of experimentation, the determinants and resources for this periodic reshaping of a class of mathematical objects could and did arise from non-mathematical contexts. Indeed for Tait, the determining context was the natural philosophical and metaphysical speculation of his day, among other factors, and for the topologists of the 1920's, it was the formalisation of mathematics and, as closer inspection reveals, indeed the larger context of intellectual modernism. Similar contextualized biographies might be given for many if not all objects of mathematical research. This should not come as a big surprise. Indeed, if Rheinberger's arguments about the interplay between epistemic objects and techniques of research are sound, they should be applicable, in suitably modified form, to all kinds of research, including mathematical research.[44] And it is to be expected that the epistemic configurations of mathematical research will turn out to be, upon closer inspection, similar hybrid units as experimental systems, albeit hybrids of a different kind.

42 For the "early modern" knot theory of the 1920's, see (Epple 1999a: ch. 10-11); for the further developments since the 1940's, see (Epple 1999b).

43 To vary a phrase coined by the collection of essays in (Daston 2000).

44 This is what I have suggested in (Epple 2004).

Cultural conjunctures of mathematical research

It remains to be discussed whether we also find in mathematics the kind of unexpected events described in the first section of this essay as *conjunctures* (in the wider sense), linking an episode of mathematical practice with an episode of another cultural and social practice.[45] Let us look, then, at a trace of such an event in the historical archive. In a remarkable monograph entitled *Das Chaos in kosmischer Auslese* (Chaos in cosmical selection), published in 1898 by an author going by the name of Paul Mongré, who obviously drew on Nietzschean ideas, we find a criticism of the Nietzschean doctrine of eternal recurrence that reads:

> "In any case, all attempts to decide, in a purely speculative way, for or against eternal recurrence will fail on the cliff which usually endangers all deductions *a priori*. Among them I also count Nietzsche's own attempt, published in his *Nachlass*, to base the necessity of recurrence on the completely untenable claim that the set of all states of the world is finite, and that the set of all moments is infinite, so that the finite content of the world must be repeated in infinite time infinitely often in order to completely fill it out. Time is linear, but the total of all states of the world could be represented even in the simplest, purely materialist conception as a point continuum of infinitely many dimensions – thus it cannot be arranged in a *continuous* manner within time. If one dispenses with the continuity of the mapping, then one indeed has to go a step further and to prove that the set of all states of the world has a higher cardinality than the linear continuum […]."[46]

The text obviously relates two topics one would not necessarily expect to be related, a then fashionable topic of Nietzscheanism, and Cantor's theory of transfinite sets. Obviously the passage tries to make use of set-theoretic notions to reject an argument sketched in Nietzsche's papers. The author of this text was Felix Hausdorff, better known to the literary public of these years – but not to us – as Paul Mongré, the Nietzschean writer. At the time he was a *Privatdozent* at Leipzig University who had written his dissertation and *Habilitation* with the astronomer Heinrich Bruns about the optics of the atmosphere, and he had not yet begun his work in set theory that would later earn him worldwide recognition. Under this pseudonym, Hausdorff experimented as an author of aphorisms and *Erkenntniskritik*.[47] The general thrust of his epistemological criticism was a fierce attack on all metaphysical assumptions about a given structure of time, space, and even the world. This "transcendent nihilism", as Hausdorff/Mongré termed his views, was coupled with the thesis that individual consciousness or activity is, in fact, the only source of the appearance of the world, for each individual, as an ordered 'cosmos'.[48] Moreover, these views were expressed not in the

---

45 That there are many examples of conjunctures of research episodes *within* mathematics in the sense that Rheinberger gave this term does not need, in my view, much further argument, once one accepts to study what I proposed to call the epistemic configurations of mathematical research; see the previous note.

46 My translation of (Hausdorff 1898: 193-194).

47 On some of Hausdorff's publications under the name Paul Mongré, see (Epple 2007) and (Epple 2009b). All of his writings have been or will be published in the volumes of Hausdorff's *Gesammelte Werke* (Heidelberg: Springer Verlag).

48 A first version of these views was published in aphoristic form in the last section *Zur Kritik des Erken-*

form of a scholarly philosophical treatise, but in a brilliant essay stretching the literary possibilities of the time, not least by using mathematical and scientific language as a rich source of literary imagination.

In pursuit of this literary activity (and not in his mathematical work of the time) Hausdorff encountered Cantor's recent work on transfinite sets. Just one year earlier, he had published another, aphoristic critique of Nietzsche's speculation on eternal recurrence that did *not* use Cantorian notions.[49] Between these two passages – and most probably motivated by his epistemological interests – Hausdorff learned about Cantor's work, and his monograph of 1898 makes extensive use of it for epistemological purposes. At this point, the two episodes of thought – Nietzschean criticism and the framing of a theory of transfinite sets – touched each other in a way that had not been anticipated. Indeed, the rich possibilities offered by Cantor's theory to imagine structures of temporal order or spatial point sets different from the usual ones obviously fascinated the young Nietzschean thinker. As it turned out, the conjuncture would also shape Hausdorff's later mathematical work. In 1901 he would be the first to ever give a lecture course on the theory of transfinite sets. At least in part, he was still motivated by epistemological concerns as another lecture course on *Zeit und Raum* taught in the winter term of 1903/1904 documents. Now, set theory was used as a tool to develop an axiomatic analysis of the possibilities of framing scientific notions of time and space.[50] More and more, however, set theory became a focus of Hausdorff's research rather than a resource for Mongré, the writer. Instead, the critical attitude of the latter became a resource for the mathematician, reversing the functional roles of the two activities.[51]

For instance, in an undated manuscript written sometime before 1908, Hausdorff addressed a curious mathematical problem and ended in a conjecture and an answer to a slightly twisted problem. The problem was the following: Suppose the 'world' to be such that space has only two points, whereas 'time' in this world can be modelled by the continuum of real numbers. Then any 'world motion' could be described by a function from the real numbers to the two space points. In this situation, can it happen that this motion (or 'world history') is *completely aperiodic* in the sense that no two intervals of time in which

---

nens of (Hausdorff 1897), then in monographic form in (Hausdorff 1898). There, the last chapter is entitled "Transcendenter Nihilismus". For a discussion of Hausdorff/Mongré's epistemological views, see (Epple 2006).

49 See (Hausdorff 1897: 349-354). That the difference between the two arguments against Nietzsche's 'proof' of eternal recurrence can be used to date Hausdorff's encounter with set theory has first been pointed out by Walter Purkert, see (Purkert 2002: 2-5).

50 The manuscript of the lecture course *Zeit und Raum* is contained in the Nachlass Hausdorff at the Universitäts- und Landesbibliothek Bonn, Abt. Handschriften und Rara, Fasz. 71. It will be published in vol. VI of Hausdorff's *Gesammelte Werke*. An epistemological analysis of space also was the topic of Hausdorff's inaugural lecture as an Extraordinarius in Leipzig (Hausdorff 1903). – Besides set theory, the axiomatic method as presented in Hilbert's *Grundlagen der Geometrie* of 1899 was another crucial resource for Hausdorff's analysis.

51 Still, Hausdorff's papers contain many sheets (most of them undated) on which he simultaneously engaged in epistemological reflection and mathematical argument. A significant amount of this boundary work will be included in vol. VI of Hausdorff's *Gesammelte Werke*.

infinitely many changes occur are mapped to the two points in a similar way?[52] The mathematical problem raised was tricky. It required proving the existence of a fairly monstrous function. Hausdorff showed in the manuscript that he could find such a function *if* a certain type of what he had baptized "gestufte Mengen" existed as a subset of the real numbers. It is not necessary to give a full definition of *gestufte Mengen* here,[53] but it should be added that Hausdorff had introduced this notion in an attempt to prove Cantor's so-called continuum hypothesis, one of the most difficult open problems in the theory of infinite sets before Paul Cohen gave an epistemologically surprising answer to it in the 1960's.[54] Moreover, Hausdorff showed that if time was taken to be a *slightly different* time continuum, not modelled by the real numbers but by a closely related, larger ordered set, then a positive answer to his problem could be given: Even with just 2 space points, a completely aperiodic 'world motion' was possible.

Hausdorff's discussion of this problem was clearly a further refinement of his earlier arguments about Nietzschean recurrence. If in such a simple model a completely aperiodic world history could be imagined, then even more so in more complex models. In other words, the intellectual link was still in place between Nietzscheanism, with all its cultural overtones, and advanced, pure mathematics at the forefront of research of the time. But now the Nietzschean motive had become a generator of a subtle mathematical problem. The intellectual *habitus* formed by working in this borderland between literature, epistemological criticism and set theory still informed Hausdorff's groundbreaking monograph *Grundzüge der Mengenlehre*, published in 1914.

Conclusion

My aim in this paper has not been to prove that experimental or mathematical knowledge is a "social" or "cultural construction", not even to show that there is a wealth of cultural determinants of scientific practice. Rather, I attempted to show why and how it is at all possible that practices of scientific knowledge generation could and can make use of cultural resources, in what forms they may be tied to cultural practices outside science. In experimental, as well as in mathematical science, the condition of this possibility is, I have argued, an intrinsic fluidity of all elements of scientific practice, ranging from the materialities of research activities over concepts to the normative patterns guiding these activities.

---

52 For Hausdorff the term 'similar' had a precise mathematical meaning in this context, i.e. there had to exist an order-preserving mapping from one time interval to the other such that the values of the function were preserved on composition with this mapping.

53 A 'gestufte Menge' is an ordered set in which no two initial segments are of the same order type.

54 Hausdorff's ideas were first published in (Hausdorff 1901). For a discussion of Hausdorff's early papers on set theory and their relation to the literary production of Paul Mongré see (Scholz 1996). – In 1963, Paul Cohen showed the continuum hypothesis to be independent of the usual (ZFC) axioms of set theory, i.e. it can neither be proved nor be disproved on their basis. This and similar results led to the insight that set theory (and mathematics) cannot be said to have *one* axiomatic foundation but that in fact many possible axiomatisations of slightly different set theories need to be studied – a result that is epistemologically similar to the 19th century acceptance of *many* systems of geometry in place of just one. Hausdorff would most probably have welcomed this conclusion, see (Koepke 1996).

On the one hand, this fluidity is due to the fact that scientific activities are, simply, just one of many cultural activities, which of course *all* share this variability. On the other hand, and this has been most strongly pointed out by Rheinberger, the epistemic and technological objects of both experimental research and mathematical argument are, and must be, fluid if research, by definition an activity of innovation, shall be possible. This indeed implies that the boundaries between scientific knowledge production, i.e. between the epistemic core of scientific activity, and culture at large are open boundaries. If there is no complete 'internal' (experimental or mathematical) determination of what a 'vacuum' or a 'knot' is, there is room for all kinds of other factors orienting scientific thought, enabling and inspiring scientific creativity, and helping material practices of research.

If this is so, my arguments also indicate a path along which more historical insight into the ways in which the generation of scientific knowledge is a cultural practice can be found. Localized, detailed investigations of the epistemic processes at these open boundaries of scientific research, of knowledge in the making, might not only lead to further unexpected insights into scientific creativity, but also to a better understanding of the social and cultural preconditions and effects of particular episodes of scientific activity. The most interesting episodes are those which redefined the style or content of experimentation or of the conception of mathematics, steps of mathematisation which changed the meaning of doing mathematics or concerned the uses of mathematics in socially and culturally important new areas of science.

What is needed for pursuing these ideas is a reordering of the historical archive. Rather than grouping its material around people, concepts, discoveries, or institutions, one would have to place traces of intellectual links at the center – traces of cultural, social, political, material resources involved in the production of specific scientific knowledge, traces of cultural norms regulating scientific research, or traces of unexpected conjunctures between scientific and cultural or social practices. The understanding of scientific knowledge production that would emerge from a reconstruction of the fabric of events around such links might differ substantially from the one we have today.

List of contributors

Lorraine Daston is one of the directors of the Max Planck Institute for the History of Science in Berlin and has published on a wide range of topics in the history of science, including the history of probability and statistics, wonders in early modern science, the emergence of the scientific fact, scientific models, objects of scientific inquiry, the moral authority of nature, and the history of scientific objectivity. Among her recent books is *Objectivity* (coauthored with Peter L. Galison, New York: Zone Books, 2007).

Sven Dierig works at the Institute for Nanotechnology Karlsruhe and teaches history of science at Technische Universität Berlin. Recent publication: *Wissenschaft in der Maschinenstadt. Emil Du Bois-Reymond und seine Laboratorien in Berlin* (Göttingen: Wallstein, 2006).

Moritz Epple teaches history of science at Johann Wolfgang Goethe University, Frankfurt am Main. His research concerns the history of the mathematical sciences since the 18th century, in particular in relation with technological, political, and philosophical issues. A recent publication is the exhibition catalogue *Jüdische Mathematiker in der deutschsprachigen akademischen Kultur* (coedited with Birgit Bergmann, Heidelberg: Springer, 2009).

Evelyn Fox Keller is professor emerita of the history and philosophy of science at the MIT. Keller's research focuses on the history and philosophy of modern biology and on gender and science. She is the author of several books, including *Making Sense of Life: Explaining Biological Development with Models, Metaphors, and Machines* (Cambridge, Mass.: Harvard University Press, 2002).

Mary-Joe Nye is a historian of science and retired faculty member of the History Department at Oregon State University. She is author and editor of several books, including: *Blackett. Physics, War, and Politics in the Twentieth Century* (Cambridge, Mass.: Harvard University Press, 2004).

Dominique Pestre, historian of science and technology, is Directeur d'études at the Ecole des Hautes Etudes en Sciences Sociales in Paris. He has worked on the relationship between physics and technology in the 20th century, and on the influence of war on scientific methods such as game theory, system analysis, or operative research. He is the author of the volume *Sience in the XXth Century* (coedited with John Krige, Amsterdam: Harwood Academic Publishers, 1997).

Hans-Jörg Rheinberger is one of the directors of the Max Planck Institute for the History of Science in Berlin. The main focus of his research lies in the history and epistemology of experimentation in the life sciences. Among his recent publications is *Epistemologie des Konkreten: Studien zur Geschichte der modernen Biologie* (Frankfurt am Main: Suhrkamp, 2006).

Friedrich Steinle teaches history of science at Technische Universität Berlin and is doing research on experiments in physics from a historical and philosophical perspective. His publications include *Explorative Experimente: Ampère, Faraday und die Ursprünge der Elektrodynamik* (Stuttgart: Franz Steiner, 2005).

Catherine Wilson is professor of philosophy at City University of New York. Her research interests concern the history and philosophy of science of the 17th and 18th centuries. A recent publication is *Moral Animals: Ideals and Constraints in Moral Theory* (Oxford: Oxford University Press, 2004).

M. Norton Wise is a professor of the history of science at the University of California in Los Angeles. He works on topics from the late 18th century to the present, specializing in the history of physics, and has been contributing to the historiographical development of a 'cultural history of science'. Presently, he is writing a book on *Bourgeois Berlin and Laboratory Science.*

Claus Zittel works at the Kunsthistorisches Institut / Max Planck Institute in Florence and teaches philosophy and German literature at the universities of Frankfurt am Main and Olsztyn (Poland). He is doing research, inter alia, on the historical epistemology of scientific images. A recent publication is *Theatrum philosophicum: Descartes und die Rolle ästhetischer Formen in der Wissenschaft* (Berlin: Akademie Verlag, 2009).

Bibliography

Archival Sources

Andrew 1903. Letter to Rivers, 13 May, in Cambridge University Library Haddon MSS envelope 12040.

Bois-Reymond, Emil du 1845. Certificate of membership in Berlin Physical Society. Staatsbibliothek Preußischer Kulturbesitz, Handschriftenabteilung. Sammlung Darmstädter, Emil du Bois-Reymond, 3 k 1841 (3), Blatt 59.

Bois-Reymond, Emil du 1853. Letter to Jeannette Claude, 10 March, Staatsbibliothek Preußischer Kulturbesitz, Handschriften Abteilung, Dep. 5, K.11, Nr. 5 [letter no. 66] (Nachlass Runge-du Bois-Reymond).

Bois-Reymond, Emil du 1924-1926. Laborheft no 1, Staatsbibliothek Preußischer Kulturbesitz, Handschriftenabteilung, acc. Darmstadt.

Bourne, Alfred Gibbs, Thurston, Edgar 1897. Administration report of the Madras Government Museum (1896-1897), Madras Education Proceedings 553, India Office Records P5278.

Christie, William 1898. Report on Indian Observatories with reference to the proposed scheme of reorganization, India Office Records L/E/7/425, Revenue and Statistics Department file 3225/99, fols. 51-59.

Eliot, John 1896. Letter to Ibbetson, 29 May and Moos to Ibbetson, 28 November 1896, in India Office Records L/E/7/425, Revenue and Satistics 3225/99, fols. 91-106, 111.

Eliot, John 1902. Letter to Christie, 23 January, Royal Greenwich Observatory MSS 7/158.

Evershed, John 1918. Letter to Fowler, 24 September, Royal Astronomical Society Evershed MSS 1 (1) no.4.

Evershed, John 1923. Letter to Turner, 19 and 22 December, Royal Astronomical Society Evershed MSS 1 (5) nos. 20-23.

Evershed, John 1930. Letter to Burns, 12 April, Royal Astronomical Society Evershed MSS 1 (5) nos. 20-23 ibid. 1 (1) no.15.

Frazer, James G. 1906. Letter to Rivers, Cambridge University Museum of Archaeology and Anthropology, Rivers MSS.

Gaertner, Eduard 1830. *Die Klosterstrasse*, Staatliche Museen zu Berlin – Preußischer Kulturbesitz, Nationalgalerie, Inv.-Nr. A II 736.

Gait 1910. Letter to Rivers, 25 March, Cambridge University Library Haddon MSS envelope 12031.

Hallman, Anton 1830. *Schattenstudien und Studien: Die Muskeln des Menschen*, Bayerische Staatsbibliothek München, Abteilung für Handschriften und Seltene Drucke, Ana 355. A. I. B.

Hufbauer, Karl 1998. *Kuhn's Discovery of History* (unpublished ms. 1998).

Hummel, Johann Erdmann 1831. *Das Schleifen der Granitschale*, Staatliche Museen zu Berlin – Preußischer Kulturbesitz, Nationalgalerie, Inv.-Nr. A I 884.

Hummel, Johann Erdmann 1834. *Die Entstehung der Zeichenkunst*, Staatliche Museen zu Berlin – Preußischer Kulturbesitz, Kupferstichkabinett, SZ Kat. 39.

Larmor, Joseph 1902. Letter to Christie, 26 February, Royal Greenwich Observatory MSS 7/158.

Lockyer, J. Norman 1873. Letter to Stokes. Cambridge University Library MSS Add 7656 L590.

Lockyer, J. Norman 1877. Letter to Salisbury, 26 June, reproduced in J. Norman Lockyer, Report on Indian Observatories and their Organization (1898), Indian Office Records L/E/7/425, fols. 24-49, on fol. 36.

Lockyer, J. Norman 1898. Report on Indian Observatories and their Organization. Indian Office Records L/E/7/425, fols. 24-49.

Moos, Nanabhoy Ardeshir Framji 1896. Letter to Ibbetson, 28 November, in India Office Records L/E/7/425, Revenue and Satistics 3225/99, fols. 91-106, 111.

Anonymus no date. Über die Förderung der Präzisionsmechanik durch den Staat. *Nationalzeitung*. Nachlass Emil du Bois-Reymond: Staatsbibliothek zu Berlin - Preußischer Kulturbesitz, Sammlung Darmstaedter.

Protocoll der Physikalischen Gesellschaft zu Berlin 1845. Archiv der Deutschen Physikalischen Gesellschaft, Nr. 10001, 7 March 1845.

Risley, Herbert 1908. Memorandum on the Ethnographic Survey of India (23 October) in India Office MSS Eur E.295 / 18.

Roscoe 1902. Letter to Haddon, 12 December, Cambridge University Library Haddon MSS envelope 3.

Samuel, P. 1903. Letter to Rivers, 20 May, Cambridge University Library Haddon MSS envelope 12040.

Smith, Charles Michie 1895, 1898. Letters to Christie, 25 April, 24 March, 12 May, Royal Observatory MSS 7 / 158.

Thurston, Edgar 1896a. *The Madras Museum as an aid to general and technical education* (31 October 1895). In Administration Report of the Madras Government Museum (1895-1896), Madras Education Proceedings (1896) no.454, India Office Records P5043, appendix E.

Thurston, Edgar 1904. Letter to Haddon, 18 May, Cambridge University Library, Haddon MSS envelope 3.

Thurston, Edgar 1905. Letter to Rivers, 9 August, in Cambridge University Library Haddon MSS envelope 12040.

Turner, Herbert 1923. Letter to Evershed, 31 December, Royal Astronomical Society Evershed MSS 1 (5) nos. 20-23.

Internet Sources

Dovers, Stephen R., Handmer, John W. 1992. *Contradictions in Sustainability*. URL: http://www.nmsu.edu/~iirm/sustlogos/dovers.html

Jen, Erica 2003. *Stable or Robust? What's the Difference?* URL: http://www.santafe.edu/~erica/stable.pdf.

Luntz Research Company. *The Environment: A Cleaner, Safer, Healthier America*. Document posted on the www.luntzspeak.com website (May 2005). This site is not active anymore. See similar informations on http://www.guardian.co.uk/environment/2003/mar/04/usnews.climatechange.

Locke, John. 1841 [1690]. *An Essay Concerning Human Understanding*. University of Adelaide: The University of Adelaide Library. URL: http://ebooks.adelaide.edu.au/l/locke/john/l81u/index.html.

Monographs, Essays and Anthologies

'Fighting obesity in schools' 2004. *Herald Tribune*. January 17-18.

Abelès, Marc 1999. Pour une exploration des sémantiques institutionnelles. *Ethnologie Française* 29: 501-511.

Affichard, Joëlle, de Foucauld, Jean-Baptiste, eds. 1995. *Pluralisme et équité, La justice sociale dans les démocraties*. Paris: Esprit.

Agar, Jon, Smith, Crosbie, eds. 1998. *Making space for science*. London: Macmillan.

Agassi, Joseph 1963. Towards an Historiography of Science. *History and Theory*, Beiheft 2.

Albertus Magnus 1971. De natura locorum. In Tilmann, Jean Paul, *An Appraisal of the Geographical Works of Albertus Magnus and his Contribution to Geographical Thought*. Ann Arbor: Dept. of Geography, University of Michigan.

Alderson, David, Willinger, Walter 2005. A contrasting look at self-organization in the Internet and next-generation communication networks. *Communications Magazine, IEEE*, 2005. URL: ieeexplore.ieee.org

Alembert, Jean le Rond d' 1751. Discours préliminaire. In *Encyclopédie, ou dictionnaire raisonné des sciences, des arts et des métiers*. Vol. 1. Paris: Briasson et al., i-xlv.

Alembert, Jean le Rond d' 1757. Géomètre. In *Encyclopédie, ou dictionnaire raisonné des sciences, des arts et des métiers*. Vol. 7. Paris: Briasson et al., 627-629.

Alembert, Jean le Rond d' 1767a [1753]. Essai sur la société des gens de lettres et des grands. In Alembert, Jean le Rind d', *Mélanges de littérature, d'histoire et de philosophie*. Quatrième édition. Vol. 1. Amsterdam: Zacharie Chatelain & Fils, 325-412.

Alembert, Jean le Rond d' 1767b [1759]. Essai sur les éléments de philosophie. In Alembert, Jean le Rond d', *Mélanges de littérature, d'histoire et de philosophie*. Quatrième édition. Vol. 4. Amsterdam: Zacharie Chatelain & Fils, 1-298.

Allen, Richard, ed. 1997. *Michael Polanyi, Society, Economics and Philosophy*. New Brunswick, New Jersey, London: Transaction Publishers.

Amann, Klaus 1994. Menschen, Mäuse und Fliegen. Eine wissenssoziologische Analyse der Transformation von Organismen in epistemische Objekte. In Hagner, Michael, Rheinberger, Hans-Jörg, Wahrig-Schmidt, Bettina, eds., *Objekte, Differenzen und Konjunkturen. Experimentalsysteme im historischen Kontext*. Berlin: Akademie Verlag, 259-289.

Andersen, Hanne 2001. *On Kuhn*. Belmont, CA.: Wadsworth.

Andresen, Jensine 1999. Crisis and Kuhn. *Isis* 90: 43-67.

Angell, Marcia, Relman, Arnold S. 2002. Patents, profits and American medicine: conflicts of interests in the testing and marketing of new drugs. *Daedalus* spring: 102-111.

Anstey, Peter 2002. Boyle on Seminal Principles. *Studies in History and Philosophy of Biological and Biomedical Sciences* 33: 597-630.

Aris, Rutherford et al., eds. 1983. *Springs of Scientific Creativity*. Minneapolis: University of Minnesota Press.

Ayyar, A.A. Narayana 1915. On the displacement at the Sun's limb of lines sensitive to pressure and density. *Bulletin of the Kodaikanal Observatory* 44: 113-116.

Babich, Babette 2003a. From Fleck's Denkstil to Kuhn's Paradigm: Conceptual Schemes and Incommensurability. *International Studies in the Philosophy of Science* 17: 75-92.

Babich, Babette 2003b. Paradigms and Thoughtstyles: Incommensurability and its Cold War Discontents from Kuhn's Harvard to Fleck's Unsung Lvov. *Social Epistemology* 17: 97-107.

Bachelard, Gaston 1969 [1938]. *La formation de l'esprit scientifique*. Paris: Vrin.

Bacon, Francis 1825-1834. Of Nature in Men, Essays Civil and Moral. In Montagu, Basil, ed., *Lord Bacon's Works*, 16 vols. London: William Pickering, vol. 1.

Bak, Per 1995. *How Nature Works: The Science of Self-Organized Criticality*. Oxford: Oxford University Press.

Bak, Per, Sneppen, Kim 1993. Punctuated equilibrium and criticality in a simple model of evolution. *Physical Review Letters* 71: 4083-86.

Bak, Per, Tang, Chao, Wiesenfeld, Kurt 1987. Self-Organized Criticality: An Explanation of 1/f Noise. *Physical Review Letters*. 59: 381-384.

Ballet, Jérôme, de Bry, Françoise 2002. Investissement socialement responsable et éthique, l'avenir de l'entreprise? *Cahiers Français* 309: 83-88.

Balogh, Brian 1991. Reorganizing the Organizational Synthesis: Federal – Professional Relations in Modern America. *Studies In American Political Development* 5: 119-172.

Baltas, Aristides, Gavroglu, Kostas, Kindi, Vassiliki 2000. A Discussion with Thomas S. Kuhn. In Conant, James, Haugeland, John, eds., *The Road Since Structure*. Chicago: University Press, 255-323.

Banthien, Henning, Jaspers, Michael, Renner Andreas 2003. *Final Report on Governance of the European Research Area: The Role of Civil Society*. Bensheim, Berlin, Brussels: European Commission.

Bayly, Christofer Alan 1996. *Empire and information: intelligence gathering and social communication in India 1780-1870*. Cambridge: Cambridge University Press.

Bayly, Susan 1999. *Caste, society and politics in India*. Cambridge: Cambridge University Press.

Beck, Hanno 1959. *Alexander von Humboldt*, 2 vols. Wiesbaden: Steiner.

Beck, Ulrich 2002. *Macht und Gegenmacht im globalen Zeitalter*. Frankfurt am Main: Suhrkamp.

Beer, Gillian 1996. *Open fields: science in cultural encounter*. Oxford: Oxford University Press.

Bellier, Irène 1999. L'invention de la'culture de Lomé. *Ethnologie Française* 29: 529-541.

Benda, Julien 1969. *The Treason of the Intellectuals*, trans. Aldington, Richard. New York: W. W. Norton. Originally *La Trahison des Clercs* (Paris: Grasset, 1927); English translation William Morrow and Co.: 1928.

Bergdoll, Barry 1994. *Karl Friedrich Schinkel: An Architecture for Prussia*. New York: Rizzoli.

Bernard, Claude 1961 [1865]. *Einführung in das Studium der Medizin*. Leipzig: Barth.

Bernoulli, Jakob 1713. *Ars conjectandi*. Basel: Impensis Thurnisiorum, fratrum.

Bessel, Friedrich Wilhelm 1815. Einleitung. *Astronomische Beobachtungen auf der Sternwarte zu Königsberg* 1: 9.

Beurton, Peter et al., eds. 2000. *The concept of the gene in development and evolution: historical and epistemological perspectives*. Cambridge: Cambridge University Press.

Biagioli, Mario 1993. *Galileo, Courtier: The Practice of Science in the Culture of Absolutism*. Chicago: University of Chicago Press.

Biagioli, Mario 1999a. *Galilei, der Höfling. Entdeckung und Etikette. Vom Aufstieg der neuen Wissenschaft*. Frankfurt am Main: Fischer.

Biagioli, Mario, ed. 1999b. *The Science Studies Reader*. New York: Routledge.

Biagioli, Mario 2006. *Galileo's Instruments of Credit: Telescopes, Images, Secrecy*. Chicago: University of Chicago Press.

Biagioli, Mario 2009. Postdisciplinary Liaisons: Science Studies and the Humanities. *Critical Inquiry* 35: 816-835.

Bialostocki, Jan 1986. *Dürer and his Critics, 1500-1971: Chapters in the History of Ideas, Including a Collection of Texts*. Baden-Baden: Koerner.

Biermann, Kurt-R. 1959. *Johann Peter Gustav Lejeune Dirichlet, Dokumente für sein Leben und Wirken*. Berlin: Akademie-Verlag.

Biermann, Kurt-R., ed. 1982. *Briefwechsel zwischen Alexander von Humboldt und Peter Gustav Lejeune Dirichlet*. Berlin: Akademie-Verlag.

Biermann, Kurt-R., ed. 1985. *Alexander von Humboldt. Vier Jahrzehnte Wissenschaftsförderung. Briefe an das preußische Kultusministerium, 1818-1859*. Berlin: Akademie-Verlag.

Bigg, Charlotte 2007. The Panorama, or La Nature à Coup d'Oeil. In Fiorentini, Erna, ed., *Observing Nature – Representing Experience. The Osmotic Dynamics of Romanticism 1800–1850*. Berlin: Reimer Verlag, 73–95.

Bigo, Didier 1998. Sécurité et immigration: vers une gouvernementalité par l'inquiétude? *Cultures et Conflits* 31-32 (automne-hiver): 13-38. URL: http://conflits.revues.org/article

Bilikiewicz, Tadeusz 1932. *Die Embryologie im Zeitalter des Barock und des Rokokko*. Leipzig: Thieme.

Bilikiewicz, Tadeusz 1939a. Odpowiedź na replikę Ludwika Flecka. *Przegląd Współczesny* 8-9: 175-176 (Reprinted in: Fleck 2007: 287-288).

Bilikiewicz, Tadeusz 1939b. Uwagi nad artykułem Ludwika Flecka "Nauka a środowisko". *Przegląd Współczesny* 18: 157-167, (Reprinted in: Fleck 2007: 271-281).

Bilski, Emily D. 1999. *Berlin Metropolis: Jews and the New Culture 1890-1918*. Berkeley, Los Angeles, London: University of California Press.

Birch, Thomas 1744. Life of the Honourable Robert Boyle. In Boyle, Robert, 1744 *The works of the Honourable Robert Boyle. In five volumes. To which is prefixed the life of the author.* 5 vols., edited by Birch, Thomas, vol. I. London: printed for A. Millar.

Bird, Alexander 2001. *Thomas Kuhn*. Princeton: Princeton University Press.

Blondiaux, Loïc, Sintomer, Yves 2002. L'impératif délibératif. *Politix* 15 (57): 17-35.

Boesinger, J. 1872. The total eclipse as seen at Ootacamund. *Nature* 5: 300-301.

Bohannan, Paul, Dalton, George 1965. Karl Polanyi 1886-1964. *American Anthropologist* 67: 1508-1511.

Bohr, Nils 1928. Das Quantenpostulat und die neuere Entwicklung der Atomistik. *Die Naturwissenschaften* 16 (35): 245-257.

Bois Reymond, Estelle du 1918. *Jugendbriefe von Emil du Bois-Reymond an Eduard Hallmann*. Berlin: Reimer.

Bois-Reymond, Emil du 1848-1884. *Untersuchungen über thierische Elektricität*, 2 vols. [vol.I, 1848; vol.II(1), 1849; vol. II(2), 1860; II(3), 1884]. Berlin: Reimer.

Bois-Reymond, Emil du 1862. *Über das Barrenturnen und die sogenannte rationelle Gymnastik*. Berlin: Reimer.

Bois-Reymond, Emil du 1863. *Hr. Rothstein und der Barren: Eine Entgegnung*. Berlin: Reimer.

Bois-Reymond, Emil du 1890. Nachruf auf Johann Georg Halske. *Verhandlungen der Physikalischen Gesellschaft zu Berlin* 9 (7): 39-44.

Bois-Reymond, Emil du 1912. Naturwissenschaft und bildende Kunst. In Bois-Reymond, Estelle du, ed., *Reden von Bois-Reymond, Emil du*. 2 vols, vol. II. Leipzig: Veit, 390-425.

Bois-Reymond, Emil du 1912a. Über tierische Bewegung. Im Verein für wissenschaftliche Vorträge zu Berlin am 22. Februar 1851 gehaltene Rede. In Bois-Reymond, Estelle du, ed., *Reden von Emil du Bois-Reymond*. Leipzig: Veit, vol. 2, 27-50.

Bois-Reymond, Emil du 1912b. Der physiologische Unterricht sonst und jetzt. Bei Eröffnung des neuen physiologischen Instituts der Berliner Universität am 6. November 1877 gehaltene Rede. In Bois-Reymond, Estelle du, ed., *Reden von Emil du Bois-Reymond*. Leipzig: Veit, Vol. I, 630-653.

Bois-Reymond, Emil du 1912c. Gedächtnisrede auf Johannes Müller. Gehalten in der Leibniz-Sitzung der Akademie der Wissenschaften am 8. Juli 1858. In Bois-Reymond, Estelle du, ed., *Reden von Emil du Bois-Reymond*. Leipzig: Veit, vol. 1, 135-317.

Bois-Reymond, Emil du 1912d. Naturwissenschaft und bildende Kunst. Zur Feier der Leibniz-Sitzung der Akademie der Wissenschaften zu Berlin am 3. Juli 1890 gehaltene Rede. In Bois-Reymond, Estelle du, ed., *Reden von Emil du Bois-Reymond*. Leipzig: Veit, vol. II, 390-425.

Bois-Reymond, Emil du 1912e . Goethe und kein Ende. In der Aula der Berliner Universität am 15. Oktober 1882 gehaltene Rektoratsrede. In Bois-Reymond, Estelle du, ed., *Reden von Emil du Bois-Reymond*. Leipzig: Veit, vol. II, 157-184.

Bois-Reymond, Emil du 1912f. Über die Übung. Zur Feier des Stiftungsfestes der militärärztlichen Bildungsanstalten am 2. August 1881 gehaltene Rede. In Bois-Reymond, Estelle du, ed., *Reden von Emil du Bois-Reymond*. Leipzig: Veit, vol. II., 99-141.

Bois-Reymond, Emil du 1912g. Über die wissenschaftlichen Zustände der Gegenwart. In der Sitzung der Akademie der Wissenschaften zur Geburtstagsfeier des Kaisers und Königs am 23. März 1882 gehaltene Rede. In Bois-Reymond, Estelle du, ed., *Reden von Emil du Bois-Reymond*. Leipzig: Veit, vol. II, 141-156.

Bois-Reymond, Emil du 1912h [1877]. Kulturgeschichte und Naturwissenschaft. In Bois-Reymond, Estelle du, ed., *Reden von Emil du Bois-Reymond*. Leipzig: Veit, vol. I, 577-629.

Bois-Reymond, Emil du 1912i. Die Humboldt-Denkmäler vor der Berliner Universität. In der Aula der Berliner Universität am 3. August 1883 gehaltene Rektoratsrede. In Bois-Reymond, Estelle du, ed., *Reden von Emil du Bois-Reymond*. Leipzig: Veit, vol. II, 249-285.

Bois-Reymond, Emil du 1918. Letter to Hallmann [May 1841]. In Bois-Reymond, Estelle du, ed., *Jugendbriefe von Emil du Bois-Reymond an Eduard Hallmann*. Berlin: Reimer, 93.

Bois-Reymond, Emil du 1927 [January 1841]. Letter to Ludwig. In Bois-Reymond, Estelle du, ed., *Zwei große Naturforscher des 19. Jahrhunderts. Ein Briefwechsel zwischen Emil du Bois-Reymond und Karl Ludwig*. Leipzig: J. A. Barth, 5.

Bollenbeck, Georg 1994. *Bildung und Kultur: Glanz und Elend eines deutschen Deutungsmusters*. Frankfurt am Main: Suhrkamp.

Boltanski, Luc, Chiapello, Eve 2000a. *Le nouvel esprit du capitalisme*. Paris: Gallimard.

Boltanski, Luc, Chiapello, Eve 2000b. The New Spirit of Capitalism. Paper presented to the Conference of Europeanists, March 14-16, Chicago.

Boltanski, Luc, Thévénot, Laurent 1991. *De la justification, Les économies de la grandeur*. Paris: Gallimard.

Bonah, Christian 2002. Experimental rage: the development of medical ethics and the genesis of scientific facts. Ludwik Fleck: an answer to the crisis of modern medicine in interwar Germany? *Social History of Medicine* 15 (2): 187-207.

Borck, Cornelius 2004. Message in a bottle from 'the crisis of reality': on Ludwik Fleck's interventions for an open epistemology. *Studies in History and Philosophy of Biological and Biomedical Sciences* [Special Issue – Ludwik Fleck: Epistemology and Biomedical Sciences] 35C (3): 447-464.

Börsch-Supan, Helmut 1997. Adolph Menzel und der Berliner Kunst. In Keisch, Claude, Riemann-Reyher, Marie Ursula, eds., *Adolph Menzel, 1815-1905: Das Labyrinth der Wirklichkeit*. Berlin: Nationalgalerie und Kupferstichkabinett, Staatliche Museen zu Berlin – Preußischer Kulturbesitz, 381-392.

Börsch-Supan, Helmut, ed. 1971. *Die Kataloge der Berliner Akademie-Ausstellungen 1786-1850*, 2 vols. & Registerband. Berlin: Bruno Hessling.

Börsch-Supan, Helmut, Grisebach, Lucius, eds. 1981. *Karl Friedrich Schinkel: Architektur, Malerei, Kunstgewerbe*. Berlin: Verwaltung der Staatlichen Schlösser und Gärten, Nicolaische Verlagsbuchhandlung.

Boruttau, Heinrich 1922. *Emil Du Bois-Reymond*. Wien, Leipzig, München: Rikola.

Bos, Henk 2001. *Redefining Geometrical Exactness. Descartes' Transformation of the Early Modern Concept of Construction*. New York: Springer.

Bose, Georg Matthias 1744. *Tentamina electrica in academiis regiis Londinensi et Parisana primum habita, omni studio repetita*. Wittenberg: Johann Ahlfeld.

Boudia, Soraya 1997. The Curie laboratory: radioactivity and metrology. *History of technology* 13: 249-265.

Boyle, James 2002. Fencing off ideas. *Daedalus* Spring: 13-25.

Boyle, Robert 1646-1647. Letter to Francis Tallents 20 Feb 1646-7. In (Boyle 1744), vol. I.

Boyle, Robert 1659. Some Motives and Incentives to the Love of God, in HD, vol. 1.

Boyle, Robert 1663. The Sceptical Chymist. In HD, vol. 2.

Boyle, Robert 1664a. Experiments and Considerations Touching Colours. In HD, vol 4.

Boyle, Robert 1664b. Occasional Reflections. In HD, vol. 5.

Boyle, Robert 1671. Excellency of Theology. In HD, vol. 8.

Boyle, Robert 1674. About the Excellency and Grounds of the Mechanical Hypothesis. In HD, vol. 8.

Boyle, Robert 1675a. On the Reconcileableness of Reason and Religion. In HD, vol. 8.

Boyle, Robert 1675b. Some Physico-theological Considerations about the Possibility of the Resurrection. In HD, vol 8.
Boyle, Robert 1675c. Some Considerations about the Reconcileableness of Reason and Religion. In (Boyle 1772), vol. 4.
Boyle, Robert 1686. A Free Inquiry into the Received Notion of Nature. In HD, vol. 10.
Boyle, Robert 1690. The Christian Virtuoso. In HD, vol. 11.
Boyle, Robert 1744. *The works of the Honourable Robert Boyle*. In five volumes. To which is prefixed the life of the author. 5 vols., edited by Birch, Thomas. London.
Boyle, Robert 1744a. An Account of Philaretus during his Minority. In (Boyle 1744), vol. I.
Boyle, Robert 1744b. Christian Virtuoso I, Appendix and II. In HD, vol. 12.
Boyle, Robert 1772. *The works of the Honourable Robert Boyle*. In six volumes. To which is prefixed The life of the author. 6 vols., edited by Thomas Birch, London.
Boyle, Robert 1966 [1772]. *The works*. Ed. Birch, Thomas, 6 vols. Reprint Hildesheim: Olms.
Boyle, Robert 2000. *The Works of Robert Boyle*. 14 vols., edited by Hunter, Michael, Davis, Edward B., London: Pickering and Chatto, (cited as "HD").
Brain, Robert 1998. Standards and semiotics. In Lenoir, Timothy, ed., *Inscribing Science*. Stanford: Stanford Universitiy Press, 249-284.
Brain, Robert M., Wise, M. Norton 1994. Muscles and Engines: Indicator Diagrams in Helmholtz's Physiology. In Krüger, Lorenz, ed., *Universalgenie Helmholtz: Rückblick nach 100 Jahren*. Berlin: Akademie Verlag, 124-145. (reprinted in Brain, Robert M., Wise, M. Norton 1999. Muscles and Engines: Indicator Diagrams in Helmholtz's Physiology. In Biagioli, Mario, ed., *The Science Studies Reader*. New York: Routledge, 51-66.)
Brandom, Bob 1994. *Making it explicit. Reasoning, representing, and discursive commitment*. Cambridge, Mass.: Harvard Universtity Press
Braufman, Nathalie, Foucart, Stéphane 2004. L'Etat veut diminuer ses dépenses informatiques grace aux logiciels libres, *Le Monde*, June 22.
Bredekamp, Horst 2007. *Galilei der Künstler. Die Zeichnungen, der Mond, die Sonne*. Berlin: Akademie Verlag.
Brian, Eric 1994. *La mésure de l'état: Administrateurs et géomètres au XVIIIéme siécle*. Paris: Albin Michel.
Brücke, Ernst 1892. *Schönheit und Fehler der menschlichen Gestalt*. Wien, Leipzig: Braumüller.
Brücke, Ernst Theodor 1928. *Ernst Brücke*. Wien: Springer.
Brudny, Michelle-Irène 2002. *Karl Popper: un philosophe heureux*. Paris: Grasset.
Buffon, Georges 1749. Premier discours. In Buffon, Georges, *Histoire naturelle*. Vol. I. Paris: Imprimerie Royale, 1-62.
Buomann, Johann 1932. *Der Barrenstreit und das Stützproblem*. Dissertation: Universität München.
Burg, Meno ^{3}1916 [1847]. *Geschichte meines Dienstlebens*. Leipzig: Kaufmann.
Busch, Werner 1985. *Die notwendige Arabeske: Wirklichkeitsaneignung und Stilisierung in der deutschen Kunst des 19. Jahrhunderts*. Berlin: Mann.
Busch, Werner, Beyrodt, Wolfgang, eds. 1982. *Kunsttheorie und Kunstgeschichte des 19. Jahrhunderts in Deutschland*: Texte und Dokumente, vol. I: Kunsttheorie und Malerei; Kunstwissenschaft. Stuttgart: Reclam.

Buse, Kent, Walt, Gill 2002. The World Health Organization and Global Public-Private Health Partnerships: In Search of 'Good' Global Health Governance. In Reich, Michael R., ed., *Public-Private Partnerships for Public Health*. Cambridge, Mass.: Harvard University Press, 169-195.

Caesarani, David 1998. *Arthur Koestler: The Homeless Mind*. New York: The Free Press.

Callon, Michel, Lascoumes, Pierre, Barthe, Yannick 2001. *Agir dans un monde incertain, essai sur la démocratie technique*. Paris: Seuil.

Calvin, Melvin 1991-1992. Memories of Michael Polanyi in Manchester. *Tradition and Discovery* 18: 40-42.

Camerini, Jane Rouder 1987. *Darwin, Wallace, and Maps*. University of Wisconsin-Madison Ph.D. Diss.

Cannon, Susan Faye 1978. *Science in Culture: The Early Victorian Period*. New York: Dawson and Science History Publications.

Casaubon, Meric 1669. *Letter of Meric Casaubon to Peter du Moulin concerning natural experimental philosophie*. Cambridge: Printed for William Mordgen.

Cash, John M. 1977. *Guide to the Papers of Michael Polanyi*. The Joseph Regenstein Library. Department of Special Collections. Chicago: University of Chicago Press.

CCE International (la revue des Conseillers du Commerce Extérieur de la France), 514, Juin-Juillet 2004, special issue on ONG – Entreprises, la nouvelle dynamique.

Charleton, Walter 1657. *The Immortality of the Soul*. London: Printed by William Wilson for Henry Herringman.

Choluj, Bozena, ed. 2007. *Von der wissenschaftlichen Tatsache zur Wissensproduktion. Ludwik Fleck und seine Bedeutung für die Diskussion um Wissen, Kultur und Macht*. Lang: Frankfurt am Main.

Christie, William, Turner, Herbert 1898. Report on the expedition to Sahdol. *Monthly Notices of the Royal Astronomical Society* 58: appendix, 1-21.

Clarke, Edwin, Jacyna, L. Stephen 1987. *Nineteenth-Century Origins of Neuroscientific Concepts*. Berkeley, Los Angeles, London: University of California Press.

Clay, John 1999. Robert Boyle: A Jungian Perspective. *British Journal for the History of Science* 32: 285-298.

Cohen, H. Floris 1994. *The Scientific Revolution: A Historiographical Inquiry*. Chicago: University of Chicago Press.

Cohen, Robert, Schnelle, Thomas, eds. 1986. *Cognition and Fact. Materials on Ludwik Fleck*. Boston: Reidel Publishing.

Cohn, Bernard 1990. *An anthropologist among the historians and other essays*. Delhi: Oxford University Press.

Cohn, Bernard 1996. *Colonianism and its Forms of Knowledge: The British in India*. Princeton: Princeton University Press.

Collingwood, Robin G. 1960. *The Idea of Nature*. Oxford: Oxford University Press.

Collins, Harry 1974. The TEA Set: Tacit Knowledge and Scientific Networks. *Science Studies* 4: 165-186.

Comenius, Amos 1657. Pansophia. De rebus omnibus in unum, perpetuum, lucidum, immotumque ordinem, et sub aeternis veritatis leger reducendis consultatur. In Comenius, Johann Amos, *Opera didactica*. Vol. 1, Amsterdam: De Geer.

Conant, James Bryant 1947. *On Understanding Science: An Historical Approach*. New Haven: Yale University Press.

Conant, James, Haugeland, John, eds. 2000. *The Road Since Structure: Thomas S. Kuhn*. Chicago: University of Chicago Press.

Condorcet, Marie Jean Antoine Nicolas 1828 [1817]. Commentaire sur le vingt-neuvième livre de l'Esprit des lois. In Destutt de Tracy, Antoine-Louis, *Commentaire sur l'Esprit des Lois de Montesquieu*. Paris: Dalibon, 380–381.

Congdon, Lee 1991. Exile and Social Thought: Hungarian Intellectuals in Germany and Austria, 1919-1933. Princeton: Princeton University Press.

Corry, Leo 2008. Number Crunching vs. Number Theory: Computers and FLT, from Kummer to SWAC (1850-1960), and beyond. *Archive for History of Exact Sciences* 62 (1): 393-455.

Crombie, Alistair C., ed. 1963. *Scientific Change: Historical Studies in the Intellectual, Social, and Technical Conditions for Scientific Discovry and Technical Invention, from Antiquity to the Present*. New York: Basic Books, originally published in London: Heinemann, 1961.

Czermak, Johann Nepomuk 1879. Über das physiologische Privat-Laboratorium an der Universität Leipzig. Rede, gehalten am 24. December 1872 bei Gelegenheit der Eröffnung seines Amphitheaters. In Czermak, Johann Nepomuk, *Gesammelte Schriften*, 2 vols., Leipzig: Engelmann, vol. II(2), 118-144.

Dahan, Amy, Pestre, Dominique, eds. 2004. *Les sciences pour la guerre*. Paris: EHESS.

Dąmbska, Izydora 1937. Czy intersubiektywne podobieństwo wrażeń zmysłowych jest niezbędnym założeniem nauk przyrodniczych? *Przegląd Filozoficzny* 40 (3): 288-294, (Reprinted in Fleck 2007: 251-258).

Danto, Arthur C. 1965. *Analytical Philosophy of History*. Cambridge: Cambridge History of Science.

Darrigol, Olivier 2005. *Worlds of Flow: A History of Hydrodynamics from the Bernoullis to Prandtl*. Oxford: Oxford University Press.

Dasgupta, Partha 2000. Science as an Institution: Setting Priorities in a New Socio-economic Context. *World Conference on Science, Science for the 21st century, A New Commitment*. Paris: UNESCO, 264-271.

Daston, Lorraine 1988. *Classical Probability in the Enlightenment*. Princeton: Princeton University Press.

Daston, Lorraine 1991. The Ideal and Reality of the Republic of Letters in the Enlightenment. *Science in Context* 4 (2): 367-338.

Daston, Lorraine 2007. Condorcet and the Meaning of Enlightenment. *Proceedings of the British Academy* 151: 113-134.

Daston, Lorraine, ed. 2000. *Biographies of Scientific Objects*. Chicago: University of Chicago Press.

Daston, Lorraine, Galison, Peter 1992. The Image of Objectivity. *Representations* 40: 81-128.

Daston, Lorraine, Galison, Peter 2007. *Objectivity*. New York: Zone Books.

Davis, Mike 2001. *Late Victorian holocausts: el Niño famines and the making of the Third World*. London: Verso.

Davies, Paul Charles William 1989. The physics of complex organisation. In Goodwin Brian, Saunders Peter, eds., *Theoretical Biology: Epigenetic and Evolutionary Order from Complex Systems*. Edinburgh (Scotland, United Kingdom): Edinburgh University Press, 101-111.

Dear, Peter 1995a. Cultural History of Science: An Overview with Reflections. *Science, Technology, and Human Values* 20: 150-170.

Dear, Peter 1995b. *Discipline and Experience: The mathematical way in the scientific revolution*. Chicago: University of Chicago Press.

Dejours, Christophe 1998. *Souffrance en France, La banalisation de l'injustice sociale*. Paris: Seuil.

Demeritt, David 2001. The Construction of Global Warming and the Politics of Science. *Annals of the Association of American Geographers* 91 (2): 307-337.

Derham, William 1716. *Physico-Theology*. 4th rev. ed. London: W. Innys.

Descartes, René 1954 [1637]. *La géométrie*. Leiden: Ian Maire. Reprinted and translated into English by David Eugene Smith and Marcia L. Latham 2007 as: *The Geometry of René Descartes, With a Facsimile of the First Edition*. New York: Dover.

DeRose, Keith 1995. Contextualism and Knowledge Attributions. *Philosophy and Phenomenological Research* 52: 913-929.

Detienne, Marcel, Vernant, Jean-Pierre 1978. *Cunning Intelligence in Greek Culture and Society*. Hassocks: Harvester.

Dettelbach, Michael 1999. The Face of Nature: Precise Measurement, Mapping, and Sensibility in the Work of Alexander von Humboldt. *Studies in History and Philosophy of Biological and Biomedical Sciences* 30: 473-504.

Dhombres, Nicole, Dhombres, Jean 1989. *Naissance d'un pouvoir: Sciences et savants en France*, 1793-1824. Paris: Payot.

Dickinson, Henry Winram, Jenkins, Rhys reprint 1981 [1927]. *James Watt and the Steam Engine*. Derbyshire: Moorland.

Diderot, Denis 1875 [1753]. *Pensées sur l'interpretation de la nature*. In Diderot, Denis, *Œuvres complètes*, ed. J. Assézat. Vol. 2. Paris: Garnier Frères.

Dierig, Sven 2001. Consordino for Piano and Brain. Bohemian Neuroscience in a 1900 Cultural Metropolis. *Configurations* 9: 413-440.

Dierig, Sven 2003. Engines for Experiment – Laboratory Revolution and Industrial Labor in the Nineteenth-Century City. *Osiris* 18: 116-134.

Dierig, Sven 2004. Die Kunst des Versuchens. In Schmidgen, Henning; Geimer, Peter, Dierig, Sven, eds., *Kultur im Experiment*. Berlin: Kadmos, 123-146.

Dierig, Sven 2006. *Wissenschaft in der Maschinenstadt*. Göttingen: Wallstein.

Dierig, Sven 2006a. Science and Craftsmanship: The art of Experiment and Instrument Making. *Comptes rendus Biologies* 329: 348-353.

Dierig, Sven 2006b. *Wissenschaft in der Maschinenstadt: Emil du Bois-Reymond und seine Laboratorien in Berlin*. Göttingen: Wallstein.

Digby, Kenelm 1643. *Observations upon Religio Medici*. London: Printed by R.C. for Lawrence Chapman and Daniel Frere.

Dijksterhuis, Eduard Jan 1956. *Die Mechanisierung des Weltbildes*. Berlin: Springer.

Dirichlet, G. Lejeune 1829. Sur la convergence des séries trigonométriques qui servent a représenter une fonction arbitraire entre des limites données. In *G. Lejeune Dirichlet's Werke* 1889 & 1897, 2 vols. Berlin: Reimer, vol. I: 118-132 (Originally in Crelle's *Journal für die reine und angewandte Mathematik* 4: 157-169).

Dirichlet, G. Lejeune 1837. Ueber die Darstellung ganz willkührlicher Funktionen durch Sinus- und Cosinusreihen. *Repertorium der Physik* 1: 152-174.

Dirks, Nicholas B. 2001. *Castes of Mind: Colonialism and the Making of Modern India*. Princeton: Princeton University Press.

Dobbert, Eduard, Meyer, Alfred 1899. *Chronik der königlichen technischen Hochschule zu Berlin: 1799-1899*. Berlin: Wilhelm Ernst & Sohn.

Donagan, Alan 1964. The Popper-Hempel Theory of Historical Explanation. *History and Theory* 4: 3-26.

Doppelmayr, Johann Gabriel 1744. *Neu-entdeckte Phaenomena von bewunderungswürdigen Würkungen der Natur*. Nürnberg: Johann Joseph Fleischmann.

Dove, Heinrich Wilhelm 1826a. *De barometri mutationibus*. Dissertation, Berlin.

Dove, Heinrich Wilhelm 1826b. *De distributione caloris per tellurem*. Habilitations-Schrift, Königsberg.

Dove, Heinrich Wilhelm 1827. Einige meteorologische Untersuchungen über den Wind. *Annalen der Physik und Chemie* 11: 545-590.

Dove, Heinrich Wilhelm 1830. Correspondirende Beobachtungen über die regelmässigen stündlichen Veränderungen und über die Perturbationen der magnetischen Abweichung im mittleren und östlichen Europa; mit einem Vorwort von Alexander von Humboldt. *Annalen der Physik und Chemie* 19: 357-391.

Dove, Heinrich Wilhelm 1837. *Meteorologische Untersuchungen*. Berlin: Sander'sche Buchhandlung.

Dove, Heinrich Wilhelm 1841. Ueber das Gesetz der Stürme. *Annalen der Physik und Chemie* 52: 1-41 (first presented in the Academy of Sciences in 1840).

Dove, Heinrich Wilhelm 1869. *Gedächtnissrede auf Alexander von Humboldt*. Berlin: F. Dümmler.

Dove, Heinrich Wilhelm 1857. *Ueber das Gesetz der Stürme*. Berlin: Reimer.

Dray, William 1964. *Philosophy of History*. Englewood Cliffs, NJ: Prentice-Hall.

Dray, William 1973. The Politics of Contemporary Philosophy of History: A Reply to Hayden White. *Clio* 3: 54-76.

Dubet, François 2006. *Injustices, l'expérience des inégalités au travail*. Paris: Seuil.

Dufay, Charles François de Cisternai 1733. Quatrième mémoire sur l'électricité. De l'attraction et répulsion des corps électriques. *Histoire de l'Académie Royale des Sciences, avec les Mémoires de Mathématique & de Physique pour la même année*: 457-477.

Dufay, Charles François de Cisternai 1734. A Letter from Mons. Du Fay, F. R. S. and of the Royal Academy of Sciences at Paris, to His Grace Charles Duke of Richmond and Lenox, concerning Electricity. Translated from the French by T. S. M D. *Philosophical Transactions* 38 (431): 258-266.

Dupuy, François 2005. *La fatigue des élites, le capitalisme et les cadres*. Paris: Seuil.

Dürer, Albrecht 1808. Albrecht Dürers christlich-mythologische Handzeichnungen: Lithographien von J.N. Strixner nach den Randzeichnungen im Gebetbuch Kaiser Maximilians. München: Lithographische Presse Alois Senefelder.

Earman, John, Glymour, Clark 1980a. Relativity and eclipses. *Historical studies in physical science* 11: 49-85.

Earman, John, Glymour, Clark 1980b. The gravitational redshift as a test of general relativity: history and analysis. *Studies in History and Philosophy of Science* 11: 251-278.

Eckert, Michael 2002. Euler and the Fountains of Sanssouci. *Archive for History of the Exact Sciences* 56: 451-468.

Eco, Umberto, Sebeok, Thomas, eds. 1983. *The Sign of Three: Dupin, Holmes, Peirce.* Bloomington: Indiana University Press.

Eddington, Arthur 1917. Einstein's theory of gravitation. *Monthly Notices of the Royal Astronomical Society* 77: 377-382.

Edgerton, David 2004. The linear model' did not exist: Reflections on the history and historiography of science and research in industry in the twentieth century. In Grandin, Karl, Wombs, Nina, eds., *The Science - Industry Nexus: History, Policy, Implications.* New York: Watson, 31-57.

Edwards, Elizabeth 1990. The image as anthropological document - photographic types: the pursuit of method. *Visual anthropology* 3: 235-258.

Edwards, Elizabeth 2001. *Raw histories: photographs, anthropology and museums.* Oxford: Berg.

Egloff, Rainer, ed. 2004. *Tatsache Denkstil Kontroverse: Begegnungen mit Ludwik Fleck.* Zürich: Collegium Helveticum.

Ehrenberg, Alain 2000. *La Fatigue d'être soi, Dépression et Société.* Paris: Odile Jacob.

Eigen, Manfred 1971a. Selforganization of matter and the evolution of biological macromolecules. *Naturwissenschaften* 58 (10): 465-523.

Eigen, Manfred 1971b. Molecular self-organization and the early stages of evolution. *Quartly Revviews Biophysics.* Aug; 4 (2): 149-212.

Eliot, George 1963. *Essays of George Eliot.* Ed. Thomas Pinney, New York: Columbia University Press.

Engelhardt, Dietrich von 1981. Der Begriff der Bildung und Kultur bei du Bois-Reymond. In Mann, Gunter, ed., *Naturwissen und Erkenntnis im 19. Jahrhundert, Emil du Bois-Reymond.* Hildesheim: Gerstenberg Verlag, 173-186.

Engelmann, Wilhelm 1969 [1906]. *Daniel Chodowiecki's sämmtliche Kupferstiche,* facsimile of 2nd ed. 1906. Hildesheim: Olms.

Epple, Moritz 1998. Topology, Matter, and Space, I: Topological Notions in 19th-Century Natural Philosophy. *Archive for History of Exact Sciences* 52: 297-392.

Epple, Moritz 1999a. *Die Entstehung der Knotentheorie: Kontexte und Konstruktionen einer modernen mathematischen Theorie.* Wiesbaden: Vieweg.

Epple, Moritz 1999b. Geometric Aspects in the Development of Knot Theory. In James, Ioan M., ed., *History of Topology.* Amsterdam: Elsevier, 301-357.

Epple, Moritz 2004. Knot Invariants in Vienna and Princeton During the 1920s: Epistemic Configurations of Mathematical Research. *Science in Context* 17: 131-164.

Epple, Moritz 2006. Felix Hausdorff's Considered Empiricism. In Ferreiros, José, Gray Jeremy J., eds., *The Architecture of Modern Mathematics: Essays in History and Philosophy.* Oxford: Oxford University Press, 263-289.

Epple, Moritz 2007. An Unusual Career between Cultural and Mathematical Modernism: Felix Hausdorff, 1868-1942. In Charpa, Ulrich, Deichmann, Ute, eds., *Jews and Sciences in German Contexts: Case Studies from the 19th and 20th Centuries.* Tübingen: Mohr und Siebeck, 77-99.

Epple, Moritz 2008a. Mathematische Wissenschaften. In *Enzyklopädie der Neuzeit.* Vol. 8. Stuttgart: Metzler, 140-163.

Epple, Moritz 2008b. The Gap Between Theory and Practice: Hydrodynamical and Hydraulical Utopias in the 18th Century. In Zittel, Claus et al. eds., *Philosophies of Technology: Francis Bacon and his Contemporaries.* Leiden: Brill, 457-494.

Epple, Moritz 2009a. Kulturen der Forschung: Mathematik und Modernität am Beginn des 20. Jahrhunderts. In Fried, Johannes, Stolleis, Michael, eds., *Wissenskulturen: Über die Erzeugung und Weitergabe von Wissen*. Frankfurt am Main: Campus, 125-158.

Epple, Moritz 2009b. Spielräume des Denkens: Felix Hausdorff und Paul Mongré. In Schwarz, Astrid, Nordmann, Alfred, eds., *Das bunte Gewand der Theorie: Vierzehn Begegnungen mit philosophierenden Forschern*. Freiburg, München: Karl Alber, 235-262.

Eriksson, Gunnar 1994. Linnaeus the Botanist. In Frangsmyr, Tore, ed., *Linnaeus: The Man and His Work*. Canton, Mass.: Science History Publications, 63–109.

Erkman, Suren 1998. *Vers une écologie industrielle*. Paris: Charles Leopold Mayer.

Evershed, John 1909. Radial movements in sunspots. *Monthly Notices of the Royal Astronomical Society* 69: 454-457.

Evershed, John 1911. Autocollimating spectroheliograph of the Kodaikanal Observatory. *Monthly Notices of the Royal Astronomical Society* 71: 719-723.

Evershed, John 1912. Kodaikanal and Madras Observatories. *Monthly Notrices of the Royal Astronomical Society* 72: 286-287.

Evershed, John 1913a. Kodaikanal and Madras Observatories. *Monthly Notices of the Royal Astronomical Society* 73: 258-259.

Evershed, John 1913b. The displacement of the lines of the solar spectrum towards the red. *Observatory* 37: 124-128.

Evershed, John 1918. Displacement of cyanogen bands in the solar spectrum. *The Observatory* 41: 371-375.

Evershed, John 1920. Displacement of the lines in the solar spectrum and Einstein's prediction. *Observatory* 43: 153-157.

Evershed, John 1920. Kodaikanal and Madras Observatories. *Monthly Notes of the Royal Astronomical Society* 80: 395-396.

Evershed, John 1921a. The relativity shift in the solar spectrum. *The Observatory* 44: 159-160, 243-245.

Evershed, John 1921b. The Einstein Tower. *The Observatory* 44: 372-373.

Evershed, John 1955. Recollections of seventy years of scientific work. *Vistas in astronomy* 1: 33-40.

Falk, Richard 1999. *Predatory Globalization: A Critique*. Cambridge: Polity Press.

Fassin, Didier, Memmi, Dominique, eds. 2004. *Le gouvernement des corps*. Paris: EHESS.

Fauser, Winfried, ed. 1982. *Die Werke des Albertus Magnus in ihrer handschriftlichen Überlieferung*. Münster: Aschendorff.

Fehlemann, Sabine 1974. *Der Maler-Architekt Anton Hallmann, 1812-1845. Leben und Werk mit einem Oeuvre-Verzeichnis*. München: Diss.

Fend, Mechthild 2003. *Grenzen der Männlichkeit*. Berlin: Reimer.

Festa, Egido, Roux, Sophie 2001. Le 'παρα φυσιν' et l'imitation de la nature dans quelques commentaires du prologue des Questions mécaniques. In Montesinos, José, Solís, Carlo, eds., *Longo campo di filosofare. Eurosymposium Galileo 2001*. Canary Islands: Fundación Canaria Orotavia de Historia de la Ciencia, 237–253.

Fichman, Martin 2002. *Evolutionary Theory and Victorian Culture*. Amherst, NY: Humanity Books.

Fifty Years of Chemical Dynamics 1981. *Berichte der Bunsen-Gesellschaft* 86: 348-464.

Figlio, Karl 1999. Psychoanalysis and the Scientific Mind: Robert Boyle. *British Journal for the History of Science* 32: 299-314.

Finkelstein, Gabriel W. 1996. Emil Du Bois-Reymond: The Making of a Liberal German Scientist, 1818-1851. PhD Diss.: Princeton University.

Finkelstein, Gabriel W. 2003. M. Du Bois-Reymond goes to Paris. *British Journal for the History Science* 36 (3): 261-300.

Fischer, Franz 1936. Rezension von Ludwik Flecks Entstehung und Entwicklung einer wissenschaftlichen Tatsache. *Der Nervenarzt* 9.

Fitoussi, Jean-Paul 2002. Démocratie et mondialisation. *Revue de l'OFCE* mars 2002: 7-18.

Fleck, Ludwik 1937. W sprawie artykułu p. Izydory Dąmbskiej w Przeglądzie Filozoficznym, rocz. 40 zesz. III [In der Angelegenheit des Artikels von Fr. Izydora Dąmbska] in 'Przgląd Filozoficzny', Jg. 40, Heft III]. *Przegląd Filozoficzny* 41: 192-195, (Reprinted in: Fleck 2007: 258-263).

Fleck, Ludwik 1939a. Nauka a środowisko [Wissenschaft und Umwelt]. *Przegląd Współczesny* 18 (8-9): 149-156, (Reprinted in: Fleck 2007: 264-270).

Fleck, Ludwik 1939b. Odpowiedź na uwagi Tadeusza Bilikiewcza [Antwort auf die Anmerkungen von Tadeusz Bilikiewicz]. *Przegląd Współczesny* 18: 168-174, (Reprinted in: Fleck 2007: 281-287).

Fleck, Ludwik 1983. *Erfahrung und Tatsache. Gesammelte Aufsätze*. Frankfurt am Main: Suhrkamp.

Fleck, Ludwik 2007. In der Angelegenheit des Artikels von Fr. Izydora Dąmbska in 'Przegląd Filozoficzny'. In: Choluj, Bozena, ed., *Ludwik Fleck und seine Bedeutung für die Diskussion um Wissen, Kultur und Macht*. Frankfurt an der Oder, 439-473.

Fleck, Ludwik 2007. *Style myślowe i fakty. Artykuły i świadectwa*. Pod redakcją Sylwii Werner, Clausa Zittla i Floriana Schmaltza. Warszawa: Wydawnictwo Instytutu Filozofii i Socjologii Polskiej Akademii Nauk.

Fleming, James Rodger 1990. *Meteorology in America, 1800–1870*. Baltimore, London: Johns Hopkins University Press.

Fontenelle, Bernard de 1728. Préface sur l'utilité des Mathématiques et de la Physique, et sur les travaux de l'Académie des Sciences. In Fontenelle, Bernard de, *Histoire du renouvellement de l'Académie Royale des Sciences*. Paris: Jean Boudot, 1-34.

Fontenelle, Bernard de 1733. Histoire de l'Académie Royale des Sciences: Sur l'électricité. Histoire de l'Académie Royale des Sciences, avec les Mémoires de Mathématique & de Physique pour la même année: 4-13.

Forbes, Eric 1961. A history of the solar red shift problem. *Annals of science* 17: 129-164.

Forman, Paul 1971. Weimar Culture, Causality, and Quantum Theory: Adaptation by German Physicists and Mathematicians to a Hostile Environment. *Historical Studies in the Physical Sciences* 3: 1–115.

Forman, Paul 1984. Kausalität, Anschaulichkeit and Individualität, or How Cultural Values Prescribed the Character and Lessons Ascribed to Quantum Mechanics. In Stehr, Nico, Meja, Volker, eds., *Society and Knowledge*. New Brunswick: Transaction Books, 333–347.

Forman, Paul 1987. Behind Quantum Electronics: National Security as Basis for Physical Research in the United States, 1940–1960. *Historical Studies in the Physical Sciences* 18: 149–229.

Forman, Paul 1997. Recent science: Late-modern and post-modern. In Söderqvist, Thomas, ed., *The Historiography of Contemporary Science and Technology*. Amsterdam: Harwood Academic Publishers, 179-213.

Forman, Paul, Sánchez-Ron, José M., eds. 1996. *National Military Establishments and the Advancement of Science and Technology: Studies in Twentieth Century History*. Dordrecht: Kluwer.

Foucault, Michel 1969. *L'Archéologie du savoir*. Paris: Gallimard.

Foucault, Michel 1994. *Dits et Ecrits, IV*. Paris: NRf Gallimard.

Fox Keller, Evelyn 2002. *Making Sense of Life: Explaining Biological Development with Models, Metaphors, and Machines*. Cambridge: Harvard University Press.

Fox Keller, Evelyn 2005. Revisiting "Scale-Free" Networks. *BioEssays* 27 (1): 1060-1068.

Fox Keller, Evelyn, Segel, Lee A. 1970. Slime Mold Aggregation viewed as an Instability. *Journal of Theoretical Biology* 26: 399-415.

Francis, W. 1906. *Gazetteer of the Madura district*. Madras: Government Press.

Frazer, James 1907-1915. *The Golden Bough*, 11 volumes. London: Macmillan.

Frercks, Jan 2004. Disziplinenbildung und Vorlesungsalltag. Funktionen von Lehrbüchern der Physik um 1800 mit einem Fokus auf die Universität Jena. *Berichte zur Wissenschaftsgeschichte* 27: 27-52.

Fritsch, Gustav 1910. Das physiologische Institut. In Lenz, Max, ed., *Geschichte der Königlichen Friedrich-Wilhelms-Universität zu Berlin*. Halle: Verlag der Buchhandlung des Waisenhauses, Bd. 3: Wissenschaftliche Anstalten, 154-163.

Fuchs, Thomas 1992. *Die Mechanisierung des Herzens. Harvey und Descartes – Der vitale und der mechanische Aspekt des Kreislaufs*. Frankfurt am Main: Suhrkamp.

Fuller, Steve 2000. *Thomas Kuhn: A Philosophical History for Our Times*. Chicago: University of Chicago Press.

Fuller, Steve 2004. *Kuhn vs. Popper: The Struggle for the Soul of Science*. New York: Columbia University Press.

Gale, Ernest F., Folkes, Joan P. 1954. Effect of nucleic acids on protein synthesis and amino-acid incorporation in disrupted staphylococcal cells. *Nature* 173: 1223-1227.

Galen 1968. *Galen on the Usefulness of the Parts of the Body. De usu partium..* Tr. Margaret Tallmadge May, 2 vols., Ithaca: Cornell University Press.

Galilei, Galileo 1974 [1638]. *Two New Sciences*, trans. Stillman Drake. Madison: University of Wisconsin Press.

Galison, Peter 2003. *Einstein's clocks, Poincaré's maps: empires of time*. New York: Norton.

Galison, Peter L., Gordin, Michael, Kaiser, David, eds. 2001. *Science and Society: The History of Modern Physical Science in the Twentieth Century*. 4 vols. New York: Routledge.

Galison, Peter, Thompson, Emily, eds. 1999. *The architecture of science*. Cambridge, Mass: MIT Press.

Galton, Francis 1888. Personal identification and description – II. *Nature* 28 June: 201-202.

Galton, Francis, 1908. *Memories of my Life*. London: Methuen.

Gardner, Patrick, ed. 1974. *The Philosophy of History*. Oxford: Oxford University Press.

Garnier, Jean-Pierre 2001. La critique radicale a-t-elle encore droit de Cité? *Espaces et Sociétés* 100-102: 35-59.

Garson, John George 1900. The metric system of identification of criminals as used in Great Britain and Ireland. *Journal of the Anthropological Institute* 30: 161-198.

Gaukroger, Stephen 2008. The Académie des Sciences and the Republic of Letters: Fontenelle's Role in the Shaping of a New Natural-Philosophical Persona, 1699-1734. *Intellectual History Review* 18: 385-402.

Geertz, Clifford 1973. Thick Description. Toward an Interpretive Theory of Culture. In Geertz, Clifford, ed., *The Interpretation of Cultures. Selected Essays*. New York: Basic books, 3-30.

Genz, Axel 1976. *Die Emanzipation der naturwissenschaftlichen Physiologie in Berlin*. Diplom-Arbeit: Universität Magdeburg.

Gerhart, John, Kirschner, Marc 2007. The theory of facilitated variation. *Proceedings of the National Academy of Sciences* 104 (suppl.): 8582-8589.

Gerth, Hans H., Mills, C. Wright, trans. and ed. 1948. *From Max Weber: Essays in Sociology*. London: Routledge and Kegan Paul.

Ghurye, Govind Sadasiv 1932. *Caste and Race in India*. New York: Kegan Paul, Trench, Trubner & Co.

Giere, Ronald N. 1988. *Explaining Science: a Cognitive Approach*. Chicago: University of Chicago Press.

Ginsborg, Hannah 2001. Kant on Understanding Organisms as Natural Purposes. In Watkins, Eric, ed., *Kant and the Sciences*. Oxford: Oxford University Press, 231-258.

Ginzburg, Carlo 1990. *Myths, Emblems, Clues*. London: Hutchinson.

Ginzburg, Carlo 1993. Microhistory: two or three things that I know about it. *Critical Inquiry* 20: 10-35.

Ginzburg, Carlo, Poni, Carlo 1991. The name and the game: unequal exchange and the historiographic marketplace. In Muir, Edward, Ruggiero, Guido, eds., *Microhistory and the Lost Peoples of Europe*. Baltimore: Johns Hopkins University Press, 1-11.

Godfrey-Smith, Peter 2003. *Theory and Reality: An Introduction to the Philosophy of Science*. Chicago: University of Chicago Press.

Goethe, Johann Wolfgang von [12]1981 [1827]. Hoffnung. In Goethe, Johann Wolfgang von, *Goethes Werke, Hamburger Ausgabe in vierzehn Bände*. Ed. Erich Trunz, vol. 1, *Gedichte und Epen I*. München: Beck, 131.

Goethe, Johann Wolfgang von 1965 [1800]. Flüchtige Uebersicht über die Kunst in Deutschland. *Propyläen* III, 2: 807-810. Reprint Stuttgart: Cotta.

Goldman, Michael 2001. The Birth of a discipline. Producing authoritative green knowledge, World Bank-style. *Ethnography* 2 (2): 191-217.

Goldman, Michael 2005. *The World Bank and Struggles for Social Justice in the Age of Globalization*. Yale: University Press.

Goldstein, Catherine 2001. L'expérience des nombres de Bernard Frenicle de Bessy. *Revue de synthèse, 4e série* 2-3-4: 425-454.

Goldstein, Catherine 2003. Weder Öffentlich Noch Privat: Mathematik im Frankreich des frühen 17. Jahrhunderts. In Wobbe, Theresa, ed., *Zwischen Vorderbühne Hinterbühne. Beiträge zum Wandel der Geschlechterbeziehungen in der Wissenschaft vom 17. Jahrhundert bis zur Gegenwart*. Bielefeld: Transcript, 41-72.

Goldstein, Catherine 2005. L'honneur de l'esprit: de la République des mathématiques. In Cosandey, Fanny, ed., *Dire et vivre l'ordre social en France sous l'Ancien Régime*. Paris: Ed. de l'EHESS, 191-230.

Goldstein, Catherine, Gray, Jeremy J., Ritter, James, eds. 1996. *L'Europe mathématique: Mythes, histoires, identités*. Paris: Editions de la Maison des sciences de l'homme.

Golinski, Jan 1998. *Making Natural Knowledge. Constructivism and History of Science*. Cambridge: Cambridge University Press.

Gonthier, Georges 2008. Formal Proof – The Four-Color Theorem. *Notices of the American Mathematical Society* 55 (11): 1382-1393.

Goodenough, Ward 1996. Navigation in the western Carolines: a traditional science. In Nader, Laura, ed., *Naked Science: anthropological inquiries into boundaries, powerand knowledge*. London: Routledge, 29-42.

Gradmann, Christoph 1993. Naturwissenschaft, Kulturgeschichte und Bildungsbegriff bei Emil Du Bois-Reymond: Anmerkungen zu einer Sozialgeschichte der Ideen des deutschen Bildungsbürgertums in der Reichsgründungszeit. *Tractrix* 5: 1-16.

Graham, Loren 1998. *What Have We Learned about Science and Technology from the Russian Experience?* Stanford: Stanford University Press.

Gralath, Daniel 1747. Geschichte der Electricität. *Versuche und Abhandlungen der Naturforschenden Gesellschaft zu Danzig* 1: 175-304.

Grattan-Guinness, Ivor 1970. *The development of the foundations of mathematical analysis from Euler to Riemann*. Cambridge: MIT Press.

Griesecke, Birgit 2005. Was machen normale Menschen, wenn sie nicht schlafen. Ludwik Fleck, Izydora Dąmbska und die ethnografische Herausforderung der frühen Wissenschaftssoziologie. In Egloff, Rainer, ed., *Tatsache – Denkstil – Kontroverse: Auseinandersetzungen mit Ludwik Fleck*. Zürich: Collegium Helveticum, 21-28.

Groeben, Christiane, Hierholzer, Klaus, eds. 1985. *Emil du Bois-Reymond (1818-1986) – Anton Dohrn (1818-1909) Briefwechsel*. Berlin: Springer.

Guicciardini, Niccolò 2009. *Isaac Newton on Mathematical Certainty and Method*. Cambridge, Mass.: MIT Press.

Gurova, Lilia 2003. Philosophy of science meets cognitive science: The categorization debate. In Ginev, Dimitri, ed., *Bulgarian studies in the philosophy of science*. Dordrecht: Kluwer Publishing, 141-162.

Guts Muths, Johann Christoph Friedrich 1804. *Gymnastik für die Jugend*. Schnepfenthal: Buchhandlung der Erziehungsanstalt.

Guttstadt, Albert 1886. *Die naturwissenschaftlichen und medicinischen Staatsanstalten Berlins*. Festschrift für die 59. Versammlung deutscher Naturforscher und Aerzte. Berlin: Hirschwald.

Hacking, Ian 1975. *The emergence of probability*. Cambridge: Cambridge University Press.

Hacking, Ian 1983. *Representing and Intervening*. Cambridge: Cambridge University Press.

Hacking, Ian 1990. *The taming of chance*. Cambridge: Cambridge University Press.

Hacking, Ian 1992. 'Style' for Historians and Philosophers. *Studies in History and Philosophy of Science* 23: 1-20.

Hacohen, Malachai Haim 2000. *Karl Popper: The Formative Years 1902-1945*. Cambridge: Cambridge University Press.

Haddon, Alfred 1896. Contributions to the anthropology of British India. *Nature* 54 (27. August): 404-405.

Haddon, Alfred 1903. Report of the Council. *Journal of Anthropological Institute* 33: 2-8.

Haddon, Alfred 1906. Plea for an expedition to Melanesia. *Nature* 74: 187-188.

Haken, Hermann 1977. *Synergetics, An Introduction: Nonequilibrium Phase-Transitions and Self-Organization in Physics, Chemistry and Biology*. Berlin: Springer-Verlag.

Haken, Hermann, Graham, Robert 1971. Synergetik – Die Lehre vom Zusammenwirken. *Umschau in Wissenschaft und Technik* 6: 191.

Haken, Hermann, Wagner, Max, eds. 1973. *Synergetics - Towards a New Discipline in Cooperative Phenomena*. Berlin: Springer Verlag.
Halffman, Willem 2003. *Boundaries of Regulatory Science. Eco/toxicology and aquatic hazards of chemicals in US, England and the Netherlands, 1970-1995*. Albatros: Boechout.
Hall, A. Rupert 1961. Comment. In Crombie, Alistair C., ed., *Scientific Change*. London: Heinemann, 370-375.
Hall, Peter, Soskice, David 2001. *Varieties of Capitalims. The Institutional Foundations of Comparative Advantage*. Oxford: University Press.
Hankins Thomas L. 1970. *Jean d'Alembert: Science and the Enlightenment*. Oxford: Oxford University Press.
Haraway, Donna 1991. Situated Knowledge. The Science Question in Feminism and the Priviledge of Partial Perspective. In Haraway, Donna, ed., *Simians, Cyborgs, and Women: The Reinvention of Nature*. New York: Routledge, 183-201.
Harrington, Anne 1996. *Reenchanted science: Holism in German culture from Wilhelm II to Hitler*. Princeton: Princeton University Press.
Harrison, E. Bruce 1993. *Going Green*. Homewood: Business One Irwin.
Harten, Ulrike 2000. *Die Bühnenentwürfe*. In Börsch-Supan, Helmut, Riemann, Gottfried, eds., *Karl Friedrich Schinkel, Lebenswerk*, vol. XVII. München & Berlin: Deutscher Kunstverlag.
Harvard University 1945. Committee on the Objectives of a General Education in a Free Society. *General Education in a Free Society: Report of the Harvard Committee*. Cambridge, Mass.: Harvard University Press.
Harwood, John T. 1991. *Introduction to The Early Essays on Ethics of Robert Boyle*. Carbondale IL: Southern Illinois University Press.
Harwood, Jonathan 1986. Ludwik Fleck and the Sociology of Knowledge. *Social Studies of Science* 16: 173-187.
Hatchuel, Armand 2000. Agir public et conception collective: l'expertise comme processus démocratique. In Heurgon, Edith, Landrieu, Josée, *Prospective pour une gouvernance*. Paris: Editions de l'Aube.
Hatfield, Gary 1993. Helmholtz and Classicism: The Science of Aesthetics and the Aesthetics of Science. In Cahan, David, ed., *Hermann Helmholtz and the Foundations of Nineteenth-Century Science*. Berkeley, Los Angeles: Univ. California Press, 522-558.
Haus, Andreas 2001. *Karl Friedrich Schinkel als Künstler*. München, Berlin: Deutscher Kunstverlag.
Hausdorff, Felix 1901. Über eine gewisse Art geordneter Mengen. Berichte über die Verhandlungen der Königlichen Sächsischen Gesellschaft der Wissenschaften zu Leipzig, *Mathematisch-physikalische Classe* 53: 460-475.
Hausdorff, Felix 1903. Das Raumproblem. *Ostwalds Annalen der Naturphilosophie* 3: 1-23.
Hausdorff, Felix 2004a [1897]. *St. Ilario: Gedanken aus der Landschaft Zarathustras*. Leipzig: C. G. Naumann. Wiederabdruck in Hausdorff, Felix, *Gesammelte Werke*. Vol. VII. Heidelberg: Springer.
Hausdorff, Felix 2004b [1898]. *Das Chaos in kosmischer Auslese*. Leipzig: C. G. Naumann. Wiederabdruck in Hausdorff, Felix, *Gesammelte Werke*. Vol. VII. Heidelberg: Springer.
Heering, Peter, Rieß, Falk, Sichau, Christian, eds. 2000. *Im Labor der Physikgeschichte: Zur Untersuchung historischer Experimentalpraxis*. Oldenburg: BIS.

Heilbron, John L. 1979. *Electricity in the 17th and 18th centuries*. Berkeley: University of California Press.

Heilbron, John L. 1986. *The Dilemmas of an Upright Man: Max Planck as Spokesman for German Science*. Berkeley: University of California Press.

Heilbron, John L. 1998. Thomas Samuel Kuhn. 18 July 1922-17 June 1996. *Isis* 89: 505-515.

Heinemann, Felix 1965. *Nomos und Physis. Herkunft und Bedeutung einer Antithese im griechischen Denken des 5. Jahrhunderts*. Darmstadt: Wissenschaftliche Buchgesellschaft.

Helmholtz, Hermann 1850a. Messungen über den zeitlichen Verlauf der Zuckung animalischer Muskeln und die Fortpflanzungsgeschwindigkeit der Reizung in den Nerven. In (Helmholtz 1882-1895: vol. II, 764-843).

Helmholtz, Hermann 1850b. Ueber die Methoden, kleinste Zeittheile zu messen, und ihre Anwendung für physiologische Zwecke (1850b). In (Helmholtz 1882-1895: vol. II, 862-880).

Helmholtz, Hermann 1852. Messungen über die Fortpflanzungsgeschwindigkeit der Reizung in den Nerven (1852). In (Helmholtz 1882-1895: vol. II, 844-861).

Helmholtz, Hermann 1856. Ueber Combinationstöne. *Annalen der Physik und Chemie* 99: 497-540. Reprinted in (Helmholtz 1882).

Helmholtz, Hermann von 1882-1895. *Wissenschaftliche Abhandlungen von Hermann Helmholtz*. Leipzig: Barth.

Helmholtz, Hermann 1902-1903 [1848]. Probevortrag. In Koenigsberger, Leo *Hermann von Helmholtz*, 3 vols, vol I. Braunschweig: Vieweg, 95-105.

Helmholtz, Hermann von 1903a [1862]. Ueber das Verhältnis der Naturwissenschaften zur Gesammtheit der Wissenschaft. In *Vorträge und Reden*, 1. vol. Braunschweig: Friedrich Vieweg und Sohn, 158–185.

Helmholtz, Hermann von 1903b [1892]. Goethe's Vorahnungen kommender naturwissenschaftlicher Ideen. *Vorträge und Reden*, 2. vol. Braunschweig: Friedrich Vieweg und Sohn, 335–361.

Helmholtz, Hermann von 1903c [1869]. Über das Ziel und die Fortschritte der Naturwissenschaft. In *Vorträge und Reden von Hermann von Helmholtz*. Braunschweig: Vieweg & Sohn, vol. 1, 367-400.

Helmholtz, Hermann von 1903d [1871]. Zum Gedächtnis an Gustav Magnus (1802-1870). In *Vorträge und Reden von Hermann von Helmholtz*. Braunschweig: Vieweg & Sohn, vol. 2, 33-52.

Henderson, J. R. 1923. C. Michie Smith – obituary. *Proceedings of the Royal Society of Edinburgh* 43: 253-254.

Henkel, Arthur, Schöne, Albrecht, eds. 1978. *Emblemata. Handbuch zur Sinnbildkunst des 16. und 17. Jahrhunderts*. Stuttgart: Metzler.

Hensel, Wilhelm 1823. *Die Lebenden Bilder und pantomimischen Darstellungen bei dem Festspiel Lalla Rukh*. Berlin: Wittich.

Hentschel, Klaus 1993a. The conversion of St John: a case study on the interplay of theory and experiment. *Science in context* 6: 137-194.

Hentschel, Klaus 1993b. The discovery of the redshift of the solar Fraunhofer lines by Rowland and Jewell in Baltimore around 1890. *Historical studies in physical science* 23: 219-277.

Hentschel, Klaus 1997. An unwelcome discovery: the pole effect in the electric arc: a threat to early twenteith century precision spectrometry. *Archive for history of exact sciences* 51: 199-271.
Hentschel, Klaus 1999. Photographic mapping of the solar spectrum 1864-1900. *Journal for the history of astronomy* 30: 93-119, 201-224.
Herle, Anita, Rouse, Sandra, eds. 1998. *Cambridge and the Torres Strait*. Cambridge: Cambridge University Press.
Hermitte, Marie-Angèle 1998. Pour une agence de l'expertise scientifique. *La Recherche* 309: 95-97.
Hermitte, Marie-Angèle, Kahn, Philippe 2004. *Les ressources génétiques végétales et le droit dans les rapports Nord-Sud*. Bruxelles: Bruyland.
Herodotus 1995. *Histories*, trans. A. D. Godley, 2 vols., Loeb edition. Cambridge, Mass.: Harvard University Press.
Hershberg, James G. 1993. *James Conant and the Birth of Nuclear Age: From Harvard to Hiroshima*. New York: Alfred A. Knopf.
Hessenbruch, Arne 2000. Calibration and work in the X ray economy 1896-1928. *Social studies of science* 30: 397-420.
Hickey, Sam, Mohan, Giles 2003. Relocating participation within a radical politics of development: citizenship and critical modernism. draft working paper, 27-28 February, taken on the web.
Hills, Richard Leslie, Pacey, Arnold 1972. The Measurement of Power in Early Steam-driven Textile Mills. *Technology and Culture* 13: 25-43.
Hippocrates 1995. *Airs Waters Places*. In. Jones, W.H.S. trans., Loeb edition, 2 vols. Cambridge, Mass.: Harvard University Press [1923], vol. 1, 70–137.
Hoagland, Mahlon B. 1990. *Toward the Habit of Truth. A Life in Science*. New York, London: W. W. Norton & Company.
Hockings Paul 1989. British society in the Company, Crown and Congress Eras. In Hockings, Paul, ed., *Blue Mountains: the ethnography and biogeography of a south Indian region*. Delhi: Oxford University Press, 334-359.
Hockings, Paul 1992. The Yellow Bough: Rivers's use of photography in The Todas. In Edwards, Elizabeth, ed., *Anthropology and photography 1860-1920*. New Haven: Yale University Press, 179-186.
Hoffmann, Christoph, Schickore, Jutta 2001. Secondary Matters: On Disturbances, Contamination, and Waste as Objects of Reserach. *Perspectives on Science* (Special Issue) 9 (2): 123-125.
Hollinger, David 1996. *Science, Jews, and Secular Culture: Studies in Mid-Twentieth Century American Intellectual History*. Princeton: Princeton University Press.
Holmes, Frederic L, Olesko, Kathryn M. 1995. The Images of Precision: Helmholtz and the Graphical Method in Physiology. In Wise, M. Norton, ed., *The Values of Precision*. Princeton: Princeton University Press, 198-221.
Holmes, Frederic L., Levere, Trevor, eds. 2000. *Instruments and Experimentation in the History of Chemistry*. Cambridge: MIT Press.
Hook, Ernest B., ed. 2002. *Prematurity and Scientific Discovery*. Berkeley: University of California Press.
Hooykaas, Reijer 1997. *Robert Boyle: A Study in Science and Christian Belief*. Lanham, MD: University Press of America.

Hopkinson, Joseph ^{2}1857. *The Working of the Steam Engine Explained by the use of the Indicator*. London: Weale.

Houck, Oliver 2003. Tales from a Troubled marriage: Science and Law in Environmental policy. *Science* 302: 1926-1929.

Hoyningen-Huene, Paul 1993. *Reconstructing Scientific Revolution: Thomas S. Kuhn's Philosophy of Science,* trans. Alexander T. Levine. Chicago: University of Chicago Press.

Hufbauer, Karl 1991. *Exploring the Sun: solar science since Galileo*. Baltimore: Johns Hopkins University Press.

Humboldt, Alexander von 1810. *Vues des cordillières*. Paris: F. Schoell.

Humboldt, Alexander von 1817. Des lignes isotherme et de la distribution de la chaleur sur le globe. *Mémoires de physique et de chimie de la Société d'Arcueil* 3: 462-602.

Humboldt, Alexander von 1829. Ueber die Mittel, die Ergründung einiger Phänomene des tellurischen Magnetismus zu erleichtern. *Annalen der Physik und Chemie* 91: 319-336.

Humboldt, Alexander von 1987 [1806]. *Ideen zu einer Physiognomik der Gewächse*. Darmstadt: Wissenschaftliche Buchgesellschaft. Republished (with extensive notes) in Humboldt's *Ansichten der Natur*, 3rd ed. (1849), reprinted in (Humboldt 1989-1997, V: *Ansichten der Natur*).

Humboldt, Alexander von 1989-1997. *Alexander von Humboldt: Studienausgabe*, 7 vols. Ed. by Beck, Hanno. Darmstadt: Wissenschaftliche Buchgesellschaft.

Humboldt, Alexander von 1991 [1811]. Essai politique sur le royaume de la Nouvelle-Espagne (1808-1814). German in (Humboldt 1989-1997, IV: *Mexico-Werk: Politische Ideen zu Mexico, Mexicanische Landeskunde*, with plates from the *Atlas géographique et physique du royaume de la nouvelle-espagne* [*Mexico Atlas*]).

Humboldt, Alexander von 2004a [1807]. *Ansichten der Natur, mit wissenschaftlichen Erläuterungen*. Frankfurt am Main: Eichborn.

Humboldt, Alexander von 2004b [1845–1862]. *Kosmos. Entwurf einer physischen Weltbeschreibung*. Ed. Ette, Ottmar, Lubrich, Oliver. Frankfurt am Main: Eichborn.

Humboldt, Alexander von, Bonpland, Aimé 1989 [1805-1807]. Ideen zu einer Geographie der Pflanzen nebst einem Naturgemälde der Tropenländer. In (Humboldt 1989-1997, I: Schriften zur Geographie der Pflanzen, 43-161).

Humboldt, Alexander von, Lichtenstein, Heinrich 1829. *Amtlicher Bericht über die Versammlung deutscher Naturforscher und Ärzte zu Berlin im September* 1828. Berlin: Trautwein.

Hummel, Georg 1954. *Der Maler Erdmann Hummel: Leben und Werk*. Leipzig: E.A. Seemann.

Hummel, Johann Erdmann 1842. *Geometrisch-praktische Construction der Schatten für Architekten und andere zeichnende Künstler*. Berlin: Herbig.

Hummel, Johann Erdmann ^{2}1833. *Die freie Perspektive erläutert durch praktische Ausgaben and Beispiele, hauptsächlich für Maler und Architekten* (1823, 1825), 2 vols. Berlin: Herbig.

Hunter, Michael, ed. 1994. *Robert Boyle: By Himself and Friends*. London: Pickering & Chatto.

Hunter, Michael 1999. Robert Boyle: A Suitable Case for Treatment? *British Journal for the History of Science* 32: 261-275.

Hunter, Michael 2000. *Robert Boyle: Scrupulosity and Science*. Woodbridge: Boydell.

Huntington, Samuel Peter 1997. *The Clash of Civilizations and the Remaking of the World Order*. New York: Touchstone Books.

Huntington, Samuel Peter 2004. The Hispanic Challenge. *Foreign Policy* (March-April): 30-45.

Iribarne, Philippe d' 2001. Entreprise, Ethique d'entreprise et mondialisation. In Canto-Sperber, Monique, ed., *Dictionnaire d'éthique et de philosophie morale*. Paris: Presses Universitaires de France, 532-534.

Iribarne, Philippe d' 2002. La légitimité de l'entreprise comme acteur éthique aux Etats-Unis et en France. *Revue française de gestion* 140: 23-39.

Jackson, Myles 2003. Harmonious Investigators of Nature: Music and the Persona of the German Naturforscher in the Nineteenth Century. *Science in Context* 16: 121-145.

Jackson, Myles 2006. *Harmonious Triads: Physicists, Musicians, and Instrument Makers in Nineteenth-Century Germany*. Cambridge, Mass: MIT Press.

Jacob, François 1976 [1970]. *The Logic of Life*. New York: Pantheon.

Jacob, François 1987. *La statue intérieure*. Paris: Editions Odile Jacob.

Jahnke, Hans Niels 2003. Algebraic Analysis in the 18th Century. In Jahnke, Hans Niels, ed., *A History of Analysis*. Providence, Rhode Island: American Mathematical Society, 105-153.

James, Susan. *Realizing Rights*. To be published.

Jamison, Andrew 1996. Ecological Modernisation as Cultural Politics. In Lash, Scott, Szerszynski, Bronislaw, Wynne, Brian, eds., *Risk, Environment & Modernity. Towards a New Ecology*. London: Sage, 246-268.

Johnson, Monte Ransome 2003. Was Gassendi an Epicurean? *History of Philosophy Quarterly* 20: 339-359.

Johnston, Sean 2001. *History of light and colour measurement: science in the shadows*. Bristol: Institute of Physics.

Joly, Pierre-Benoît 1999. Besoin d'expertise et quête d'une légitimité nouvelle: quelles procédures pour réguler l'expertise? *Revue Française des Affaires Sociales* 1: 45-52.

Jones, Greta 1988. *Science, Politics and the Cold War*. London: Routledge.

Kahr, Brett 1999. Robert Boyle: A Freudian Perspective. *British Journal for the History of Science* 32: 277-284.

Kalaora, Bernard 1999. Global experts: la religion des mots. *Ethnologie Française 29:* 513-527.

Kant, Immanuel 1913 [1790]. *Kritik der Urteilskraft*. In Königlich Preußische Akademie der Wissenschaften, ed., *Kant's gesammelte Schriften*, vol. V. Berlin: Georg Reimer.

Kant, Immanuel 1993a [1788]. *Kritik der praktischen Vernunft*. Ed. Vorländer, Karl. Hamburg: Felix Meiner.

Kant, Immanuel 1993b [1790]. Critique of Judgement. In *Great Books* 39, Chicago: Encyclopedia Brittanica.

Kargon, Robert H. 1964. Walter Charleton, Robert Boyle and the acceptance of Epicurean atomism in England. *Isis* 55: 184-192.

Kauffman, Stuart 1993. *The Origins of Order*. Oxford: Oxford University Press.

Kauffman, Stuart 1995. *At Home in the Universe: The Search for the Laws of Self-Organization and Complexity*. New York: Oxford University Press.

Kauffman, Stuart 2003. Molecular autonomous agents. *Philosophical Transactions of the Royal Society*. A 361: 1089-1099.

Kaufmann, Doris ed. 2000. *Geschichte der Kaiser-Wilhelm-Gesellschaft im Nationalsozialismus: Bestandsaufnahme und Perspektiven der Forschung.* 2 vols. Göttingen: Wallstein.

Kekulé, Reinhard 1877. *Über die Entstehung der Götterideale der Griechischen Kunst.* Stuttgart: Verlag von W. Spemann.

Kennedy, Dane 1996. *The magic mountains: hill stations and the British Raj.* Berkeley: University of California.

Kenny, Judith 1997. Claiming the high ground: theories of imperial authority and the British hill station in India. *Political geography* 16: 655-673.

Kirsten, Christa et. al., eds. 1986. *Dokumente einer Freundschaft: Briefwechsel zwischen Hermann von Helmholtz und Emil du Bois-Reymond, 1846-1894.* Berlin: Akademie Verlag.

Klein, Ursula 1994. Origin of the Concept of Chemical Compound. *Science in Context* 7 (2): 163-204.

Klein, Ursula 1996. The Chemical Workshop Tradition and the Experimental Practice-Discontinuities within Continuities. *Science in Context* 9: 251-287.

Koch, Heinrich 1998. Gustav Peter Lejeune Dirichlet. In Begehr, Heinrich G. W. et. al., eds., *Mathematics in Berlin.* Berlin: Birkhauser, 33-48.

Kochhar, Rajesh K. 1991. The growth of modern astronomy in India. *Vistas in astronomy* 34: 69-105.

Koepke, Peter 1996. Metamathematische Aspekte der Hausdorffschen Mengenlehre. In Brieskorn, Egbert, ed., *Felix Hausdorff zum Gedächtnis: Aspekte seines Werkes.* Braunschweig, Wiesbaden: Vieweg, 71-106.

Koerner, Joseph Leo 1993. *The Moment of Self-Portraiture in German Renaissance Art.* Chicago: University of Chicago Press.

Koerner, Lisbet 1999. *Linnaeus: Nature and Nation.* Cambridge, Mass.: Harvard University Press.

Kohler, Robert 1991. Systems of production: Drosophila, Neurospora, and biochemical genetics. *Historical Studies in the Physical and Biological Sciences* 22: 87-130.

Kohut, Adolf 1885. Emil du Bois-Reymond. *Westermann's illustrierte Deutsche Monatshefte für das gesamte geistige Leben der Gegenwart* 29 (342): 803-819.

Kommission der europäischen Gradmessung 1881. *Verhandlungen der europaïschen Gradmessung.* Berlin: Georg Reimer.

Kragh, Helge 2002. The Vortex Atom: A Victorian Theory of Everything. *Centaurus* 44: 32-114.

Krinsky, Valentin I., ed. 1984. *Self-organization: autowaves and structures far from equilibrium*: proceedings of an international symposium, Pushchino, USSR, July 18-23 1983. Berlin: Springer.

Kroh, Otto 1936. Rezension von Ludwik Flecks 'Entstehung und Entwicklung einer wissenschaftlichen Tatsache'. *Zeitschrift für Psychologie Leipzig*: 1-3.

Krohn, Wolfgang, Küppers, Günter, Nowotny, Helga, eds. 1990. *Self-Organization: Portrait of a Scientific Revolution.* Dordrecht: Kluwer Academic Publishing.

Krüger, Lorenz et al. 1987. *The probabilistic revolution.* 2 vols. Cambridge, Mass.: MIT Press.

Kuhn, Thomas S. 1949. *The Cohesive Energy of Monovalent Metals as a Function of Their Atomic Quantum Defects.* Harvard University Ph.D. Thesis.

Kuhn, Thomas S. 1957. *The Copernican Revolution: Planetary Astronomy in the Development of Western Thought*. Cambridge, Mass.: Harvard University Press.

Kuhn, Thomas S. 1962. *The Structure of Scientific Revolutions*. Chicago: University of Chicago Press.

Kuhn, Thomas S. 1963. The Function of Dogma in Scientific Research. In Crombie, Alistair C., ed., *Scientific Change*. London: Heinemann, 347-369.

Kuhn, Thomas S. 1978. *Die Struktur wissenschaftlicher Revolutionen*. Frankfurt: Suhrkamp

Kuklick, Henrika 1991. *The savage within: the social history of British anthropology 1885-1945*. Cambridge: Cambridge University Press.

Kumar, Deepak 1995. *Science and the Raj 1857-1905*. New Delhi: Oxford University Press.

Kummer, Eduard 1897. Gedächtnissrede auf Gustav Peter Lejeune Dirichlet. In *G. Lejeune Dirichlet's Werke*, 2 vols. Berlin: Reimer, vol. II, 310-344.

Kutzbach, Gisela 1987. Concepts of monsoon physics in historical perspective: the Indian Monsoon. In Fein, Jay, Stephens, Pamela, eds., *Monsoons*. New York: John Wiley, 159-207.

Lagarde, Paul Anton de 1903 [1885]. Über die Klage, daß der deutschen Jugend der Idealismus fehle. In Lagarde, Paul de, *Deutsche Schriften*. 4ed. Göttingen: Becker and Eidner, 373-384.

Lammel, Gisold 1993. *Adolph Menzel und seine Kreise*. Dresden: Verlag der Kunst.

Lampe, Emil 1906. Dirichlet als Lehrer der Allgemeinen Kriegsschule. *Naturwissenschaftliche Rundschau* 21: 482-485.

Langham, Ian 1981. *The building of British social anthropology: W.H.R.Rivers and his Cambridge disciples in the development of kinship studies, 1898-1931*. Dordrecht: Reidel.

Lash, Scott, Szerszynski, Bronislaw, Wynne, Brian, eds. 1996. *Risk, Environment & Modernity, Towards a New Ecology*. London: Sage.

Latour, Bruno 1987. *Science in Action. How to Follow Scientists and Engineers through Society*. Cambridge: Harvard University Press.

Latour, Bruno 1990. Drawing things together. In Lynch, Michael, Woolgar, Steve, eds., *Representation in scientific practice*. Cambridge MA.: MIT, 19-68.

Latour, Bruno 1999. *Politiques de la nature. Comment faire entrer les sciences en démocratie*. Paris: La Découverte.

Laurance, Jeremy 2004. Pharmaceutical Companies Accused of Manipulating Drug Trials for Profit. *The Independent* (April 23).

Lavater, Johann Caspar 1968-1969 [1775-1778]. *Physiognomische Fragmente zur Beförderung der Menschenkenntnis und Menschenliebe*. Illustr. Chodowiecki, Daniel, 4 vols. Zürich: Füssli.

Lazzarato, Maurizio 2000. Du biopouvoir à la biopolitique. *Multitudes* 1. URL: multitudes.samizdat.net (May 2005).

Le Ru, Veronique 1999. L'aigle à deux têtes de l'Encyclopédie: accords et divergences de Diderot et de d'Alembert de 1751 à 1759. *Recherches sur Diderot et sur l'Encyclopédie* 26: 17-26.

Lenoir, Timothy 1983. Politik im Tempel der Wissenschaft: Emil du Bois-Reymond und die Institutionalisierung der Experimentalphysiologie in Berlin. In Wapnewski, Peter, ed., *Wissenschaftskolleg Berlin, Jahrbuch 1983/84*. Berlin: Siedler, 195-206.

Lenoir, Timothy 1986. Models and Instruments in the Development of Electrophysiology. *Historical Studies in the Physical and Biological Sciences* 17: 1-54.

Lenoir, Timothy 1988. Science for the Clinic: Science Policy and the Formation of Carl Ludwig's Institute Leipzig. In William, Coleman, Holmes, Frederic L., eds., *The Investigative Enterprise. Experimental Physiology in Nineteenth-Century Medicine*. Berkeley, Los Angeles, London: University of California Press, 139-178.

Lenoir, Timothy 1992. Laboratories, medicine and public life in Germany 1830-1839. Ideological Roots of the Institutional Revolution. In Cunningham, Andrew, Williams, Perry, eds., *The Laboratory Revolution in Medicine*. Cambridge: Cambridge University Press, 14-71.

Lenoir, Timothy 1997. The Politics of Vision: Optics, Painting, and Ideology in Germany, 1845-95. In Lenoir, Timothy, *Instituting Science. The Cultural Production of Scientific Disciplines*. Stanford, CA: Stanford University Press, 131-178.

Lessig. Lawrence 2004. *Free Culture, How Big Media Uses Technology and the Law to Lock Down Culture and Control Creativity*. London: The Penguin Press.

Levi, Primo 1986. *The Periodic Table*. Harmondsworth: Sphere Books.

Levine, George 2002. *Dying to Know: Scientific Epistemology and Narrative in Victorian England*. Chicago: University of Chicago Press.

Lévy-Strauss, Claude 1973. *Das wilde Denken*. Frankfurt am Main: Suhrkamp.

Lewis, Frank L. 1992. *Applied Optimal Control and Estimation*. New York: Prentice-Hall.

Li, Lun, Alderson, David, Doyle, John, Willinger, Walter 2004. A first-principles approach to understanding the Internet's router-level topology. Proc ACM SIGCOMM, Portland, Oregon, Aug 30–Sept 2.

Li, Tien-Yien, Yorke, James A. 1975. Period Three Implies Chaos. *Amer. Math. Monthly* 82: 985-992.

Livingstone, David 2003. *Putting science in its place*. Chicago: Chicago University Press.

Lochard, Yves, Simonet-Cusset, Maud, eds. 2003. *L'expert associatif, le savant et le politique*. Paris: Syllepse.

Lockyer, J. Norman 1869. Spectroscopic observations of the Sun. *Proceedings of the Royal Society* 17: 350-356.

Lockyer, J. Norman 1873. *Solar Physics*. London: Macmillan.

Lockyer, J. Norman, Hunter, W.W. 1877. Sunspots and famines. *Nineteenth century* 2 (November): 583-602.

Lockyer, Mary, Lockyer, Winifred 1928. *Life and work of Sir Norman Lockyer*. London: Macmillan.

Lohff, Brigitte 1981. Emil Du Bois-Reymonds Theorie des Experiments. In Mann, Gunter, ed., *Naturwissen und Erkenntnis im 19. Jahrhundert: Emil Du Bois-Reymond*. Hildesheim: Gerstenberg Verlag, 117-128.

Lohff, Brigitte 1992. Johannes Müller und das physiologische Experiment. In Hagner, Michael, Wahrig-Schmidt, Bettina, eds., *Johannes Müller und die Philosophie*. Berlin: Akademie Verlag, 105-124.

Lommel, Eugen, ed. 1892. *Gesammelte Abhandlungen von Georg Simon Ohm*. Leipzig: Johann Ambrosius Barth.

Lorenz, Edward N. 1963. Deterministic nonperiodic flow. *Journal of Atmospheric Sciences* 20: 130-141.

Lowood, Henry E. 1990. The Calculating Forester: Quantification, Cameral Science, and the Emergence of Scientific Forestry Management in Germany. In Frangsmyr, Tore, Heilbron, John, Rider, Robin E., eds., *The Quantifying Spirit in the Eighteenth Century*. Berkeley: University of California Press, 315-342.

Löwy, Ilana 1990. *The Polish School of Philosophy of Medicine: From Tytus Chałubiński (1820-1889) to Ludwik Fleck (1896-1961)*. Dordrecht: Kluwer Academic Publishing.

Löwy, Ilana 2000. *Medical acts and medical facts: the Polish tradition of practice-grounded reflections on medicine and science, from Tytus Chałubiński to Ludwik Fleck*. Krakau: L'Academie Polonaise des Sciences.

Löwy, Ilana 2004a. Introduction: Ludwik Fleck's epistemology of medicine and biomedical sciences. In *Studies in History and Philosophy of Science* Part C: Biological and Biomedical Sciences 3: 437-445.

Löwy, Ilana 2004b. 'A River That Is Cutting It's Own Bed': The Sereology of Syphillis between Laboratory, Society and the Law. *Studies in History and Philosophy of Biological and Biomedical Sciences* 35: 509-524.

Lubbers, Eveline, ed. 2002. *La grande mascarade, Ces multinationales qui lavent plus vert*. Paris: Parangon.

Lucretius, Titus Carus 1992. *De rerum natura*. Tr. W.H. Rouse, revised Martin Ferguson Smith, Cambridge, Mass: Harvard University Press.

Ludwig, Carl 1852. *Lehrbuch der Physiologie des Menschens*. 2 vols, vol. I. Heidelberg: Winter.

Lützen, Jesper 2003. The Foundation of Analysis in the 19th Century. In Jahnke, Hans Niels, ed., *A History of Analysis*. Providence, Rhode Island: American Mathematical Society, 155-195.

MacIntosh, Jack J. 1991. Robert Boyle on Epicurean atheism and atomism. In Osler, Margaret J., ed., *Atoms, Pneuma, and Tranquillity: Epicurean and Stoic Themes in European Thought*. Cambridge, Cambridge University Press, 197-219.

MacIntosh, Jack J. 1996. Animals, Morality, and Robert Boyle. *Dialogue* 35: 435-472.

MacKenzie, Donald 1981. *Statistics in Britain, 1865-1930: The Social Construction of Scientific Knowledge*. Edinburgh: Edinburgh University Press.

MacKenzie, Donald 1990. *Inventing Accuracy: A Historical Sociology of Nuclear Missile Guidance*. Cambridge, Mass.: MIT Press.

MacKenzie, Donald 1999. Slaying the Kraken: The Sociohistory of a Mathematical Proof. *Social Studies of Science* 29: 7-60. Erratum. Ibid., 317-318. Corrigendum. *Social Studies of Science* 30 (2000): 317-318.

MacKenzie, Donald 2006. *An Engine, not a Camera: How Financial Models Shape Markets*. Cambridge, Mass.: MIT Press.

Maddison, Robert E.W. 1963. Studies in the life of Robert Boyle, 6: the Stalbridge period, 1645-1655, and the invisible college. *Notes and records of the Royal Society* 18: 104-124.

Madras Mail (16 December 1902). Interview with W.H.R. Rivers, 12.

Magee, Bryan 1973. *Karl Popper*. New York: Viking.

Mahias, Marie-Claude 1998. Le soleil noir des Nilgiri: l'astronomie, la photographie et l'anthropologie physique en Inde du Sud. *Gradhiva* 24: 45-56.

Main, Thomas John, Brown, Thomas 1847. *The Indicator and Dynamometer, with their Practical Applications*. London: Hebert.

Mamdami, Mahmood 2000. *Beyond Rights Talk and Culture Talk. Comparative Essays on the Politics of Rights and Culture*. New York: St. Martin's Press.

Mamou, Yves 2004. La communication des laboratoires est remise en question. *Le Monde* 22 juin.

Mandelbaum, David 1980. The Todas in time perspective. *Reviews in Anthropology* 7: 279-302.

Manent, Pierre 2001. *Cours familier de philosophie politique*. Paris: Fayard.

Manin, Bernard 2005. Democratic Deliberation: why we should promote debate rather than discussion. Paper for the Program in Ethics and Public Affairs Seminar. Princeton: Princeton University Press.

Marcum, James A. 2005. *Thomas Kuhn's Revolution: An Historical Philosophy of Science*. New York: Continuum International Publishing.

Marois, Maurice, ed. 1969. *Theoretical Physics and Biology: Proceedings of The First International Conference on Theoretical Physics and Biology, Versailles, 26-30 June 1967*. Basel: S. Karger.

Marois, Maurice, ed. 1973. *From Theoretical Physics to Biology: Proceedings of The Third International Conference on Theoretical Physics and Biology, Versailles, 21-26 June 1971*. Basel: S. Karger.

Matschoss, Conrad 1911. Geschichte der Königlich Technischen Deputation für Gewerbe. Zur Erinnerung an das 100jährige Bestehen. 1811-1911. *Beiträge zur Geschichte der Technik und Industrie. Jahrbuch des Vereines deutscher Ingenieure* 3: 239-275.

Matschoss, Conrad 1921. *Preußens Gewerbeförderung und ihre grossen Männer, dargestellt im Rahmen der Geschichte des Vereins zur Beförderung des Gewerbefleißes 1821 – 1921*. Berlin: Verein Deutscher Ingenieure.

Mattick, John 2004. RNA regulation: a new genetics? *Nature Reviews Genetics* 5: 316-323.

Maunder,Walter, ed. 1899. *The Indian Eclipse 1898*. London: Hazell, Watson and Viney.

Mayer, Anna-K. 2000. Setting Up a Discipline: Conflicting Agendas of the Cambridge History of Science Committee, 1936-1950. *Studies in the History and Philosophy of Science* 31: 665-689.

McClellan III, James E. 1985. *Science Reorganized: Scientific Societies in the Eighteenth Century*. New York: Columbia University Press.

Menand, Louis 2001. *The Metaphysical Club: A Story of Ideas in America*. New York: Farrar, Straus and Giroux.

Mende, Matthias, Hebecker, Inge, eds. 1973. *Das Dürer Stammbuch von 1828*. Nürnberg: Carl.

Merton, Robert K., Gaston, Jerry, eds. 1977. *The Sociology of Science in Europe*. Carbondale: S. Illinois University Press.

Milani, Carlos, Arturi, Carlos, Solinis, Germàn 2001. *Démocratie et gouvernance mondiale, Quelles régulations pour le XXIe siècle?* Paris: UNESCO & Karthala.

Mitchell, Nora 1972. *The Indian Hill-Station: Kodaikanal*. Chicago: University of Chicago Department of Geography.

Montesquieu, Charles de Secondat de 1845 [1748]. *Esprit des lois, avec les notes de l'auteur et un choix des observations*. Paris: Firmin-Didot Frères.

Montucla, Jean Etienne 1799 [An VII]. *Histoire des Mathematiques*. Tome Second. Nouvelle édition, considérablement augmentée. Paris: Henri Agasse.

Moreau Défarges, Philippe 2001. Gouvernance. *Le Débat* 115: 165-172.

Morrison, Charles 1984. Three styles of imperial anthropology: British officials as anthropologists of India. *Knowledge and society* 5: 141-169.

Mosse, George 1996. *The Image of Man: The Creation of Modern Masculinity*. New York: Oxford University Press.

Mouffe, Chantal 1999. Deliberative Democracy or Agonistic Pluralism? *Social Research* 66 (3): 745-758.

Muecke, Frances 1999. 'Taught by Love': The Origin of Painting Again. *Art Bulletin* 81: 297-302.

Müller, Friedrich 1878. *Lectures on the origin and growth of religion*. London: Longman.

Musschenbroek, Petrus van, Gottscheden, Johann Christoph 1747. *Hrn. Peters von Muschenbroek Grundlehren der Naturwissenschaft*, nach der 2. lat. Ausg., nebst einigen neuen Zusätzen des Verfassers, ins Deutsche übers. Mit einer Vorrede ans Licht gestellt von Johann Christoph Gottscheden. Leipzig: Kiesewetter.

Musschenbroek, Pieter van 1739. *Essai de physique - Beginsels der natuurkunde*. Leyden: Luchtmans.

Muthmann, Friedrich 1955. *Alexander von Humboldt und sein Naturbild im Spiegel der Goethezeit*. Zürich: Artemis.

Myers, Greg 1989. Nineteenth-century popularizations of thermodynamics and the rhetoric of social prophecy. In Brantlinger, Patrick, ed., *Energy and Entropy: Science and Culture in Victorian Britain*. Bloomington: Indiana University Press, 307-338.

Naoroji, Dadabhai 1901. *Poverty and un-British rule in India*. London.

Needham, Joseph, Pagel, Walter, eds. 1938. *Background to Modern Science: Ten Lectures at Cambridge arranged by the History of Science Committee*. New York: Macmillan and Cambridge: Cambridge University Press.

Nersessian, Nancy J. 2000. Nomic concepts, frames, and conceptual change. *Philosophy of Science* 67: 224-241.

Neumann, Hans 1925. *Heinrich Wilhelm Dove: Eine Naturforscher-Biographie*. Liegnitz: Krumbhaar.

Neureuther, Eugen 1977. *Randzeichnungen zu Goethes Balladen und Romanzen: Baierische Gebirgslieder* (1829-1839; 1855), facsimile of 2nd ed. Unterschneidheim: Alfons Uhl.

Nickles, Thomas 1998. Kuhn, Historical Philosophy of Science, and Case-Based Reasoning. *Configurations* 6: 51-85.

Nickles, Thomas, ed. 2002. *Thomas Kuhn*. Cambridge: Cambridge University Press.

Nicolis, Grégoire, Baras, Florence, eds. 1984. *Chemical instabilities: applications in chemistry, engineering, geology, and materials science*. (NATO ASI series. Series C, Mathematical and physical sciences, 120) Dordrecht: Kluwer Academic Publishers.

Nicolis, Grégoire, Prigogine, Ilya 1977. *Self Organization in Non-Equilibrium Systems: From Dissipative Structures to Order through Fluctuations*. New York: John Wiley & Sons.

Nicolson, Malcolm 1990. Alexander von Humboldt and the Geography of Vegetation. In Cunningham, Andrew, Jardine, Nicholas, eds., *Romanticism and the Sciences*. Cambridge: Cambridge University Press, 169–185.

Nietzsche, Friedrich 1988. *Sämtliche Werke*. 15 vols. Ed. by Colli, Giorgio; Montinari, Mazzino. München, Berlin, New York: De Gruyter.

Nye, Mary Jo 2000. Laboratory Practice and the Physical Chemistry of Michael Polanyi. In Holmes, Frederic L., Levere, Trevor, eds., *Instruments and Experimentation*, 367-400.

Nye, Mary Jo 2001. At the Boundaries: Michael Polanyi's Work on Surfaces and the Solid State. In Reinhardt, Carsten, ed., *Chemical Sciences in the Twentieth Century*, 246-257.

Nye, Mary Jo 2002a. Michael Polanyi (1891-1976). *HYLE* 8: 123-127.

Nye, Mary Jo 2002b. Michael Polanyi's Theory of Surface Adsorption: How Premature? In Hook, Ernest B., ed., *Prematurity and Scientific Discovery*, 151-163.

Nye, Mary Jo 2004. *Blackett: Physics, War, and Politics in the Twentieth Century*. Cambridge, Mass.: Harvard University Press

Nye, Mary Jo 2007. Working Tools for Theoretical Chemistry: Polanyi, Eyring, and Debates over the 'Semi-Empirical Method.' *Journal of Computational Chemistry* 28: 98-108.

O'Connell, Joseph 1993. Metrology: the creation of universality by the circulation of particulars. *Social studies of science* 23: 129-173.

Ohm, Georg Simon 1843. Ueber die Definition des Tones, nebst daran geknüpfter Theorie der Sirene und ähnlicher tonbildender Vorrichtungen. *Annalen der Physik und Chemie* 59: 513-565.

Ohm, Georg Simon 1844. Noch ein Paar Worte über die Definition des Tones. *Annalen der Physik und Chemie* 62: 1-18.

Ohm, Georg Simon 1989 [1827]. *Die galvanische Kette, mathematisch bearbeitet*. Berlin: Riemann. Reprint with commentary by Dunsch, Lothar. Berlin: Verlag der Wissenschaften.

Olesko, Kathryn M., Holmes, Frederic L. 1993. Experiment, Quantification, and Discovery: Helmholtz's Early Physiological Researches, 1843-1850. In Cahan, David, ed., *Hermann von Helmholtz and the Foundations of Nineteenth-Century Science*. Berkeley: University of California Press, 50-109.

Olesko, Kathryn. *Precision in German Society, 1648-1989*. (in preparation)

Onana, Janvier. *ONGéisation de l'action publique: vers l'avènement de l'ère post-étatique en Afrique subsaharienne*. Work in progress.

Osler, Margaret 2003. Early Modern Uses of Hellenistic Philosophy: Gassendi's Epicurean Project. In Miller, Jon, Inwood, Brad, eds., *Hellenistic and Early Modern Philosophy*. Cambridge: Cambridge University Press, 30-44.

Osmont, Annick 1998. La 'Governance': concept mou, politique ferme. *Les Annales de la Recherche Urbaine* 80-81: 19-26.

Oster, Malcolm 1989. The Beame of Divinity. *British Journal for the History of Science* 22: 151-180.

Oster, Malcolm 1993. Biography, Culture, and Science: The Formative Years of Robert Boyle. *History of Science* 31: 177-226.

Paczuski, Maya, Maslov, Sergei, Bak, Per 1996. Avalanche Dynamics in Evolution, Growth, and Depinning Models. *Physical Review* E53: 414-443.

Paczuski, Maya, Bak, Per 1999. Self-Organization of Complex Systems. *Proceedings of 12th Chris Engelbrecht Summer School*. Berlin: Springer.

Padioleau, Jean-Gustave 2001. La Gouvernance ou comment s'en débarasser. (Stratégie de corruption). *Espaces et Sociétés* 101-102: 61-73.

Palladino, Paolo, Worboys, Michael 1993. Science and imperialism. *Isis* 84: 91-102.

Palló, Gábor 1996. Polányi Mihály és a kriptonlámpa. *Fizikai Szemle* 9: 311-316.

Palló, Gábor 1998. Michael Polányi's Early Years in Science. *Bulletin for the History of Chemistry* 21: 39-43.

Pang, Alex 2002. *Empire and the sun: Victorian solar eclipse expeditions*. Stanford: Stanford University Press.
Panofsky, Erwin ²1945. *Albrecht Dürer*. 2 vols. Princeton: Princeton University Press.
Paret, Peter 1980. *The Berlin Secession. Modernism and Its Enemies in Imperial Germany*. Cambridge, MA, London: Harvard University Press - Belknap.
Park, Katharine 1997. The Meanings of Natural Diversity: Marco Polo on the 'Division' of the World. In Sylla, Edith, McVaugh, Michael, eds., *Texts and Contexts in Ancient and Medieval Science*. Studies on the Occasion of John E. Murdoch's Seventieth Birthday. Leiden: Brill, 134-147.
Pestre, Dominique 2003. *Science, argent et politique, un essai d'interprtation*. Paris: INRA, coll. Sciences en Question.
Pestre, Dominique 2004. Thirty Years of Science Studies: Knowledge, Society and the Political. *History and Technology* 20 (4): 351-369.
Pestre, Dominique 2005. The Technosciences between Market, Social Worries and the Political. How to Imagine a Better Future. In Nowotny, Helga, Pestre, Dominique, Schmidt-Aßmann, Eberhardt, Schulze-Fielitz, Helmut, Trute, Hans-Heinrich, eds., *The Public Nature of Science under Assault*. Berlin: Springer, 29-52.
Pestre, Dominique 2006. *Introduction aux Science Studies*. Paris: La Découverte.
Petersen, Hans-Christian 1936. Ludwig Fleck's Lehre vom Denkstil und dem Denkkollektiv. *Klinische Wochenschrift* 15: 239-242.
Pickering, Andrew 1992. From science as Knowledge to science as practice. In Pickering, Andrew, ed., *Science as Practice and Culture*. Chicago: University of Chicago Press, 1-28.
Pickering, Andrew 1995. *The Mangle of Practice: Time, Agency, and Science*. Chicago: University of Chicago Press.
Pickering, Andrew, Stephanides, Adam 1992. Constructing Quaternions: On the Analysis of Conceptual Practice. In Pickering, Andrew, ed., *Science as Practice and Culture*. Chicago: University of Chicago Press, 139-167.
Pieper, Jan 1994. Die Maschine im Interieur: Ludwig Persius' Dampfmaschine im Babelsberger Park. *Daidalos* 53: 104-115.
Pinney, Christopher 1990. Classification and fantasy in the photographic construction of caste and tribe. *Visual anthropology* 3: 259-288.
Pinney, Christopher 1997. *Camera Indica: the social life of Indian photographs*. London: Reaktion Books.
Planck, Max 1944. *Wege zur physikalischen Erkenntnis*. 3rd ed. Leipzig: Hirzel.
Płonka-Syroka, B. 2004. Ludwik Fleck und die innere Emigration. In Scholz, Abrecht, Caris-Heidel, Petra, eds., *Emigrantenschicksale*. Frankfurt: Mabuse-Verlag, 347-357.
Polanyi, Michael 1928. Geheimrat Fritz Haber vollendet am 9. Dezember sein 60. Lebensjahr. *Metallwirtschaft* 7: 1316-1317.
Polanyi, Michael 1930. Förderung der Wissenschaft. *Der deutsche Volkswirt*, 23 May: 1149-1151.
Polanyi, Michael 1945. *Rights and Duties of Science*. Society for Freedom in Science Occasional Pamphlet.
Polanyi, Michael 1946. *The Planning of Science*. Society for Freedom in Science Occasional Pamphlet 4.

Polanyi, Michael 1947. *The Foundations of Academic Freedom*. Society for Freedom in Science Occasional Pamphlet 6.

Polanyi, Michael 1948. *Full Employment and Free Trade*. Cambridge: Cambridge University Press.

Polanyi, Michael 1951. *The Logic of Liberty*. London: Routledge and Kegan Paul.

Polanyi, Michael 1958. *Personal Knowledge: Towards a Post-Critical Philosophy*. Chicago: University of Chicago Press.

Polanyi, Michael 1963. Comment by Michael Polanyi. In Crombie, Alistair C., ed., *Scientific Change*, 375-380.

Polanyi, Michael 1964 [1946]. *Science, Faith, and Society*. Chicago: University of Chicago Press. Originally Oxford: Oxford University Press.

Polanyi, Michael 1966. *The Tacit Dimension*. Garden City: Doubleday.

Polanyi, Michael 1975 [1940]. *The Contempt of Freedom: The Russian Experiment and After*. New York: Arno Press. Originally London: Watts.

Polanyi, Michael 1997a. New Scepticism, trans. Nagy, Endre J., 29-32. In Allen, Richard, Polanyi, Michael, eds., *Society, Economics and Philosophy*, 29-32. Originally published in *Szabadgondolat (Free Thought)*, 1 (February 1919), 53-56.

Polanyi, Michael 1997b. The Struggle between Truth and Propaganda. In Allen, Richard, Polanyi, Michael, eds., *Society, Economics, and Philosophy*, 47-60. Originally in *Manchester School of Economic and Social Studies*, 7 (1936), 96-116.

Polanyi, Michael. 1997c. To the Peacemakers: Views on the Prerequisites of War and Peace in Europe, trans. Nagy, Endre J. In Allen, Richard, Polanyi, Michael, eds., *Society, Economics and Philosophy*, 15-28. Originally in Hungarian as a pamphlet of 1917 and republished in *Huszadik Század (Twentieth Century)*, 1917, 165-176.

Popper, Karl 1959 [1935]. *The Logic of Scientific Discovery*. New York: Routledge. 1st English ed., Hutchinson. Original edition, *Logik der Forschung*. Wien: Julius Springer.

Popper, Karl 1976 [1974]. *Unended Quest: An Intellectual Biography*. La Salle and London: Open Court. First published as "Autobiography of Karl Popper" in *The Philosophy of Karl Popper* in *The Library of Living Philosophers*, ed. Paul Arthur Schilpp. Chicago: Open Court.

Popper, Karl 2004 [1957]. *The Poverty of Historicism*. London and New York: Routledge. Originally published by Routledge and Kegan Paul.

Popper, Karl Raimund 1934. *Logik der Forschung*. Wien: Julius Springer.

Porter, Theodore M. 1986. *The Rise of Statistical Thinking, 1820–1900*. Princeton: Princeton University Press.

Porus, Vladimir 1983. Ludwik Fleck's epistemological conception and contemporary discussion on the nature and dynamics of scientific knowledge. *Kwartalnik Historii Nauki i Techniki* 28 (3-4): 545-560.

Posner, Richard A. 2002. The law and economics of intellectual property. *Daedalus* spring: 5-12.

Poulin, Richard 2005. *Mondialisation des industries du sexe*. Paris: Imago.

Prakash, Gyan 1999. *Another reason: science and the imagination of modern India*. Princeton: Princeton University Press.

Priestley, Joseph 1767. *The history and present state of electricity, with original experiments*. London: Dodsley.

Prigogine Ilya, Lefever Robert, Goldbeter Albert, Herschkowitz-Kaufman, Marcelle 1969. Symmetry breaking instabilities in biological systems. *Nature* 223 (209): 913-916.
Prigogine, Ilya. 1969 [1967]. Structure, Dissipation and Life. In Marois, Maurice, ed., *Theoretical Physics and Biology*, Versailles, June. Amsterdam: North-Holland Publishing, 23-52.
Prigogine, Ilya., Lefever, Robert 1975. Stability and Self-Organization in Open Systems. *Advances in Chemical Physics* 29: 1-28.
Prigogine, Ilya., Nicolis, Grégoire 1967. On Symmetry-Breaking Instabilities in Dissipative Systems. *The Journal of Chemical Physics* 46 (9): 3542-3550.
Principe, Lawrence 1994. Style and Thought of the Early Boyle: the Discovery of the 1648 Manuscript of Seraphic Love. *Isis* 85: 247-260.
Prinz, Jesse J. 2002. *Furnishing the Mind: Concepts and Their Perceptual Basis. Representation and Mind.* Cambridge, Mass.: MIT Press.
Purkert, Walter 2002. Grundzüge der Mengenlehre – Historische Einführung. In Hausdorff, Felix, *Gesammelte Werke*. Vol. II. Heidelberg: Springer, 1-89.
Pyenson, Lewis 1990. Why science may serve political ends: cultural imperialism and the mission to civilize. *Berichte zur Wissenschaftsgeschichte* 13: 69-81.
Rabinow, Paul 1996. *Essays on the Anthropology of Reason*. Princeton: Princeton University Press.
Rabinow, Paul 2004. *Anthropologie der Vernunft*. Frankfurt am Main: Suhrkamp.
Raheja, Gloria Goodwin 1999. The illusion of consent: language, caste and colonial rule in India. In Pels, Peter and Salemnik, Oscar, eds., *Colonial subjects: essays on the practical history of anthropology*. Ann Arbor: Michigan University Press, 117-152.
Raj, Kapil 1997. Les hommes-instruments de sa Majesté – des jambes et des moulins à prières pour cartographier le Transhimalaya. *La Recherche* 298 (May): 74-80.
Rancière, Jacques 2004. Les philosophes sans porte-voix. *Libération* (June 19 – 20).
Rapaport, Anatol, Horvath, Willian J. 1959. Thoughts on Organization Theory and a Review of Two Conferences. *General Systems* IV: 87-93.
Rave, Paul Ortwin 1962. *Berlin, dritter Teil: Bauten für Wissenschaft, Verwaltung, Heer, Wohnbau und Denkmäler*. Berlin: Deutscher Kunstverlag.
Ravetz, Jerome R. 1993. The Sin of Science: Ignorance of Ignorance. *Knowledge (Creation-Diffusion-Utilization)* 15 (12 December): 157-165.
Ravetz, Jerome R. 2001. Safety in the globalizing knowledge economy: an analysis by paradoxes. *Journal of Hazardous Materials* 86: 1-16.
Ravignan, Antoine de 2002. Nike ou le difficiel passage des principes aux pratiques. *Cahiers Français* 309: 88.
Reich, Michael R., ed. 2002. *Public-Private Partnerships for Public Health*. Boston: Harvard University Press.
Reidemeister, Kurt 1932. *Knotentheorie*. Berlin: Springer.
Reimer, Doris 1999. *Passion & Kalkül. Der Verleger Georg Andreas Reimer (1776-1842)*. Berlin, New York: Walter de Gruyter.
Reinhardt, Carsten, ed. 2001. *Chemical Sciences in the Twentieth Century*. Berlin: Wiley-VCH.
Remmert, Volker 1998. *Ariadnefäden im Wissenschaftslabyrinth. Studien zu Galilei: Historiographie – Mathematik – Wirkung*. Bern: Peter Lang.

Reports of the Cambridge Anthropological Expedition to Torres Straits. Vol. 2: *Physiology and Psychology* 1901, 1903. Cambridge: Cambridge University Press.

Reuleaux, Franz 1875. *Theoretische Kinematik: Grundzüge einer Theorie des Maschinenwesens*. Braunschweig: Vieweg und Sohn.

Rheinberger, Hans-Jörg 1992. *Experiment, Differenz, Schrift. Zur Geschichte epistemischer Dinge*. Marburg: Basiliskenpresse.

Rheinberger, Hans-Jörg 1997. *Toward a History of Epistemic Things. Synthesizing Proteins in the Test Tube*. Stanford: Stanford University Press.

Rheinberger, Hans-Jörg 2000. Beyond nature and culture: Modes of reasoning in the age of molecular biology and medicine. In Lock, Margaret, Young, Allan, Cambrosio, Alberto, eds., *Living and Writing with the New Medical Technologies*. Cambridge: Cambridge University Press, 19-30.

Rheinberger, Hans-Jörg 2001. Putting isotopes to work: Liquid scintillation counters, 1950-1970. In Joerges, Bernward, Shinn, Terry, eds., *Instrumentation Between Science, State and Industry*. Dordrecht: Kluwer Academic Publishers, 143-174.

Rheinberger, Hans-Jörg 2007. *Historische Epistemologie zur Einführung*. Hamburg: Junius.

Ricœur, Paul 1991. Postface au Temps de la Responsabilité. In Ricœur, Paul, *Lectures 1, Autour du politique*. Paris: Seuil, 270-293.

Ricœur, Paul 1995. La place du politique dans une conception pluraliste de la justice. In Affichard, Joëlle, Foucauld, Jean-Baptiste de, eds., *Pluralisme et équité, La justice sociale dans les démocraties*. Paris: Esprit, 71-84.

Riemann, Bernhard 1990 [1854]. Ueber die Darstellbarkeit einer Function durch eine trigonometrische Reihe (Habilitationschrift). *Abhandlungen der Königlichen Gesellschaft der Wissenschaften zu Göttingen* 13. In Weber, Heinrich, Dedekind, Richard, eds., *Bernhard Riemann: Gesammelte mathematische Werke* (reedited by Raghavan Narasimhan). Berlin: Springer, 227-264.

Riezler, Kurt 1927. Die Krise der Wirklichkeit. *Die Naturwissenschaften* 16: 705-712.

Risley, Herbert 1909. *Ethnology and caste. Imperial Gazetteer of India*. Oxford: Clarendon Press.

Rivers, William Halse R. 1903. The funeral of Sinerani. *The Eagle* 24: 337-343.

Rivers, William Halse R. 1905. Observations on the senses of the Todas. *British Journal of Psychology* 1: 321-396.

Rivers, William Halse R. 1906. *The Todas*. London: Macmillan.

Rivers, William Halse R. 1908. *The Influence of Alcohol and other Drugs on Fatigue*. London: Arnold.

Rivers, William Halse R. 1913. Report on anthropological research outside America. Reports upon the present condition and future needs of the science of anthropology. *Carnegie Institution Report* (Washington) 200: 5-28.

Rivers, William Halse R. 1914. *The History of Melanesian Society*, 2 vols. Cambridge: Cambridge University Press.

Robertson, Neil, Sanders, Daniel P., Seymour, Paul, Thomas, Robin 1997. The Four-Colour Theorem. *Journal of Combinatorial Theory, Series B* 70 (1): 2-44.

Roman, Stéphane 2000. La 'bonne gouvernance', pierre philosophale du développement économique? *L'économie politique* 6: 39-50.

Romi, Raphaël. Codex Alimentarius: de l'ambivalence à l'ambiguïté. *Revue juridique de l'environnement* 2: 201-214.

Rooksby, Richard Lionel. 1971. W.H.R.Rivers and the Todas. *South Asia* 1: 109-121.

Rorty, Richard 1993. Ist Naturwissenschaft eine natürliche Art? In Rorty, Richard, *Eine Kultur ohne Zentrum*. Stuttgart: Reclam, 13-47.

Rosenblum, Robert 1957. The Origin of Painting: A Problem in the Iconography of Romantic Classicism. *Art Bulletin* 39: 279-290.

Rothschuh, Karl E. 1964. Emil Du Bois-Reymond und die Elektrophysiologie der Nerven. In Rothschuh, Karl, ed., *Von Boerhaave bis Berger. Die Entwicklung der kontinentalen Physiologie im 18. und 19. Jahrhundert*. Stuttgart: Fischer, 85-105.

Rowell, Andy 2002a. La mascarade verte. In Lubbers, Eveline, eds., *La grande mascarade, Ces multinationales qui lavent plus vert*. Paris: Parangon, 15-24.

Rowell, Andy 2002b. Le dialogue: diviser pour mieux régner. In Lubbers, Eveline, ed., *La grande mascarade, Ces multinationales qui lavent plus vert*. Paris: Parangon, 34-47.

Ruse, Michael 1999. *Mystery of Mysteries: Is Evolution a Social Construction?* Cambridge, Mass.: Harvard University Press.

Sachse, Carola, Walker, Mark 2005. Introduction: A Comparative Perspective. Politics and Science in Wartime: Comparative International Perspectives on the Kaiser Wilhelm Institute. *Osiris* 20: 1-20.

Sachse, Carola, Walker, Mark, eds. 2005. Politics and Science in Wartime: Comparative International Perspectives on Kaiser Wilhelm Institutes. (= *Osiris*, Volume 20.) Chicago: University of Chicago Press

Sandwith, Fleming Mant1907. Hill stations and health resorts in the British tropics. *Journal of Tropical Medicine and Hygiene* 10: 361-370.

Saunders, Frances Stonor 1999. *The Cultural Cold War: The CIA and the World of Arts and Letters*. New York: New Press.

Schaanning, Espen 1997. *Vitenskap som skapt viten: Foucault og historisk praksis*. Oslo: Spartacus.

Schadow, Johann Gottfried 1890 [1801]. Ueber einige in den Propyläen abgedruckte Sätze Goethes. In Julius Friedländer, ed., *Gottfried Schadow: Aufsätze und Briefe*. Stuttgart: Ebner & Seubert, 45-55.

Schaffer, Simon 1992. Late Victorian Metrology and its Instrumentation: A Manufactory of Ohms. In Budd, Robert, Cozzens, Susan E., eds., *Invisible Connections: Instruments, Institutions, and Science*. Bellingham, Washington: SPIE Press, 23-56.

Schiebinger, Londa, Swan, Claudia, eds. 2004. *Colonial Botany: Science, Commerce, and Politics in the Early Modern World*. Philadelphia: University of Pennsylvania Press.

Scholz, Erhard 1996. Logische Ordnungen im Chaos: Hausdorffs frühe Beiträge zur Mengenlehre. In Brieskorn, Egbert, ed., *Felix Hausdorff zum Gedächtnis: Aspekte seines Werkes*. Braunschweig, Wiesbaden: Vieweg, 107-134.

Schrödinger, Erwin 1994 [1932]. Ist die Naturwissenschaft milieubedingt? In Meyenn, Karl von, ed., *Quantenmechanik und Weimarer Republik*. Braunschweig, Wiesbaden: Vieweg, 295-332.

Schuster, Peter-Klaus 1991. *Melencolia I: Dürer's Denkbild*. Vol. I, Berlin: Gebr. Mann.

Schwarz, Ingo, Wenig, Klaus, eds. 1997. *Briefwechsel zwischen Alexander von Humboldt und Emil du Bois-Reymond*. Berlin: Akademie Verlag.

Scott, Michael T. 1983. Michael Polanyi's Creativity in Chemistry. In Aris, Rutherford, et al., eds., *Springs of Scientific Creativity*. Minneapolis: University of Minnesota Press, 279-307.

Scott, William T., Moleski, Martin X. 2005. *Michael Polanyi: Scientist and Philosopher.* Oxford: Oxford University Press.
Seiler, Michael 1989. *Das Palmenhaus auf der Pfaueninsel: Geschichte seiner baulichen und gärtnerischen Gestaltung.* Berlin: Haude & Spener.
Sellars, Wilfried 1997 [1956]. *Empiricism and the Philosophy of Mind.* Cambridge, Mass.: Harvard University Press.
Sengoopta, Chandak 2003. *Imprint of the Raj: how Fingerprinting was born in colonial India.* London: Macmillan.
Sennett, Richard 2006. *La culture du nouveau capitalisme.* Paris: Albin Michel.
Shapin, Steven 1993. Personal development and intellectual biography: the case of Robert Boyle. *British Journal for the History of Science* 26: 335-345.
Shapin, Steven, Schaffer, Simon 1985. *Leviathan and the Air-Pump: Hobbes, Boyle, and the Experimental Life.* Princeton: Princeton University Press.
Shinn, Terry 2002. The transverse science and technology culture: Dynamics and roles of research-technology. *Social Science Information* 41: 207-251.
Sibum, Otto 1995. Reworking the Mechanical Value of Heat: Instruments of Precision and Gestures of Accuracy in Early Victorian England. *Studies in the History and Philosophy of Science* 26: 73-106.
Sibum, Heinz Otto 2003. Experimentalists in the Republic of Letters. *Science in Context* 16 (1/2): 89-120.
Siemens, Werner 1847. Ueber Geschwindigkeitsmessung. *Fortschritte der Physik im Jahre 1845* 1: 46-72.
Siemens, Werner 1889 [1845]. Anwendug des elektrischen Funkens zur Geschwindigkeitsmessung. In *Wissenschaftliche und technische Arbeiten von Werner Siemens.* 2 vols. Berlin: Springer, vol. I, 8-14.
Siemens, Werner 1891 [1845]. Beschreibung des Differenz-Regulators der Gebrüder Werner und Wilhelm Siemens. In *Wissenschaftliche und technische Arbeiten von Werner Siemens*, 2 vols. Berlin: Springer, vol.II, 2-11.
Simon, Herbert 1996. *The Sciences of the Artificial*, 3rd edition. Cambridge, Mass: MIT Press.
Sintomer, Yves 2001. Délibérer, participer, représenter. Vers une sociologie de la justification politique. Paris: mémoire d'habilitation, université Paris V.
Slobodin, Richard 1997. *W.H.R.Rivers.* Revised ed.. Stroud: Sutton.
Smith, Alice Kimball 1965. *A Peril and a Hope: The Scientists' Movement in America, 1945-1947.* Chicago: University of Chicago Press.
Smith, Charles Michie 1900. *Report on the Kodaikanal and Madras Observatories 1899-1900.* Madras: Asylum Press
Smith, Charles Michie 1904. Kodaikanal Observatory. *Monthly Notices of the Royal Astronomical Society* 64: 333-334.
Smith, Crosbie, Wise, M. Norton 1989. *Energy and Empire: A Biographical Study of Lord Kelvin.* Cambridge: Cambridge University Press.
Smith, Edward E., Medin, Douglas L. 1981. *Categories and concepts.* Cambridge, Mass.: Harvard University Press.
Smith, John 1660. *Select Discourses.* London: Printed by J. Flesher, for W. Morden.
Smolin, Lee 1997. *The Life of the Cosmos.* New York: Oxford University Press.
Soper, Kate 1995. *What Is Nature?* Oxford: Blackwell.

Spencer, Joseph E., Thomas, William L. 1948. The hill stations and summer resorts of the Orient. *Geographical Review* 38: 637-651.

Sponsel, Alastair 2002. Constructing a revolution in science. *British journal for the history of science* 35: 439-468.

Sprat, Thomas 1667. *History of the Royal Society*. London: Printed by T. R. for J. Martyn [...], and J. Allestry.

Srinivasan, Amriit 1998. The subject in fieldwork. In Thapan, Meenakshi, ed., *Anthropological journeys: reflections on fieldwork*. Hyderabad: Orient Longman, 54-82.

Staatliche Museen zu Berlin 2001. Nationalgalerie Berlin: Das XIX. Jahrhundert: Katalog der ausgestellten Werke. Berlin: E.A. Seemann.

Stachel, Peter 2004. "Was ist eine Tatsache? " Ludwik Flecks Beitrag zur Wissenschaftssoziologie und Erkenntnistheorie. *Simon Dubnow Institute Yearbook* 3: 351-382.

Stanley, Matthew 2003. An expedition to heal the wounds of war. *Isis* 94: 57-89.

Steinhausen, Georg 1931. *Deutsche Geistes- und Kulturgeschichte von 1870 bis zur Gegenwart*. Halle: Niemeyer.

Steinle, Friedrich 1991. *Newtons Manuskript "de gravitatione": Ein Stück Entwicklungsgeschichte seiner Mechanik*. Stuttgart: Franz Steiner Verlag.

Steinle, Friedrich 1996. Work, Finish, Publish? The formation of the second series of Faraday's 'Experimental Researches in Electricity'. *Physis* 33: 141-220.

Steinle, Friedrich 1997. Entering New Fields: Exploratory Uses of Experimentation. *Philosophy of Science* 64 (Supplement): 65-74.

Steinle, Friedrich 2000. Experiment, Instrument und Begriffsbildung: Ampère, das Galvanometer und der Stromkreis. In Meinel, Christoph, ed., *Instrument – Experiment: Historische Studien*. Berlin/Diepholz: GNT-Verlag, 98-108.

Steinle, Friedrich 2003. Experiments in History and Philosophy of Science. *Perspectives on Science* 10 (4): 408-432.

Steinle, Friedrich 2004a. Einleitung: Michael Faraday und seine "Experimental Researches in Electricity". *Experimental-Untersuchungen über Elektricität von Michael Faraday, in drei Bänden*. Erster Band. Frankfurt: Verlag Harri Deutsch, iv-xxxii.

Steinle, Friedrich 2004b. Exploratives Experimentieren. Charles Dufay und die zwei Elektrizitäten. *Physik Journal* 3 (6): 47-52.

Steinle, Friedrich 2005. *Explorative Experimente. Ampère, Faraday und die Ursprünge der Elektrodynamik*. Stuttgart: Franz Steiner Verlag.

Steinle, Friedrich 2006. Concept formation and the limits of justification. "Discovering" the two electricities. In Schickore, Jutta., Steinle, Friedrich, eds., *Revisiting Discovery and Justification. Historical and philosophical perspectives on the context distinction*. Dordrecht: Springer, 183-195.

Steinsaltz, Rabbi Adin, Funkenstein, Amos 1997. *Sociology of ignorance*. Moscow: Stirling (to be published).

Stewart, Balfour 1873. *The Conservation of Energy*. London: King.

Stigler, Stephen 1986. *The History of Statistics: The Measurement of Uncertainty before 1900*. Cambridge, Mass.: Harvard University Press.

Stillingfleet, Edward 1662. *Origines Sacrae: A Rational account of the grounds of Christian Faith, as to the truth and divine authority of the Scriptures and the matters therein contained*. London: Printed by R.W.; for Henry Mortlock.

Stock, Wolfgang 1980. Die Bedeutung Ludwik Flecks für die Theorie der Wissenschaftsgeschichte. *Grazer Philosophische Studien* 10: 105-118.

Storer, Norman W., ed. 1973. *Robert K. Merton, The Sociology of Science: Theoretical and Empirical Investigations*. Chicago: University of Chicago Press.

Stratton, Frederick John Marrion 1957. John Evershed 1864-1956. *Biographical memoirs of Fellows of the Royal Society* 3: 40-51.

Stubbe, Henry 1670. *A Censure upon Certain Passages in the History of the Royal Society as being Destructive to Established Religion and the Church of England.* Oxford: Printed for Richard Davis.

Sweetnam, George 2000. *Command of light: Rowland's school of physics and the spectrum.* Phildaelphia: American Philosophical Society.

Syfret, Rosemary H. 1950a. Some Early Critics of the Royal Society. *Notes and Records of the Royal Society* 8: 20-64.

Syfret, Rosemary H. 1950b. Some Early Reactions to the Royal Society. *Notes and Records of the Royal Society* 7: 207-258.

Symotiuk, Stefan 1983. Two Sociologies of Knowledge. L. Fleck – T. Bilikiewicz. *Kwartalnik Historii Nauki i Techniki* 3-4: 569-582.

Szapor, Judit 1997. Laura Polanyi, 1882-1957: Narrative of a Life. *Polanyiana* 6: 43-54.

Tait, Peter Guthrie 1877. Note on the Measure of Beknottedness. *Proceedings of the Royal Society of Edinburgh* 9: 289-298.

Tapsell, Paul 2003. Beyond the frame. In Peers, Laura, ed., *Museums and Source Communities*. London: Routledge, 242-251.

Technische Deputation für Gewerbe [P. C. W. Beuth and K. F. Schinkel, eds.] 1821. *Vorbilder für Fabrikanten und Handwerker*. Berlin: [Prêtre].

Terhune, Chad 2004. Coca-Cola's New Chief offers Obesity Defence. *Wall Street Journal Europe,* June 18.

Teja-Bach, Friedrich 1996. *Struktur und Erscheinung: Untersuchungen zu Dürers graphischer Kunst*. Berlin: Gebrüder Mann.

The Lancet 2001. Editorial, 'The Tightening Grip of Big Pharma', April 14.

Thompson, D'Arcy Wentworth 1917. *On Growth and Form* Cambridge, UK: Cambridge University Press, 1917; second edition, 1942; reprinted 1952 et al.

Thorpe, Charles 2001. Science against Modernism: The Relevance of the Social Theory of Michael Polanyi. *British Journal of Sociology* 52: 19-35.

Thurston, Edgar 1896b. Anthropology of the Todas and Kotas of the Nilgiri Hills. *Bulletin of the Madras Government Museum* 1 (4): 141-229.

Thurston, Edgar 1898. Anthropology in Madras. *Nature* 58 (26 May): 82-84.

Thurston, Edgar 1901. Todas of the Nilgiris. *Bulletin of the Madras Government Museum* 4 (1): 1-21.

Tieck, Ludwig 1996 [1836]. *Der junge Tischlermeister*. Frankfurt am Main: Ullstein.

Tilt, Edward 1875. *Health in hot countries for British women.* Fourth ed.. London: Churchill.

Todes, Daniel P. 2002. *Pavlov's Physiology Factory: Experiment. Interpretation, Laboratory Enterprise*. Baltimore, London: Johns Hopkins University Press.

Toulmin, Stephen 1963. *Foresight and Understanding. An Enquiry into the Aims of Science.* New York: Harper.

Turing, Alan Mathison 1952. The Chemical Basis of Morphogenesis. *Philosophical Transactions of the Royal Society*. Band 237: 37-72.

Turnbull, David, Stokes, Terry 1990. Manipulable systems and laboratory strategies in a biomedical institute. In Le Grand, Homer E., ed., *Experimental Inquiries*. Dordrecht: Kluwer Academic Publishing, 167-192.

Turner, R. Steven 1977. The Ohm-Seebeck Dispute, Hermann von Helmholtz, and the Origins of Physiological Acoustics. *British Journal for the History of Science* 10: 1-24.

Union of Concerned Scientists website 2004. *Scientific Integrity in Policymaking, An Investigation into the Bush Administration's Misuse of Science*. URL: http://www.ucsusa.org/.

Union of Concerned Scientists website. Preeminent Scientists Protest Bush Administration's Misuse of Science. URL: http://www.ucsusa.org/. February 18, 2004.

United States House of Representatives, Committee on Government Reform, Minority Staff Special Investigation Division November 13, 2003. *Politics and Science in the Bush Administration*.

United States, Office of Scientific Research and Development 1945. Science, the Endless Frontier. A Report to the President by Vannevar Bush, Director of the Office of Scientific Research and Development. July 1945. Washington, D.C.: US Government Printing Office.

Urfalino, Philippe 2005. La délibération n'est pas une conversation. Délibération, décision collective et négociation. *Négociations* 2: 99-114.

Urry, James 1993. *Before social anthropology: essays on the history of British anthropology*. Churchill: Harwood Academic.

Van De Vijver, Gertrudis, ed. 1992. *New Perspectives on Cybernetics: Self-Organization, Autonomy and Connectionism*. Dordrecht: Kluwer Academic Publishing.

Veltz, Pierre 2000. *Le nouveau monde industriel*. Paris: Gallimard.

Verein Ernst Mach in Wien 1929. Wissenschaftliche Weltauffassung. Der Wiener Kreis. Wien: A. Wolf.

Verwaltung der Staatlichen Schlösser und Gärten und der Nationalgalerie SMPK, ed. 1981. *Karl Friedrich Schinkel: Architektur, Malerei, Kunstgewerbe*. Exhibition catalogue, Berlin: Nicolai.

Vogt, Adolf M. 1985. *Karl Friedrich Schinkel, Blick in Griechenlands Blüte: ein Hoffnungsbild für Spree-Athen*. Frankfurt am Main: Fischer.

Vorbilder für Fabrikanten und Handwerker. Auf Befehl des Ministers für Handel, Gewerbe und Bauwesen herausgegeben von der Königlich Technischen Deputation für Gewerbe 1863. 2d. ed. Berlin: Königliche Staatsdruckerei.

Walker, Anthony 1986. *The Toda of South India: a new look*. Delhi: Hindustan Publishing Corporation.

Walker, Anthony 1997. The western romance with the Todas. In Hockings, Paul, ed., *Blue Mountains Revisited: cultural studies on the Nilgiri Hills*. Delhi: Oxford University Press, 106-135.

Walker, Gilbert 1908. Memorandum to Director. In Evershed, John, ed., *Report of Kodaikanal Observatory*. Madras: Asylum Press.

Walker, Gilbert 1911. Memorandum to Director. In Evershed, John, ed., *Report of Kodaikanal Observatory*. Madras: Asylum Press.

Waring, Stephen 1991. *Taylorism transformed, Scientific Management Theory since 1945*. Chapel Hill: University of North Carolina Press.

Watkins, John 1997. Karl Raimond Popper. 1902-1994. *Proceedings of the British Academy* 94: 645-684. URL: http://www.britac.ac.uk/pubs/src/popper/part1.html (21 December 2006).
Weaver, Warren 1948. Science and Complexity. *American Scientist* 36: 536-544.
Weber, Max 1948. Science as a Vocation. In Gerth, H. H., Mills, C. Wright, trans. and ed., *From Max Weber*. London: Paul, trench, Trubner, 129-156.
Weingart, Peter. How robust is 'socially robust knowledge'? (to be published).
Weissberg, Alex 1952. *Conspiracy of Silence*. Trans. Fitzgerald, Edward, pref. Koestler, Arthur. London: Hamish Hamilton.
Wessely, Anna 1991.Transposing "Style" from the History of Art to the History of Science. *Science in Context* 4: 265-278.
Westfall, Richard S. 1971. Force in Newton's physic. The science of dynamics in the seventeenth century. London: MacDonald.
White House accused of siding with food industry on obesity. *Financial Times* 17 janvier 2004
White, Hayden 1973. The Politics of Contemporary Philosophy of History. *Clio* 3: 35-53.
Wigner, Eugene P., Hodgkin, Robin A. 1977. Michael Polanyi, 12 March 1891 – 22 February 1976. *Biographical Memoirs of Fellows of the Royal Society* 23: 413-448.
Wille, Hans 1960. Die Erfindung der Zeichenkunst. In Guldan, Ernst, ed., *Beiträge zur Kunstgeschichte: Eine Festgabe für H. R. Rosemann*. München: Deutscher Kunstverlag, 279-300.
Williams, Michael 1991. *Unnatural doubts. epistemological realism and the basis of scepticism*. Oxford: Blackwell
Winkler, Johann Heinrich 1745. *Die Eigenschaften der electrischen Materie und des electrischen Feuers, aus verschiedenen neuen Versuchen erkläret, und, nebst etlichen neuen Maschinen zum Electrisiren beschrieben*. Leipzig: Breitkopf.
Winner, Langdon 1980. Do artefacts have politics? *Daedalus* 109: 121-131.
Wise, M. Norton 1999. Architectures for Steam. In Galison, Peter, Thompson, Emily, eds., *The Architecture of Science*. Cambridge, Mass: MIT Press, 107-140.
Wise, M. Norton, Wise, Elaine M. 2003. Staging an Empire. In Daston, Lorraine, ed., *Things that Talk*. Cambridge: Zone Books.
Wittich, Dieter 1978. Eine aufschlußreiche Quelle für das Verständnis der gesellschaftlichen Rolle des Denkens von Thomas Kuhn. *Deutsche Zeitschrift für Philosophie* 26: 105-113.
Wittich, Dieter 1981. Rezension von: Ludwik Fleck: Genesis and development of a scientific fact. *Deutsche Zeitschrift für Philosophie* 29: 723-728.
Wittich, Dieter 1983. Das Verhältnis von Wissenschaft und Kultur in der Wissenschaftstheorie von Ludwik Fleck. *Deutsche Zeitschrift für Philosophie* 31: 852–855.
Wojcik, Jan 1997. *Robert Boyle and the Limits of Reason*. Cambridge: Cambridge University Press.
Woods, Ngaire, Narlikar, Amrita 2001. Governance and Accountability: the WTO, the IMF and the World Bank. *International Social Science Journal* 53/170: 569-583.
Woolf, Harry 1959. *The Transits of Venus: A Study of Eighteenth-Century Science*. Princeton: Princeton University Press.
World Intellectual property Organization April 28, 2003 (Geneva). Intergovernmental Committee on Intellectual Property and Genetic Resources, Traditional Knowledge and Folklore, 5th Session, WIPO/GRTKF/IC/5/8.

Wyss, Beat 1985. *Trauer der Vollendung: Von der Aesthetik des deutschen Idealismus zur Kulturkritik an der Moderne*. München: Matthes & Seitz.
Young, Iris Marion 2001. Activist challenge to deliberative democracy. *Political Theory* 29: 670-690.
Youschkevitch, Adolf P. 1976. The Concept of Function up to the Middle of the 19th Century. *Archive for the History of Exact Sciences* 16: 37-85.
Zaun, Jörg 2002. *Instrumente für die Wissenschaft: Innovationen in der Berliner Feinmechanik und Optik 1871-1914*. Berlin: Verlag für Wissenschafts- und Regionalgeschichte Engel.
Ziman, John 2000. Commentary. *Minerva* 38: 21-25.
Zittel, Claus 2002. Konstruktionsprobleme des Sozialkonstruktivismus. In Zittel, Claus, ed., *Wissen und soziale Konstruktion*. Berlin: Akademie Verlag, 87-108.
Zittel, Claus 2009. *Theatrum philosophicum: Descartes und die Rolle ästhetischer Formen in der Wissenschaft*. Berlin: Akademie Verlag.
Zubov, Vladimir Ivanovich, Liapunov, Aleksandr Mikhailovich 1961. *Methods of A. M. Lyapunov and their application*. Oak Ridge, Tenn.: US Atomic Energy Commission.

www.ingramcontent.com/pod-product-compliance
Lightning Source LLC
Chambersburg PA
CBHW080938240926
51849CB00004BA/492
9783050044071